Infinite Science
Publishing

Ulrich M. Engelmann

**Assessing Magnetic
Fluid Hyperthermia**

Magnetic Relaxation Simulation,
Modeling of Nanoparticle
Uptake inside Pancreatic Tumor
Cells and in vitro Efficacy

© 2019 Infinite Science Publishing
University Press and
Academic Printing

Imprint of Infinite Science GmbH,
Technikzentrum | MFC 1
Maria-Goeppert-Straße 1
23562 Lübeck, Germany

Cover Design: Infinite Science Publishing
Editorial and Copy Editing: Institute of Applied Medical Engineering, RWTH Aachen University
D82 (Diss. RWTH Aachen University, 2019)

Publisher: Infinite Science GmbH, Lübeck, www.infinite-science.de
Printed in Germany

ISBN Hardcover: 978-3-945954-58-4

Bibliografische Information der Deutschen Nationalbibliothek:
Die Deutsche Nationalbibliothek verzeichnet diese Publikation in der Deutschen Nationalbibliografie; detaillierte bibliografische Daten sind im Internet über http://dnb.d-nb.de abrufbar.

Author

Ulrich M. Engelmann
Institute of Applied Medical Engineering
RWTH Aachen University, Germany
E-mail: engelmann@ame.rwth-aachen.de

D82 (Diss. RWTH Aachen University, 2019)

Ulrich M. Engelmann

**Assessing Magnetic
Fluid Hyperthermia**

Magnetic Relaxation Simulation,
Modeling of Nanoparticle
Uptake inside Pancreatic Tumor
Cells and in vitro Efficacy

**Institute of Applied Medical Engineering
RWTH Aachen University**

Infinite Science
Publishing

To
David Fenimore,
who sparked my enthusiasm for physics.

Abstract

As a leading cause of death worldwide with patient-specific evolving mutations, cancer requires innovative therapies capable of individualized treatment. The use of magnetic nanoparticles (MNP) as thermal agents offers such individualized cancer therapy: After local accumulation at the tumor site, MNP can be triggered to transform the energy of an externally applied alternating magnetic field into heat via relaxation of their magnetic moments. This process named magnetic fluid hyperthermia (MFH) enables organ-confined cancer treatment by delivering therapeutic temperatures higher than $43\,°C$ inside tumors, inducing tumor cell death. In this way, MFH efficacy relies on MNP efficiency to generate such elevated temperatures in interaction with the biological environment. However, this environment imposes severe limitations to the MNP magnetic relaxation and heating behavior by restricting the MNP mobility and causing MNP agglomeration. Based on in vitro experiments, this thesis addresses the applicability of MFH to pancreatic tumor cells and discusses opportunities to optimize intracellular MNP heating for clinical application of MFH. The interaction of iron oxide MNP with pancreatic tumor cells and the MNP uptake kinetics inside these cells are investigated using transmission electron microscopy and magnetic particle spectroscopy. The impact of MNP-cell interaction on heating efficiency is quantified with inductive heating experiments and compared to artificially agglomerated and immobilized MNP, mimicking the conditions in cellular environments. Furthermore, Monte-Carlo (MC-)simulations of MNP magnetic relaxation are used to predict sets of parameters varying field amplitude and frequency as well as MNP size and magnetic properties to optimize MFH efficiency under medically tolerable field parameters.

Combined agglomeration and immobilization of MNP upon internalization inside cells decrease heating efficiency by nearly two thirds compared to freely dispersed MNP. This decrease is attributed to one half (one third of the overall heating) to the inhibition of physical rotation upon MNP immobilization, blocking the Brownian contributions of larger MNP to overall heating. The other half can be related to demagnetization effects due to increased magnetic interparticle interactions upon MNP agglomeration. Despite this decrease in MNP heating, MFH can still effectively damage cell in vitro even without a perceptible bulk temperature rise, by local heating on the cellular level. This requires a sufficiently large MNP uptake inside cells, which is reached after approx. 6 h of incubation as predicted from MNP uptake modeling. The cell damage depends on the thermal energy deposited per cell (TEC). Interestingly, healthy cells are more resistant to MFH treatment with a 50 % margin in TEC to damage healthy cells compared to pancreatic tumor cells. As a consequence, MFH therapy can be tuned to deal tumor-specific damage while healthy surrounding cells remain unharmed by controlling TEC via the intracellular uptake of MNP, the MNP heating efficiency and the duration of MFH treatment. The improvement of intracellular MNP uptake and heating efficiency, e. g. by using MNP with large particle cores ($d_C \geq 25\,nm$) as predicted by MC-simulations, will therefore remain one focus of future work. It seems most important, however, to translate MFH to in vivo experiments in the near future to establish MFH among the standard clinical cancer therapies.

Zusammenfassung

Übersetzung des englischen Originaltitels: *Evaluierung magnetischer Fluid-Hyperthermie: Simulation magnetischer Relaxation, Modellierung der Aufnahme von Nanopartikeln in Pankreas-Tumorzellen und Wirksamkeit in vitro.*

Als eine der weltweit führenden Todesursachen mit patientenspezifischen, sich entwickelnden Mutationen benötigt Krebs innovative Behandlungsmethoden, die eine individuelle Behandlung erlauben. Die Verwendung von magnetischen Nanopartikeln (MNP) als thermische Wirkstoffe bietet eine solche individualisierte Krebstherapie: Nach der lokalen Ansammlung von MNP an der Tumorort, können die MNP durch ein extern angelegtes, magnetisches Wechselfeldes zu Relaxationsprozesse angeregt werden und dadurch die Feldenergie in Wärme umwandeln. Dieser Prozess, genannt magnetische Fluid-Hyperthermie (MFH), erlaubt die lokal auf einzelne Organe beschränkte Krebsbehandlung: Hierbei werden therapeutisch wirksame Temperaturen von über 43 °C in den Tumor eingebracht und schädigen dadurch Tumorzellen irreparabel. In dieser Weise beruht die Wirksamkeit von MFH auf der Effizienz von MNP, diese erhöhten Temperaturen zu erzeugen, wobei die MNP dabei in Wechselwirkung mit der sie umgebenden biologischen Umgebung stehen. Diese Umgebung schränkt dabei das magnetische Relaxations- und Aufheizverhalten der MNP erheblich ein, indem sie die Beweglichkeit der MNP reduziert und MNP Agglomeration verursacht. Diese Arbeit befasst sich auf Basis von in vitro Experimenten mit der Evaluierung der Anwendbarkeit von MFH auf Pankreastumorzellen und diskutiert Ansätze zur Verbesserung der intrazellulären Aufheizung von MNP für die klinische Anwendung von MFH. Die Wechselwirkung von Eisenoxid-MNP mit Pankreas-Tumorzellen und die MNP-Aufnahmekinetik innerhalb dieser Zellen werden mittels Transmissionselektronenmikroskopie und Magnet-Partikel-Spektroskopie untersucht. Der Einfluss der MNP-Zell-Interaktion auf die MNP Aufheizung wird mit induktiven Aufheizexperimenten quantifiziert und mit synthetisch immobilisierten und agglomerierten MNP Modellsystemen verglichen, welche die Bedingungen der biologischen Umgebung nachahmen. Darüber hinaus werden Monte-Carlo (MC-)Simulationen der magnetischen Relaxation von MNP verwendet, um Parametersätze mit variierender Feldamplitude und -frequenz, sowie Partikeldurchmesser und magnetischen Eigenschaften zu bestimmen, die den Wirkungsgrad von MFH optimieren und zwar unter der zusätzlichen Einschränkung von medizinisch tolerierbaren Feldparametern.

Die kombinierte Agglomeration und Immobilisierung von MNP durch die Internalisierung in Zellen verringert die MNP Aufheizung um fast zwei Drittel im Vergleich zu frei dispergierten MNP. Dieser Rückgang der MNP Aufheizung wird zur Hälfte (einem Drittel der Gesamtaufheizung) auf die Hemmung der physikalischen Rotation der MNP bei ihrer Immobilisierung zurückgeführt, die die Brown'schen Beiträge großer MNP zur Gesamtaufheizung blockiert. Die andere Hälfte kann mit Entmagnetisierungseffekten in Verbindung gebracht werden, die durch erhöhte magnetische Wechselwirkung zwischen den MNP aufgrund der Agglomeration in Zellen entstehen. Trotz dieser Reduktion der Gesamtaufheizung der MNP kann MFH Zellen in vitro effektiv schädigen und das sogar ohne einen messbaren Anstieg in der globalen Temperatur der

Probe. Der Grund hierfür liegt auf einer lokalen Aufheizung auf der Zellebene, welche jedoch eine ausreichende Aufnahme von MNP in die Zellen voraussetzt. Aus der Modellierung der MNP-Aufnahme-Kinetik für Pankreastumorzellen folgt eine solche ausreichende MNP-Aufnahme nach 6 h Inkubationszeit. Die Zellschädigung hängt maßgeblich von der eingebrachten Wärmeenergie pro Zelle (EWZ) während der Anwendung von MFH ab. Interessanterweise zeigen sich gesunde Zellen resistenter gegen eine MFH-Behandlung in vitro und eine Marge von 50 % in der EWZ besteht zwischen gesunden Zellen und Tumorzellen. Dadurch kann die Tumorbehandlung durch MFH mittels des EWZ-Parameters auf eine tumorspezifische Schädigung abgestimmt werden, wobei gesunde Zellen in der Umgebung unversehrt bleiben. Die EWZ kann dabei entweder durch die Menge an intrazellulären MNP, die MNP Aufheizleistung oder die MFH-Behandlungsdauer gesteuert werden. Die Verbesserung der intrazellulären MNP Aufnahme- und Aufheizeffizienz, z B. durch den mittels MC-Simulationen vorhergesagten Einsatz von größeren Partikelkernen ($d_C \geq 25\,$nm), bildet daher einen Schwerpunkt für zukünftige Arbeiten. Für die Etablierung von MFH als klinisch anerkannte Krebstherapie muss allerdings vorrangig die Untersuchung von MFH in in-vivo Experimenten vorangetrieben werden.

Contents

Index of Abbreviations

AAm	Acrylamide		**FFP**	Field-Free Point (in MPI)
AAS	Atomic Absorption		**FHG**	Ferrohydrogels
	Spectroscopy		**FM**	Ferromagnetism
AC	Alternating Current		**FWHM**	Full-Width-Half-Maximum
AFM	Antiferromagnetism		**GA**	Glutaraldehyde
AMF	Alternating Magnetic Field		**ICDD**	International Centre
APS	Ammonium persulfate			for Diffraction Data
BIS	N,N'-methylenebisacrylamide		**ICNIRP**	International Commission on
CA	Clonogenic Assay			Non-Ionizing Radiation
CDF	Cummulative Distribution			Protection
	Function; eq. (4.5)		**ICP-OES**	Inductively Coupled Plasma
CEM	Cumulative Equivalent Minutes;			Optical Emission Spectroscopy
	eq. (8.2)		**LA**	Lauric Acid
DC	Direct Current		**ILP**	Intrinsic Loss Power; eq. (5.3)
DI-H$_2$O	Deionized water		**LLG**	Landau-Lifshitz-Gilbert
DLS	Dynamic Light Scattering			(equation); eq. (2.33)
DM	Diamagnetism		**LM**	Low-Melting (Agarose)
DMEM	Dulbecco's Modified Eagle		**LRT**	Linear Response Theory
	Medium (cell culture medium)		**LUM**	Liposomes Ultra Magnétique
DMPC	Dimyristoylphosphatidyl-			MNP sample from FFAM
	choline; phospholipid		**MC**	Monte-Carlo (Simulation)
DMPG	Dimyristoylphosphatedyl-		**MDT**	Magnetic Drug Targeting
	glycerol; phospholipid		**MFH**	Magnetic Fluid Hyperthermia
DPBS	Dulbecco's Phosphate-Buffered		**ML(1,2,3)**	Magnetoliposome MNP samples
	Saline		**MNP**	Magnetic Nanoparticles
DSMZ	Deutsche Sammlung von Mikro-		**MPI**	Magnetic Particle Imaging
	organismen & Zellkulturen		**MPS**	Magnetic Particle Spectroscopy
EELS	Electron Energy Loss		**MRI**	Magnetic Resonance Imaging
	Spectroscopy		**MTT**	3-(4,5-Dimethylthiazol-2-yl)-
ECM	Extracellular Cell Matrix			2,5-diphenyltetrazoliumbromid
FCS	Fetal Calf Serum		**MVB**	Multivesicular Bodies
FDA	U.S. Food and Drug		**NaCl**	Sodium Chloride
	Administration		**PA**	Photometric Analysis
FeM	Ferrimagnetism			(of sample iron concentration)
FFAM	FerroFluide Anionique		**PAAm**	Poly(acrylamide)
	Magnétique; MNP sample		**PBS**	Phosphate-Buffered Saline
FFLA	FerroFluid stabilized		**PDAC**	Pancreatic Ductal Adeno-
	with Lauric Acid; MNP sample			carcinoma

Index of Symbols and Constants

μ_{agar}	Weight fraction of agarose dissolved in DI-H_2O [%]		σ	Distribution width of the PDF of the log-normal distribution
μ_B	Bohr magneton: $9.274 \cdot 10^{-24}\,A \cdot m^2$		$\sigma_{\hat{\mu}}$	Standard deviation of the mean of the PDF of the log-normal distribution; eq. (3.24)
μ_m	Weight fraction [%]; eq. (7.1)		σ_s	Elastic stress [$N \cdot m^2$]
μ_r	Relative permeability		σ_{el}	Shear stress [$N \cdot m^2$]; eq. (2.74)
μ_v	Volume fraction [%]		$\boldsymbol{S}$	Spin momentum [$kg \cdot m^2/s$]
$\boldsymbol{M}$	Magnetization [kA/m]		S	Spin quantum number
$M(H)$	Magnetization curve (also referred to as hysteresis loop)		$\boldsymbol{\tau}$	(Magnetic) torque [$kg \cdot m^2/s^2$]
M_S	Saturation magnetization [kA/m]		τ	Relaxation time [s]; eq. (2.25)
M_r	Remanence magnetization [kA/m]		τ'	Autocorrelation time [ms]
$\boldsymbol{m}$	Magnetic (dipole) moment [$A \cdot m^2$] (also of MNP; eq. (2.30))		τ_B	Brownian relaxation time [s]; eq. (2.31)
m_0	Saturation mass of MNP adsorbed at the cellular membrane [pg(Fe)]		τ_N	Néel relaxation time [s]; eq. (2.32)
			τ_R	Effective relaxation time [s]; eq. (2.70)
m_{CM}	Mass of carrier matrix of MNP in solution [mg]		T	(Absolute) temperature [K] $(273.15\,K \stackrel{\wedge}{=} 0\,°C)$
m_e	Electron (rest) mass: $9.1094 \cdot 10^{-31}\,kg$		T_0	Initial temperature [°C]
$m_{Fe_3O_4}$	Mass of MNP in solution [mg]		T_B	Blocking temperature [K]
ν_{el}	Number of polymer chains per unit volume		T_{bra}	Branching temperature (where ZFC and FC-curve diverge) [K]
N	Number of turns in a coil		T_C	Curie temperature [K]
N_d	Demagnetization factor		T_{eff}	Effective temperature [°C]; eq. (8.1) (sensed by cells in MFH treatment)
N_t	Number of time steps used in MC-simulations for one $M(H)$-loop		T_{max}	Peak temperature (in ZFC-curve) [K]
$\boldsymbol{n}$	Direction of magnetic nano-particle easy axes		T_N	Néel temperature [K]
n_{cyc}	Number of full $M(H)$-loops simulated in MC-simulations		t	Time [s]
			t_{chase}	Incubation time at 37°C during chase (cell) experiments [min]
P	Number of particles used in MC-simulations		t_{exp}	Time-scale of the experiment [s]
Ξ	Dilution factor of MNP samples		t_{inc}	Incubation time [s] (of ML with cells)
ξ	Reduced field parameter ; eq. (2.65)		t_{HT}	Time at elevated temperatures during hyperthermia treatment [s]
Q	Heat [J]			
Q_{eddy}	Heat of an eddy current [J]		t_{ex}	Delay time for the onset of exocytosis [min]
$\rho_{Fe_3O_4}$	Mass density of magnetite: $5\,180\,kg/m^3$		U	Internal energy (of a MNP) [J]; eq. (2.37)
R^2_{adj}	Fitting quality parameter			
Σ	Entropy [J/K]		Φ	Magnetic flux [J/A]; eq. (2.47)

Φ_0	Magnetic flux quantum: $2.0678 \cdot 10^{-15}$ J/A	w	Peak width at FWHM [rad] (from XRD diffraction pattern)
ϕ_0	Maximum active cellular membrane surface fraction capable of endo- and exocytosis processes	χ	Magnetic (volume) susceptibility [1/m^3]
φ	Rotation angle about z-axis [°] (for XRD sample holder)	$\hat{\chi}$	Complex magnetic (volume) AC-susceptibility [1/m^3]; eq. (2.66)
V	Sample volume [m^3]	χ_0	Initial magnetic (mass) susceptibility
V_C	Particle core volume [m^3]	X	Number of repetitions of single MC-simulations
V_e	Potential difference accelerating electrons [J]	$\Psi_{\parallel}(\tau')$	Autocorrelation function; eq. (4.7)
V_H	Particle hydrodynamic volume [m^3]	ψ	Rotation angle about x-axis [°] (for XRD sample holder)
V_M	Particle magnetic volume [m^3]	∇	Nabla operator
$V(t)$	Voltage induced in MPI or MPS receive coil [V]; eq. (2.48)	ω	Rotation angle about y-axis [°] (for XRD sample holder)
V_{AMF}	Excitation signal in MPI or MPS [V]	Z_{med}	Medically tolerable limit of the magnetic field [kHz $\cdot$ mT/μ_0]
V_{MNP}	Particle signal in MPI or MPS [V]; eqs. (2.49),(2.59)	z_{avg}	Mean hydrodynamic size* [nm] (defined by ISO 22412:2017)
$\hat{V}_k$	Fourier coefficients of voltage induced in MPI or MPS receive coil [V]		
$\boldsymbol{W}(t)$	Wiener integral; eq. (3.2)		

* By definition, the particle *size* denotes the particle *diameter* throughout this thesis.

** The magnetic field strengths are reported in units of mT/μ_0 in this thesis by convention. The aim of this convention is to report the numbers on a Tesla scale, which most readers with a medical background are more familiar with (e. g. from MRI, MFH), but, on the other hand still use the correct units for the magnetic field strength.

1. Introduction

With about 14 million new cases reported in 2014 [1] and accounting for 9 million deaths in 2016 [2], cancerous deseases are one of the leading causes of death worldwide [3]. Developed and developing countries are affected alike; e.g. 25.2 % of deaths were caused by cancer in Germany in 2015 [4]. Since cancer is a complex and constantly mutating disease [5], modern medicine is invariably challenged to develop new treatment approaches. Besides improving conventional therapy approaches such as chemotherapy, radiation and resection, much promise lies in novel approaches, adjustable to the cancer-specific challenges of each patient individually. Among these individualized therapies, locally confined *magnetic fluid hyperthermia* (MFH), i.e. the overheating of cancerous tissue, has gained much attention in current research due to its potential for the individualization of therapy, as will be outlined in the following.

The MFH principle relies on the use of *magnetic nanoparticles* (MNP) to generate therapeutic heat. MNP are nanosized magnets with sizes[1] below 100 nm [6] and introduce the phenomenon of *superparamagnetism* [7]. As such, MNP display unique heating characteristics when subjected to an alternating magnetic field (AMF), potentially allowing a precise temperature control required for individual therapy [8]. MNP can also be magnetically guided to and accumulated at the tumor site after injection in the body's circulatory system. This provides a high degree of individualization, as virtually any tumor site accessible by a magnetic field can be targeted with MNP and treated with therapeutic heat.

First successful clinical trials of MFH therapy performed between 2004 and 2011 confirmed the feasibility of clincial MFH [9]; however, after 2011, these trials were discontinued and broad clinical application of MFH is still missing. This is due to the fact that the concentration of MNP necessary to induce therapeutic temperatures (T$\leq$ 43 °C [10]) used in these trials has been very high and the biocompatibility of the MNP at such high concentration is ultimately not proven yet. Furthermore, MNP unavoidably interact with biological and cellular components inside the body. E.g., MNP internalization inside cells confines the MNP arrangement and immobilizes the MNP [11], and the effects of intracellular internalization of MNP on particle heating are poorly understood at present [12]. From this, two demands can be identified to advance the clinical application of MFH: (I) MNP heating performance must be improved in order to reach therapeutic temperatures at lower MNP concentrations that are biocompatible and (II) the interaction of MNP with tumor cells and its effects on particle heating must be deciphered. These demands translate into four research questions that this thesis endeavors to answer:

1. *What MNP (i.e. which MNP properties) maximize the particle heating in MFH?*

2. *How do MNP interact with cells and how can the MNP uptake inside cell be quantified?*

---

[1]By definition, the particle *size* denotes the particle *diameter* throughout this thesis.

3. *How does particle heating change upon MNP internalization inside cells?*

4. *How efficient is intracellular MFH applied to (tumor) cells; esp. at low MNP concentrations?*

This thesis sequentially investigates these four questions on the basis of well characterized MNP systems using an experimental approach: First, the particle heating in terms of the MNP properties and AMF parameters are evaluated experimentally and the results are compared to theoretical particle heating simulations using MNP relaxation theory. In this way, optimal MNP properties and AMF parameters for maximum particle heating are identified (Chapter 5). Second, the MNP-cell interactions in vitro with pancreatic tumor cells and healthy control cells are investigated, assessing the changes in arrangement that MNP undergo upon intracellular internalization and deriving a model for predicting the uptake kinetics of MNP (Chapter 6). Third, based on the MNP-cell interaction analysis, the effects of MNP internalization on particle heating are examined using artificial in situ model systems, which mimic the states of intracellular MNP (Chapter 7). Finally, the efficacy of MFH on pancreatic tumor cells is assessed in vitro using low MNP concentrations and the main parameters contributing to MFH effectiveness are identified (Chapter 8).

Short Description of each Chapter

This thesis is divided in seven main chapters. Chapter 2 provides background knowledge on key topics covered in this thesis: First, a basic introduction to the physical principles of magnetism with special focus on superparamagnetism is given, followed by an overview of the physical concepts governing the particle relaxation and heating processes. Moreover, the state-of-the-art in applying MFH in tumor therapy is summarized, also explaining how the application of MFH can be implemented in a treatment approach for pancreatic tumor therapy. Lastly, the chapter concludes with a description of the mechanical properties of hydrogels that are later used as tissue-mimicking modeling systems to incorporate MNP (cf. Chapter 7).

Chapter 3 presents the simulations of magnetic particle relaxation used to predict optimized particle heating. It first outlines the current state of research on simulating magnetic particle heating and, second, explains the implementation of the Monte-Carlo-based (MC-)simulation employed for predicting particle heating in this thesis. The simulation results are verified and compared to common theories of particle heating. From this, general trends for particle heating in dependence of MNP properties and AMF parameters are derived, which are later used to derive the optimal MNP properties and AMF parameters for maximum particle heating (cf. Chapter 5).

Chapter 4 describes the synthesis of MNP and covers the characterization of MNP properties. Each characterization technique used is introduced briefly by explaining the experimental procedure, followed by the analysis and discussion of the characterization results and the specific

MNP properties. These MNP properties are key parameters for the investigations of particle heating, MNP-cell interactions and intracellular particle heating effects in all following chapters.

Particle relaxation and heating measurements are presented in Chapter 5. This chapter introduces magnetic particle spectroscopy (MPS) as a means of characterizing MNP relaxation processes. Following, the particle heating of each MNP system is analyzed and discussed on the basis of the predictions of particle heating in dependence of MNP properties and AMF parameters derived from MC-simulations in the previous Chapter 3. Most importantly, MC-simulation results for particle heating are validated against experimental data. These validated MC-predictions are then employed to predict the optimal MNP properties and AMF parameters for maximum particle heating, answering research question 1 from above.

Chapter 6 investigates MNP-cell interactions with special consideration of changes in MNP morphology upon internalization inside cells and the quantitative description of the MNP uptake kinetics. A mathematical model is developed and fitted to the uptake kinetics, allowing to predict the amount of intracellular MNP for arbitrary incubation times. Key factors of how MNP arrangement, mobility and magnetic properties are affected upon MNP uptake inside cells are identified and quantified: Most importantly, MNP are found to be immobilized and agglomerated intracellularly. Overall, this chapter is dedicated to answer research question 2 from above.

On the basis of MNP immobilization and agglomeration effects inside cells from the previous Chapter 6, MNP model systems are presented in Chapter 7, which allow the study of the effects of either immobilization or agglomeration on particle heating in isolation. Therefore, MNP are suspended in two hydrogels of tunable mesh size, allowing the gradual control of the degree of immobilization of the incorporated MNP. Moreover, two systems of controlled MNP agglomeration are prepared, one of which even allows the evaluation of the combined effect of immobilization and agglomeration simultaneously. This chapter answers research question 3 from above.

In the last main Chapter 8, the efficacy of MFH applied to pancreatic tumor cells after incubation with low MNP concentrations is studied and the cytotoxic effects of MFH treatment are assessed. The superior efficacy of MFH compared to standard hyperthermia is demonstrated and reveals the presence of so-called nanoheating effects on the cellular level during MFH application. Furthermore, three main factors for controlling and improving MFH efficacy are identified: The overall particle heating performance, the amount of MNP internalized inside cells and the duration of treatment. These three factors can be combined in the single parameter of the thermal energy deposited per cell (TEC), allowing to assess the efficacy of MFH for specific tumors individually. In this way, Chapter 8 provides an answer to research question 4.

2. Background

This chapter introduces to the basics of nanomagnetism, magnetic particle relaxation, imaging and heating, as well as the current status of applying particle heating in tumor therpay and the use of hydrogels as tissue-equivalent model systems in biomedical applications. In detail, the basics of the theory of magnetism precedes the current chapter in Section 2.1. This section forms the foundation for a subsequent summary on the physics of magnetic particle relaxation in Section 2.2, followed by a brief introduction in the principles of magnetic particle imaging in Section 2.3. Subsequently, the physics of magnetic particle heating are explained in Section 2.4, including the leading mathematical theories developed for describing particle heating. An overview on the present-day applications of magnetic fluid hyperthermia is given in Section 2.5, where also a treatment strategy for the example of pancratic tumors is discussed. Finally, Section 2.6 concludes the present chapter by reviewing the application and mechanical description of hydrogels as tissue-equivalent model systems.

2.1. Theory of Magnetism

Generally, the phenomena of magnetism are easy to measure experimentally, but very complex to explain theoretically. Only when applying quantum mechanics down to the atomic scale, can a satisfactory explanation be given. The following sections are compiled to explain the basic magnetic phenomena (Section 2.1.1) neccessary to understand the concepts of magnetic anisotropy (Section 2.1.2) and superparamagnetism (Section 2.1.3). To this end, this section focuses on illustrating and understanding these phenomena rather than on complex calculation and is based on the concepts from [13] and [14], if not mentioned otherwise.

2.1.1. Basics of Magnetism

Materials exposed to a magnetic field, H, acquire a magnetization, M. This magnetization originates from the alignment of the magnetic (dipole) moments, $m = g\mu_B J$, of each individual atom within the material and is defined per unit volume, V, as $M = \frac{1}{V}\sum_{n=1}^{N} m_n$. Here, $g = 2.0023$ is the Landé-factor for electrons, $\mu_B = 9.274 \cdot 10^{-24}\,\mathrm{A \cdot m^2}$ is the Bohr magneton and $J = L + S$ is the atom's total angular momentum, combining the orbital momentum L and the intrinsic atomic momentum S, denoted as spin. For most materials the magnetization aligns with an applied magnetic field, H, linearly according to

$$M = \chi \cdot H. \tag{2.1}$$

χ is the magnetic susceptibility, which is generally a three-dimensional tensor. For homogeneous and isotropic materials all non-diagonal entries vanish and χ reduces to a scalar.
The response of any material to the applied magnetic field, H, is called magnetic induction,

B, and is defined by

$$B = \mu_0(H + M),$$ (2.2)

with the magnetic constant $\mu_0 = 4\pi \cdot 10^{-7} \, \text{N A}^{-2}$. If a magnetic material is inserted into the magnetic field H in free space, a demagnetizing field, H_d, develops inside the material. H_d opposes the intrinsic magnetization of the material, M, and can be approximated for magnetic materials of ellipsoidal shape as

$$H_d = -N_d \cdot M$$ (2.3)

with the demagnetizing factor N_d, which is a function of the geometry of the material (e. g. $H_d = 1/3$ for a spherical magnet). Therefore, the field inside a magnetic material actually reads $H_{in} = H + H_d$.

By introducing the susceptibility from eq. (2.1) in eq. (2.2) one obtains $B = \mu H$, with the permeability

$$\mu = \mu_0(1 + \chi) = \mu_0\mu_r.$$ (2.4)

The relative permeability μ_r is used to classify all materials according to their magnetic behavior. Note that this classification is not without exceptions but provides a simple basis, upon which five different types of magnetism can be distinguished, divided according to the arrangement and interaction of the materials' atomic magnetic moments and the temperature dependence of the materials' magnetization:

Diamagnetism (DM)

For diamagnets $\mu_r < 1$ holds as their susceptibility is $-1 < \chi_{DM} < 0$. With no external magnetic field applied, the net magnetic moment of each individual atom of the diamagnet is zero. Typical examples are materials with completely filled electronic shells (e. g. noble gases such as argon or neon) or semiconductors with strong covalent bonding (e. g. silicon, diamond). Only when a magnetic field is applied a magnetic moment arises from the precession of the electron orbits around the direction of the field. The magnetic moment is proportional to the field strength according to eq. (2.1), but in the opposite direction of the field, with $\chi_{DM} < 0$. On a side note: Diamagnets with $\chi_{DM} = -1$ form the special class of ideal superconductors. In the superconducting state, usually accessible only at very low temperatures close to $T = 0 \, \text{K}$, these materials expel the magnetic induction B from their interior and pass electric currents with zero resistance. Note that diamagnetism is present in all known materials but is weak compared to the other types of magnetism (s. below). Hence it is usually not detectable experimentally in para- or ferromagnets as it is dominated by the response of the atomic magnetic moments to the applied field.

Paramagnetism (PM)

Paramagnets follow eq. (2.1) with $\mu_r > 1$ and thus $0 < \chi_{PM} < 1$. PM is observed in metals like aluminum, gold and copper [15]. On a microscopic scale, paramagnets differentiate from diamagnets as the magnetic moment of each atom is not equal to zero in the absence of an external magnetic field. When the magnetization in zero field is measured however, DM and PM are indistinguishable and both show $|M| = 0$. Even though paramagnets have magnetic moments for each atom, they add up to zero net magnetization in the ensemble average at room temperature, since the thermal energy is sufficiently high for the magnetic moments to orientate randomly due to thermal fluctuation. Applying a weak magnetic field aligns some of the magnetic moments, resulting in a small net magnetization. As the field is increased, more and more magnetic moments align along the direction of the field, usually following a linear increase in M (corresponding to a constant susceptibility χ, eq. (2.1)). If the temperature is increased at constant field instead, the additional thermal fluctuation causes a decrease in M and in χ. Generally, the susceptibility χ is inversely proportional to the temperature T, described by Curie's law of paramagnetism, reading:

$$\chi_{PM} = C/T, \tag{2.5}$$

with $C = \frac{\mu_0 N |m|^2}{3k_B}$, where N denotes the number of magnetic moments in the entire paramagnetic ensemble, each with the magnetic dipole moment magnitude $|m|$. Note that eq. (2.5) is only valid for $T < T_C$, where T_C is the Curie temperature, an experimentally determined parameter characteristic for a specific paramagnetic material.

Paramagnets can be described in theory by assuming equal magnetic moments for each atom that only interact with the field H, but not with each other, and have a total angular momentum quantum number J. The magnetization of such an idealized paramagnet with N atoms is mathematically described by

$$M = N \cdot g\mu_B J \cdot B_J(x) = N \cdot m \cdot B_J(x). \tag{2.6}$$

With the Brillouin function

$$B_J(x) = \frac{2J+1}{2J} \cdot \coth\left(\frac{2J+1}{2J} \cdot x\right) - \frac{1}{2J} \cdot \coth\left(\frac{x}{2J}\right) \tag{2.7}$$

and $x = \dfrac{m \cdot H}{k_B T}$.

$B_J(x)$ is a function of temperature, T, and the Boltzmann constant, $k_B = 1.3806 \cdot 10^{-23}$ J/K. For small fields $H \ll 1$, and thus $x \ll 1$, eq. (2.7) can be expanded in a Taylor series reading to first order approximation $B_J(x) \approx \frac{J+1}{3J} \cdot x + O(x^3)$. Inserting this Taylor series in eq. (2.6) yields the linear relationship generally observed for paramagnets $M \propto H$ (cf. eq. (2.1)).

Ferromagnetism (FM)

A few materials (primarily iron, nickel, cobalt and their alloys [15]) display a permanent magnetization even in the absence of an applied external field. They are called ferromagnets. With $\mu_r \gg 1$, ferromagnets show the highest magnetization of all magnetic materials, when exposed to an external field H. FM is caused by the individual atoms' spins that interact with each other via a strong but short-ranged interaction, aligning neighboring spins in parallel arrangement. This so-called exchange interaction causes a permanent spin order within ferromagnets and leads to a permanent magnetization.

Ferromagnetic Hysteresis Ferromagnetic materials do not fulfill eq. (2.1) and exhibit a previous-magnetization-dependent response when exposed to H, called *hysteresis*. A typical hysteresis curve (loop) is depicted in Fig. 2.1, where the magnetization M is plotted versus the applied external field H. The enveloping loop can be recorded by applying a sufficiently strong

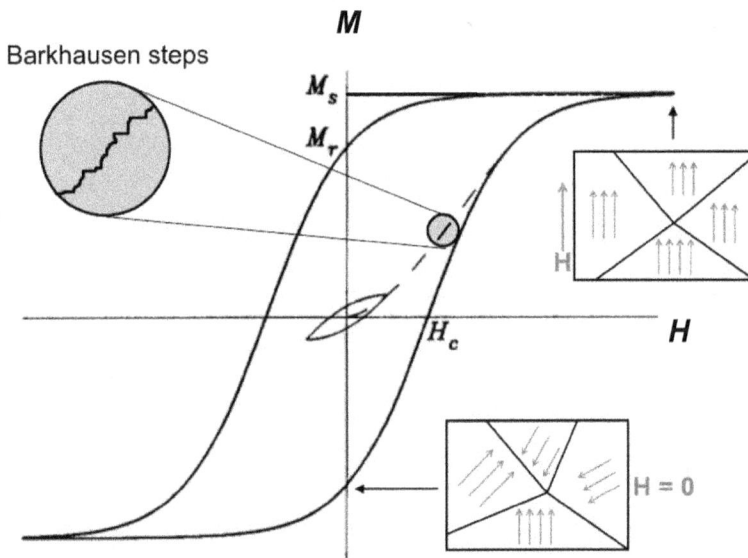

Fig. 2.1.: Schematic diagram of a hysteresis curve $M(H)$. The enveloping loop marks the major hysteresis loop, reaching saturation magnetization, M_S, for large fields $|H| > H_C$. $M = 0$ when the coercivity field, $|H| = H_C$, is applied, whereas for $H = 0$ a permanent magnetization called remanence, M_r, remains. The inner curve shows a minor loop for small H. The dashed line markes the virgin curve. The mircoscopic regions of uniform spontaneously aligned magnetization, called domains, are marked at two destinct points (at positive saturation for large H and at $H = 0$). Barkhausen steps arise from irreversible domain wall motions on the microscale (shown in zoom). Adapted from [13].

field to align all magnetic moments with the field and reach the saturation magnetization M_S. When H is reduced to zero a permanent magnetization $|M(H = 0)| = M_r$, the so-called remanence, remains within the ferromagnet. Now applying H in the opposite direction, a certain field strength is required to reset $M = 0$. This field is named coercivity H_C. Both segments of the hysteresis curve, going from $|H| = 0$ to $|H| = \pm H_C$ are denoted as demagnetizing curves. By applying an equally strong and saturating field in the opposite direction, the lower tail of the loop can be traced. The entire outer loop is named major (hysteresis) loop. The loops are actually a whole continuum of curves enveloped by the major loop. The smaller curves do not reach M_S for applied fields below coercivity, $|H| < H_C$ and are thus denoted as minor hysteresis loops. By adapting $|H|$ appropriately, every remanent magnetization value between $-M_r r < |M| < +M_r$ can be selected. Ferromagnets can lose their permanent (remnant) magnetization, meaning to set $M = 0$ for $H = 0$, either by cycling the applied field with steadily decreasing amplitude or heating the ferromagnet above the critical Curie-temperature $T > T_C$ (e. g. $T_C \approx 1\,041\,K$ for Fe). Above T_C the spin order characteristic for FM is broken by thermal fluctuation and the ferromagnet behaves like a paramagnet with $\mu_r > 1$. A third way of demagnetizing a ferromagnet is to mechanically break the spin order, e. g. by forcefully throwing it to the ground. However, this methods holds a high risk of damaging the ferromagnet permanently.

Magnetic Domain Formation A qualitative description of the magnetic properties of ferro-magnets was first introduced by Pierre-Ernst Weiss in 1907. Weiss stated that atomic magnetic moments (later named spins, as is also the denotation throughout this thesis) align against thermal fluctuation in areas within the ferromagnet and he termed those areas of aligned spins *(magnetic) domains* [16]. Within a single domain, some exchange interaction (definded below, cf. eq. (2.8)) between spins aligns them parallel to each other. Between domains, however, the direction of the magnetization vector varies, if no external field is applied. Thus, when the bulk magnetization over many domains is measured in the absence of a field, the superposition of the randomly oriented domain magnetization vectors yields $|M| < M_S$, as not all vectors are aligned. An external field H forces the domain walls to shift position. As some of these positions lead to preferred local minima in the total energy of the crystal, the random shift of the walls causes a discontinuous change in magnetization on the microscale, known as Barkhausen steps [17]. Consequently, the magnetization in a gradually increasing field occurs in discontinu-ous steps on the microscale as sketched in Fig. 2.1. A sufficiently large field ($|H| \geq H_C$) aligns all domains and the full saturation magnetization M_S is reached. When the field is reduced after saturation, the domains remain in their newly preferred orientation and the overall addition of domain magnetization vectors yields $|M| = M_r$ for $H = 0$.

The formation of domains originates from the interaction between exchange and anisotropy en-ergies within a ferromagnetic crystal, which are explained briefly in the following: The (Heisen-berg) exchange interaction is the short-ranged (between next neighbors) but strong quantum mechanical mechanism between atomic spins that aligns these spins parallel to each other. Its energy, ε_{ex}, can be summed over all atoms in the crystal pair by pair, i, j, reading

$$\varepsilon_{ex} = -2 \cdot \sum_{i \neq j} J_{ij} S_i \cdot S_j, \tag{2.8}$$

with the factors J_{ij}, the so called exchange integrals, describing the strength and range of the interaction. J_{ij} is accessible by experiment from the Curie temperature, T_C, where the magnets saturation magnetization becomes zero and therefore T_C is a natural measure for the strength of the atomic exchange interaction.

Anisotropy within ferromagnets arises from the spin-orbit interaction, which mainly couples the electronic orbits to the crystallographic structure. Further contributions to anisotropy arise from geometrical deviance from spherical symmetry and spin-canting effects at the surface of nanosized ferromagnets. These different contributions to anisotropy are discussed in detail in the next Section 2.1.2. For now, simply consider the total effective anisotropic energy ε_{ai} resulting from these contributions (cf. eq. (2.21) for a mathematical definition). The anisotropy energy is generally much lower than the exchange energy $\varepsilon_{ex} \gg \varepsilon_{ai}$. But since the exchange interaction is always isotropic in space, the direction of magnetization M in a domain is exclusively determined by anisotropy energy ε_{ai}. The preferred directions of magnetization within the crystal, i. e. those directions, where ε_{ai} is minimal, are named easy axes

Both the exchange and anisotropy energies compete with a third energy, named magnetostatic or demagnetizing energy, ε_{ms}, whose energy reads

$$\varepsilon_{ms} = -\frac{\mu_0}{2} \int M \cdot H_d \, dV. \tag{2.9}$$

with the magnetization M and the demagnetizing field H_d, while the integration is over the whole volume of the ferromagnetic body, V and the factor $\frac{1}{2}$ corrects the otherwise twofold counting of the interaction between two atoms. ε_{ms} is a complicated function of the domain geometry and therefore also denoted shape anisotropy (cf. Section 2.1.2). The magnetostatic interaction is long-ranged compared to the exchange interaction, spanning many atoms or several hundreds of nanometers. The energy ε_{ms} is stored in a magnetic stray field (s. Fig. 2.2). The

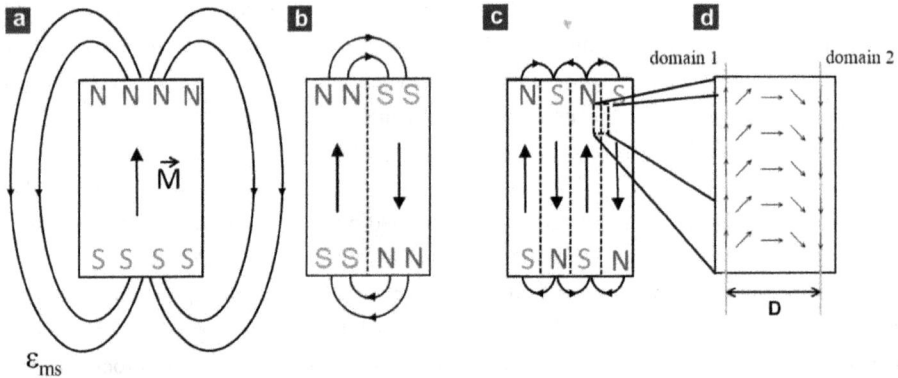

Fig. 2.2.: Magnetic domains and domain wall formation: (a) shows a single domain with the magnetization $\vec{M}$ directed from magnetic south S to magnetic north N. The stray field is large and the magnetostatic energy ε_{ms} is maximal. In comparison, in (b) two domains and even four in (c) reduce the stray field and consequently ε_{ms} remarkably. But this reduction comes at the expense of an increase in exchange energy ε_{ex}, which causes a gradual reversal of the spin direction between domain 1 and domain 2 across the length D, as shown in (d). Adapted from [18].

three energies, ε_{ex}, ε_{ai} and ε_{ms} allow for a deeper understanding of the formation of domains in a ferromagnetic crystal, since the magnetization direction, strength and area of the domains are governed by minimizing the net energy

$$\varepsilon_{net} = \varepsilon_{ex} + \varepsilon_{ai} + \varepsilon_{ms} \tag{2.10}$$

with $\varepsilon_{ex} \sim \varepsilon_{ms} \gg \varepsilon_{ai}$.

The interplay between energy minimization of ε_{ex} by aligning spins parallel and ε_{ms} by reducing the stray field causes the formation of magnetic domains, as shown in Fig. 2.2. Furthermore, ε_{ex} is responsible for the gradual change of the magnetization vector M between domains, called domain walls, cf. Fig. 2.2d. Here, ε_{ex} aims to keep the angular separation between neighboring spins minimal, which increases the wall thickness. ε_{ex} competes with the anisotropy energy ε_{ai}, to which the spins favorably align along the easy axes (directions of easiest magnetization, s. sec. 2.1.2 for details) and thus forms comparatively thinner walls. The actual wall thickness, D, is a result of the equilibrium between exchange and anisotropy energy.

Antiferromagnetism (AFM)

In antiferromagnets the neighboring spins show a strong but negative (i. e. repulsive) interation, therefore they align anti-parallel to each other. The spins, all originating from a single magnetic species, are distributed in equal numbers on two interpenetrating, antiparallel crystal lattices. Consequently, they add up to zero net magnetization in the absence of an applied field. Antiferromagnetic materials typically show paramagnetic properties at room temperature but display a complex behavior in susceptibility χ below a critical temperature, named the Néel temperature T_N: For $T < T_N$, antiferromagnetic spins align in the characteristic antiparallel spin order and χ depends on whether H is applied perpendicular or parallel to the orientation of the spins. Common AFM examples are cobalt(II)oxide (CoO), nickel(II)oxide (NiO) and wustite (FeO) [15].

Ferrimagnetism (FeM)

In materials with two or more magnetic species, their different magnetic moments and sub-lattices give rise to a strong and negative interaction between the individual species' spins, leading to an anti-parallel arrangement of these spins. In contrast to AFM, neighboring atoms have different magnitudes of magnetic moments. Therefore, ferrimagnets show a non-zero (permanent) net magnetization in zero magnetic field below a critical temperature. As this magnetization is permanent as in ferromagnets, this critical temperature is also denoted Curie temperature, T_C. For $T > T_C$ the spin order of FeM is broken by thermal fluctuations in the same way as described above for FM. Typical ferrimagnetic materials are oxides, including iron-oxides maghemite (γ-Fe_2O_3) and magnetite (Fe_3O_4), which will be used throughout this thesis.

2.1.2. Magnetic Anisotropy

Experiments reveal that the magnetization of a ferromagnet (or a antiferro- or ferrimagnet for that matter) has a preferential orientation along certain internal directions. These preferential directions depend on the crystallographic structure, the magnet geometry, mechanical stress within the magnetic material or — in the case of small magnetic objects — spin-canting at the surface. These four dependencies are referred to as *magnetocrystalline* (with energy ε_{mc}), *shape* (or magnetostatic) (ε_{ms}), *stress* (or magnetoelastic) (ε_{me}), and *surface anisotropy* (ε_{sur}), respectively. They all contribute to the net anisotropy energy, ε_{ai}, reading

$$\varepsilon_{ai} = \varepsilon_{mc} + \varepsilon_{ms} + \varepsilon_{me} + \varepsilon_{sur}. \tag{2.11}$$

and will be discussed individually briefly in the following. Note that the general description will be discussed in more detail for the special case of spherical magnetic nanoparticles, where appropriate.

Magnetocrystalline Anisotropy

Every magnet exhibits magnetocrystalline anisotropy, which arises from spin-orbit interaction: The symmetrical order of atoms in the crystal lattice generates electrostatic fields that couple with the electronic orbits, which in turn are interacting with the electronic spins. This results in preferential directions of magnetization. Directions, in which saturation magnetization is reached easiest (i.e. at lowest applied fields) are called easy axes and those directions along which magnetizing is most difficult, hard axes. Depending on the crystallographic structure of the magnetic material, generally, two different formulations of magnetocrystalline anisotropy — *uniaxial* and *cubic* anisotropy — are discussed:

Uniaxial Anisotropy In crystals with only one easy axis, e.g. hexagonal crystals such as cobalt, the magnetocrystalline energy is isotropic for any given angle, φ, in the plane perpendicular to the easy axis. The anisotropy of such crystals is therefore *uniaxial* and the magnetocrystalline anisotropy energy is given by:

$$\varepsilon_{mc}^{u} = K_u V \cdot \sin^2(\theta). \tag{2.12}$$

Where K_u denotes the temperature-dependent uniaxial anisotropy constant, which is usually determined experimentally,[1] V is the sample volume and θ describes the angle between the easy axis and the direction of magnetization. Note that eq. (2.12) is an approximation, only taking into account the dominant contribution to angle-dependent anisotropy (since the direction of magnetization inside a magnet is exclusively determined by the angular dependency of ε_{ai}, as stated in Section 2.1.1).

Assuming spherical magnetic nanoparticles with a particle magnetic moment $\boldsymbol{m}$ (equivalent to the magnet's magnetization, $\boldsymbol{M}$, mentioned above), eq. (2.12) simplifies further to:

$$\varepsilon_{mc}^{u} = K_u V \cdot (\boldsymbol{m} \cdot \boldsymbol{n})^2, \tag{2.13}$$

---

[1]Anisotropy constants of thin film materials are for example determined by angle-dependent ferromagnetic resonance (FMR) measurements [19]. However, magnetic nanoparticles, as used throughout this thesis, are very challenging to prepare for FMR measurements.

where θ marks the angle between m, and the particle easy axis, n, and further assuming that $m, n \in K$, where K is the unit sphere.

Cubic Anisotropy Cubic crystals (such as iron and nickel, but also iron-oxides as used in this thesis) posess more than one easy axis of magnetization. Therefore, their magnetocrystalline energy, ε_{mc}^c, depends on the direction that the magnetization vector M makes with the principal crystallographic axes of the crystal [14]: These directions can be expressed as the directional cosines of M, $\alpha_1 = \cos(a)$, $\alpha_2 = \cos(b)$ and $\alpha_3 = \cos(c)$, defined as shown in Fig. 2.3. Using

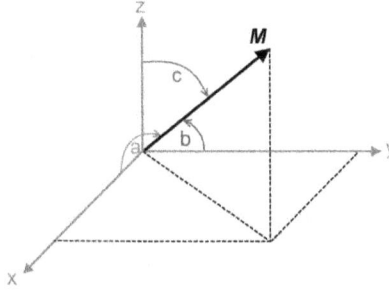

Fig. 2.3.: Sketch of the directional cosines of the magnetization M: $\alpha_1 = \cos(a)$, $\alpha_2 = \cos(b)$ and $\alpha_3 = \cos(c)$. Adapted from [14].

these directional cosines, ε_{mc}^c reads:

$$\varepsilon_{mc}^c = K_{c_1}V \cdot (\alpha_1^2\alpha_2^2 + \alpha_2^2\alpha_3^2 + \alpha_3^2\alpha_1^2) + K_{c_2}V \cdot \alpha_1^2\alpha_2^2\alpha_3^2, \tag{2.14}$$

with the sample volume V and the cubic anisotropy constants K_{c_1} and K_{c_2}. As explained for uniaxial anisotropy already, the angle-independent terms were neglected again. Typical values are $K_{c_1} \approx 48\,\mathrm{kJ/m^3}$ and $K_{c_2} \approx 5\,\mathrm{kJ/m^3}$ for iron [14] and $K_{c_1} \approx -11\,\mathrm{kJ/m^3}$ and $K_{c_2} \approx -3\,\mathrm{kJ/m^3}$ for magnetite at room temperature [20]. The directions of the easy axes in a cubic crystal depend on the sign of K_{c_1} and K_{c_2}, as depicted in Fig. 2.4. Knowing that $K_{c_1} < 0$ and $K_{c_2} < 0$ holds for magnetite at room temperature [20], one directly sees from Fig. 2.4 that the easy axis aligns along the [111]-direction in magnetite.

The Anisotropy Field The forces binding the magnetization to the easy axes can be expressed in terms of a pseudo-field, denoted *anisotropy field*, H_K. This field is parallel to the easy axes and defined by the torque, $\tau = \mu_0 M \times H_K = \frac{d\varepsilon_{mc}}{d\theta}$, on the magnetization M with the angle between easy axes and M, θ. For small angles $\sin(\theta) \approx \theta$ and with $|M| = M_S$ the saturation magnetization, a general approximation for the magnitude of the anisotropy field yields

$$|H_K| = H_K = \frac{2K_i}{\mu_0 M_S}, \tag{2.15}$$

with $i = u, c$, for uniaxial or cubic anisotropy.

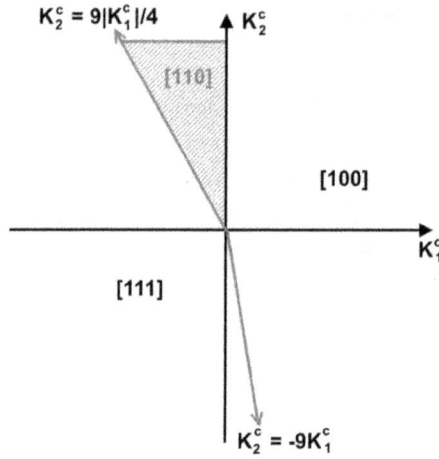

Fig. 2.4.: Directions of magnetization along easy axes for cubic crystals as a function of the anisotropy constants K_{c_1} and K_{c_2}. The direction [110] is shown in red (holds for the shaded region). Adapted from [14].

Shape Anisotropy

Shape anisotropy arises from the divergence in the magnetization at the surface of a magnet, generating a stray or demagnetizing field, as described previously in eq. 2.9. In the following, a solution to the integration for ε_{ms} (cf. eq. (2.9)) is presented for the case of a prolate spheroid: Assuming the magnetization in the x-z-plane with $M_S^x = M_S \cdot \sin(\theta)$ and $M_S^y = M_S \cdot \cos(\theta)$, the angle between the direction of magnetization and the easy axis is denoted as θ. If the dimensions of the spheroid are elongated along the z-axis, with $l_x = l_y < l_z$ and the demagnetization factors N_x, N_y and N_z are known, ε_{ms} can be expressed by

$$\varepsilon_{ms} = \frac{1}{2}\mu_0 N_z M_S^2 V + \frac{1}{2}\mu_0 (N_x - N_z) M_S^2 V \cdot \sin^2(\theta) := K_0 V + K_{sh} V \cdot \sin^2(\theta). \quad (2.16)$$

The easy axis of a prolate spheroid determined by shape anisotropy is therefore governed by the strength of the shape anisotropy constant K_{sh} with respect to the other anisotropy constants, while K_0 can be neglected as it has no angle dependency.

Stress Anisotropy

A magnetic material subjected to a sufficiently strong magnetic field changes its dimensions, which changes of overall volume, V. This places mechanical stress (or strain) on the crystal lattice, which in turn induces preferential directions for the magnetization vector known as stress or magnetoelastic anisotropy. The magnetoelastic energy of a cubic crystal under stress can be described in terms of the elastic stress σ_s and the magnetostriction constant λ_s:

$$\varepsilon_{me} = \frac{3}{2} \int \sigma_s \cdot \lambda_s dV \quad (2.17)$$

integrated over the entire magnetic volume, V. For isotropic crystals and uniform stress, eq. (2.17) simplifies to

$$\varepsilon_{me} = K_{me} \sin^2(\vartheta), \tag{2.18}$$

with ϑ the angle between magnetization, M, and the applied stress direction and $K_{me} = \frac{3}{2}\sigma_s \cdot \lambda_s$. Typical values of magnetostriction constants are $\lambda_{s_{Fe,bcc}} = -7$ for polycrystalline iron and $\lambda_{Fe_3O_4} = 40$ for polycrystalline magnetite [14].

Surface Anisotropy

Nanostructured magnetic objects (such as magnetic nanoparticles), which are characterized by a significant surface to volume ratio (and therefore a relatively large number or surface atoms), can additionally experience surface anisotropy. Here, the deviations from crystal symmetry at the surface due to structural defects, broken exchange bonds and surface strain can dominate the magnetocrystalline and shape anisotropy effects [21]. The effective anisotropy increases as the size of spherical nanomagnets decreases [22], which is described very well by an effective anisotropy constant, adding bulk (core) anisotropy, K_B, and surface anisotropy, K_S, contributions:

$$K_{eff} = K_B + \frac{6}{d}K_S, \tag{2.19}$$

where d is the particle diameter. The surface anisotropy energy arises solely from K_S and reads

$$\varepsilon_{sur} = K_S V \frac{6}{d}. \tag{2.20}$$

Total Anisotropy Energy of Spherical Nanomagnets

Throughout this thesis, spherical nanomagnets, so-called *magnetic nanoparticles*, are considered, whose high geometric symmetry influence the anisotropy contributions as follows: For ideally spherical nanoparticles, shape anisotropy contributions average to zero, therefore, $\varepsilon_{ms} = 0$. Note that the same argument should hold for surface anisotropy contributions. However, in reality, the surface roughness and spin disorder of the nanostructure of nanoparticles induce an effective surface anisotropy, $\varepsilon_{sur} \neq 0$. Lastly, the magnetoelastic anisotropy contributions are very small due to the nanoparticle small size and are therefore neglected, $\varepsilon_{me} \approx 0$. Hence, the net anisotropy energy reduces to

$$\varepsilon_{ai} \approx \varepsilon_{mc} + \varepsilon_{sur}, \tag{2.21}$$

which can be rewritten in terms of eqs. (2.13) and (2.14) (depending on uniaxial or cubic crystal structure), containing the modified effective anisotropy constants, combining bulk and surface anisotropy contributions from eq. (2.19):

$$\varepsilon_{ai}^{u} = K_{eff,u} V \cdot (\boldsymbol{m} \cdot \boldsymbol{n})^2, \quad \text{with} \quad K_{eff,u} = K_u + \frac{6}{d} K_S \tag{2.22}$$

and

$$\varepsilon_{ai}^{c} = K_{eff,c_1} V \cdot (\alpha_1^2 \alpha_2^2 + \alpha_2^2 \alpha_3^2 + \alpha_3^2 \alpha_1^2) + K_{eff,c_2} V \cdot \alpha_1^2 \alpha_2^2 \alpha_3^2, \tag{2.23}$$

$$\text{with} \quad K_{eff,c_1} = K_{c_1} + \frac{6}{d} K_S \quad \text{and} \quad K_{eff,c_2} = K_{c_2} + \frac{6}{d} K_S.$$

2.1.3. The Origin of Superparamagnetism

When decreasing the dimensions of a ferromagnet (or antiferro- or ferrimagnet as well) below a certain size, it is energetically no longer efficient to form domain walls (cf. Section 2.1). Magnetic objects below this size will be called *magnetic nanoparticles* (MNP) or simply *(magnetic) particles* throughout this thesis. This critical size is $d < 100$ nm for typical MNP materials like Fe_3O_4, FePt or nickel [23].

For a quantitative description, one assumes a simple system of only next-neighbor exchange interaction and uniaxial anisotropy, eqs. (2.8) and (2.12), in which the energy is given by

$$\varepsilon = K_u V \sin^2 \theta - |\boldsymbol{m}| |\boldsymbol{H}| \cos(\theta). \tag{2.24}$$

Equation (2.24) describes the energy landscape in dependence of θ, the angle between the direction of magnetization and the easy axis. $\varepsilon(\theta)$ is plotted in Fig. 2.5 and has two minima: one at $\theta = 0$ with $\varepsilon_1 = -|\boldsymbol{m}||\boldsymbol{H}|$ and the other one at $\theta = \pi$ with $\varepsilon_2 = +|\boldsymbol{m}||\boldsymbol{H}|$. The one maximum is at $\theta = \arccos(-|\boldsymbol{m}||\boldsymbol{H}|/(2K_u V))$ with $\varepsilon_m = K_u V \cdot (1 + (|\boldsymbol{m}||\boldsymbol{H}|/(2K_u V))^2)$. At

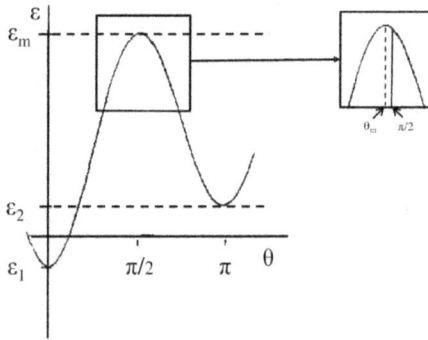

Fig. 2.5.: Energy ε for spherical magnetic particles with uniaxial anisotropy. The two minima at $\theta = 0$ and $\theta = \pi$ are separated by an energy barrier caused by anisotropy with the maximum ε_m at $\theta_m = \arccos(-|\boldsymbol{m}||\boldsymbol{H}|/(2K_u V))$ (s. inset for zoom). Adapted from [18] and inspired by [13].

thermal equilibrium the magnetization, M, resides in the vicinity of the minima $\varepsilon_{1,2}$, separated by the energy barrier $\Delta E_{1,2} = \varepsilon_m - \varepsilon_{1,2}$. The initial position of the particle (whether minimum

1 or 2) is determined by the initial direction of the magnetization. The magnetization can flip its direction if the thermal energy exceeds the energy barrier, e. g. when thermally activated. The rate of flips, ν, across the energy barrier from one minimum to the other is approximated by $\nu \propto \exp(-(K_u V(1 \pm (H_0 M_S)/(2K_u))^2)/(k_B T))$. ν is independent of the direction of the flips in the case of $H = 0$ (i. e. the energy barrier is of equal height, $\Delta E_1 = \Delta E_2$). The average time $\tau = \frac{1}{\nu}$ between one flip of M between the two minima due to thermal activation is called *relaxation time*:

$$\tau = \tau_0 \cdot e^\psi, \tag{2.25}$$

$$\text{with} \quad \tau_0 = \frac{M_S}{2K_u \gamma} \cdot \sqrt{\frac{\pi}{\psi}}, \ \psi = \frac{K_u V}{k_B T}$$

and the electron gyromagnetic ratio $\gamma_0 = 1.76 \cdot 10^{11}\,\text{rad} \cdot \text{Hz/T}$. Obviously, the relaxation time is exponentially dependent on the particle magnetic volume, $V = V_M = \frac{1}{6}\pi d_M^3$, with the particle magnetic size (diameter), d_M: Within a rather short range of decreasing d_M, the relaxation time changes rapidly from large to small values by several orders of magnitude.

Superparamagnetism can be observed if measuring the particle magnetization over a specific time t_{exp}: If $\tau \gg t_{exp}$, no change in magnetization is observed during the experiment. However, if $\tau \ll t_{exp}$ holds, the magnetization vector flips back and forth many times during the experiment, as thermal fluctuations dominate anisotropy effects. Therefore for $H = 0$, the measured value of magnetization averaged over the time t_{exp} is zero. Whereas for a sufficiently high field the magnetization M aligns with the applied field against the preferential direction defined by anisotropy, i. e. $H \geq H_K$ (cf. eq. (2.15)). As described in Section 2.1, M reaches saturation for $|H| > H_C$. The interaction energy between M and H can be described by simplifying eq. (2.24) to $\varepsilon = -|m||H|\cos\theta$ (cf. eq. (2.38) in the following Section). Hence, the magnitude of magnetization M_H parallel to an applied field is described as the average of the angle θ over many particles

$$|M_H| = |M| \cdot \langle \cos\theta \rangle = M_S \cdot L(\xi), \tag{2.26}$$

where

$$L(\xi) = \coth(\xi) - \frac{1}{\xi} \tag{2.27}$$

$$\text{with} \ \xi = \frac{m \cdot H}{k_B T} = \frac{\mu_0 V_M M_S H_0}{k_B T}, \tag{2.28}$$

with the amplitude of the applied field H_0. $L(\xi)$ is called the *Langevin function*, which is also the limit of the Brillouin function (eq. (2.7)) for high spin quantum numbers, $S \to \infty$. Note that $L(\xi)$ is valid for non-interacting magnetic moments that are assumed to be homogeniously distributed inside the magnet (isotropic distribution) [24]. Furthermore, the Langevin function is defined under the assumption that MNP are always at thermal equilibrium, i. e. that the applied field vector, $\boldsymbol{H}$ is always parallel to the magnetization vector of the MNP, $\boldsymbol{M}$. This assumption is only true for static applied fields, as otherwise, the MNP show relaxation phenomena, which is why in the next Section 2.2 the MNP relaxation including thermal fluctuations are discussed in detail.

Such a magnet behaves like a ferromagnet without hysteresis but with a high saturation magnetization, as several thousand spins are aligned as a result of the single domain nature (compared to only single spins in paramagnets). This phenomenon of "loss of ferromagnetism" in small particles in dependence of the time-scale of the experiment and the temperature is therefore known as *superparamagnetism* (SPM), with 'super' referring to the high saturation and 'paramagnetism' to the paramagnetic behavior. The respective $M(H)$-curve is depicted in Fig. 2.6a.

The Blocking Temperature Superparamagnetism may also be induced by temperature (not just by the time-scale of the experiment, as described in the paragraph above): Assuming a typical value of $t_{exp} = 100\,\mathrm{s}$ for static magnetic measurements such as SQUID [14], one can see from the exponent in eq. (2.25), that a particle of known anisotropy constant K_u and known magnetic volume, $V = V_M$, shows superparamagnetic behavior with $\tau \ll t_{exp}$ for temperatures T high enough to induce thermal fluctuations of the particle's magnetic moment. Decreasing the temperature, superparamagnetism is lost and the particle's magnetic moment will be blocked for $\tau \geq t_{exp}$. The temperature where $\tau = t_{exp}$ holds is denoted as the *blocking temperature*, $T = T_B$. By measuring the sample magnetization as a function of temperature in zero-field-cool (ZFC) experiments, the characteristic blocking temperature of the whole particle size distribution can be estimated. Furthermore, the mean anisotropy energy barrier, $\Delta E = K_{eff} V_M$, can be estimated and from this the effective anisotropy constant K_{eff} can be calculated — provided that the particle magnetic volume V_M is known. Comparing T_B of MNP of nominally equivalent particle magnetic size can also give a qualitative interpretation of the effect of magnetic dipole-dipole particle interactions. For a ZFC measurement, the sample is cooled from its superparamagnetic state with zero field applied, freezing all particles' magnetic moments in random orientation. At low temperature (e.g. $T = 5\,\mathrm{K}$), a small magnetic field, e.g. $H = 1\,\mathrm{mT}/\mu_0$ ($\approx 800\,\mathrm{A/m}$), is applied and the sample magnetization, M is recorded under gradual temperature increase. As T increases, M increases as well, as more and more particles are passing over their blocking temperature and their magnetic moments align with the applied field. Above a certain temperature, $T > T_{max}$, M decreases again, as thermal fluctuations dominate the applied field, randomly orienting the particle magnetic moments again. From the peak in temperature, T_{max}, one can estimate the mean blocking temperature, T_B. In the complement field-cooled (FC) experiment, the small field, H, is applied at room temperature and the sample is consecutively cooled. As T decreases, M increases as the thermal fluctuations lessen and the particles' magnetic moments align with H upon blocking (freezing). M will continue to increase up to its saturation temperature, T_{sat}. For temperatures below the so-called branching temperature, $T < T_{bra}$, the ZFC and FC magnetization curves branch off, while for $T > T_{bra}$ the ZFC and FC curves superimpose (i.e. $M_{ZFC} = M_{FC}$), as all particles relax driven by thermal fluctuation. An exemplary ZFC-FC curve is depicted in Fig. 2.6b.

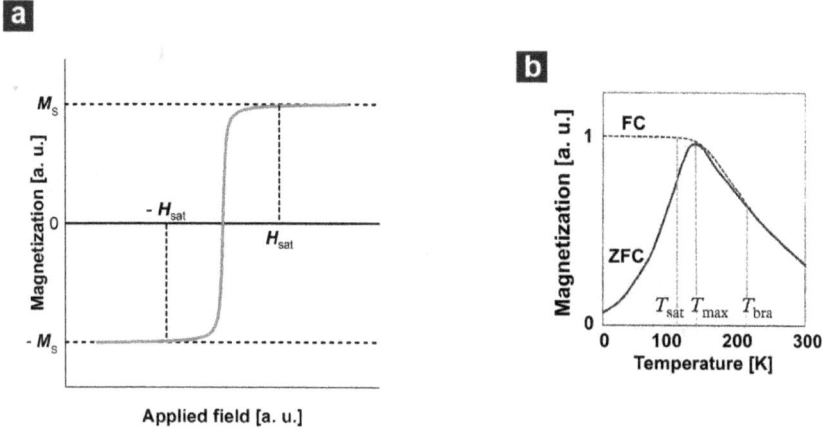

Fig. 2.6.: (a) magnetization curve of a superparamagnetic material; no hysteresis is measured. The saturation magnetization, M_S, is reached for applied fields $|H_0| > H_{sat}$. (b) Zero-field-cooled/field-cooled (ZFC-FC) magnetization curves; the ZFC and FC curves separate for $T \leq T_{bra}$, the ZFC curve peaks at $T = T_{max}$, and the FC curve saturates for $T \leq T_{sat}$. (b) adapted from [14].

2.2. Physics of Magnetic Particle Relaxation

When MNP are exposed to an external alternating magnetic field (AMF),

$$H_{ac}(t) = H_0 \cdot \cos(2\pi f \cdot t), \tag{2.29}$$

with the field amplitude, H_0, the frequency, f, and the time of exposure, t, they begin to rotate with the magnetic field. At the same time, their internal magnetic moment m with

$$|m| = \mu_0 M_S V_M, \tag{2.30}$$

where M_S is the saturation magnetization, $V_M = \frac{\pi}{6} d_M^3$ the particle magnetic volume and μ_0 the permeability of free space, flips with a certain probability dependent on thermal energy $\varepsilon_{therm} = k_B \cdot T$. This process of simultaneous rotation and flipping of the particle's magnetic moment is denoted as *relaxation*. In other words, there two ways for the magnetic moment to relax in a particle: *Brownian relaxation*, which describes the rotation of the entire particle relative to its surrounding and *Néel relaxation*, describing the internal rotation of the magnetic moment, while the particle itself remains stationary [23, 25] (s. Fig. 2.7). At zero-field, their contributions can be described via the characteristic Brownian and Néel relaxation times, τ_B and τ_N, respectively, which are given by:

$$\tau_B = \frac{\pi \eta}{2 k_B T} \cdot d_H^3 \tag{2.31}$$

$$\tau_{\text{N}} = \tau_0 \cdot \exp(\frac{\Delta E}{k_{\text{B}} T}) \tag{2.32}$$

with the anisotropy energy barrier ΔE (which is typically $\Delta E = K_{\text{eff}} \cdot V_{\text{M}}$, cf. eq. (2.25)), the viscosity of the carrier liquid η, and the particle hydrodynamic size d_{H}. As shown in the

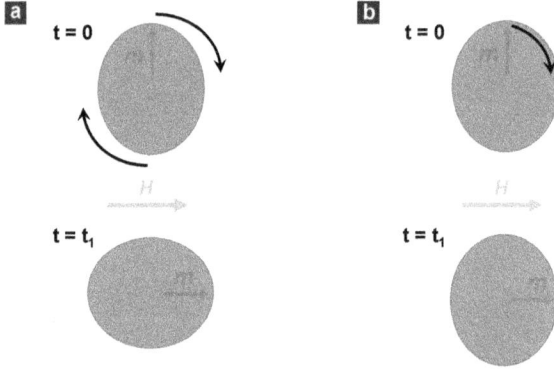

Fig. 2.7.: Schematic diagram of magnetic particle relaxation processes in the presence of a magnetic field $\boldsymbol{H}$, which is switched on at time $t = 0$: (a) particle rotation by revolving the entire particle relative to its surrounding while the magnetic moment $\boldsymbol{m}$ remains stationary, called Brownian relaxation. (b) relaxation of the magnetic moment $\boldsymbol{m}$, while the particle remains stationary relative to its surrounding, denoted as Néel relaxation. Note that $\boldsymbol{m}$ is aligned with $\boldsymbol{H}$ for both cases at time $t = t_1$. Note further that particles are drawn as prolate spheroids for emphasis of the relaxation mechanism; in reality, MNP are assumed to be spherical.

previous Section 2.1.3, the magnetization dynamics of MNP can be described in terms of the Langevin function $L(\xi)$, Eq. (2.27). However, $L(\xi)$ assumes an isotopic distribution of spins within MNP [24] and therefore does not account for magnetic anisotropy. Furthermore, it is valid only for thermal equilibrium and does not consider particle relaxation processes. Only if the change in the applied field is slow enough for the magnetization of the MNP to follow the field, i. e. the relaxation time τ is much smaller than applied field frequency, $\tau \ll 1/f$, relaxation effects can be neglected and $L(\xi)$ accurately describes the magnetization of the MNP [24]. Applications such as magnetic fluid hyperthermia and magnetic particle imaging, however, rely on relatively high field amplitudes $H_0 \sim 10\,\text{mT}$ and frequencies $f \sim 100\,\text{kHz}$, that force the MNP response to the applied field well into the non-equilibrium regime, where $\tau \sim 1/f$. Here, particle relaxation is driven by the applied field $\boldsymbol{H}$ and furthermore influenced by magnetic anisotropy as well as by thermal fluctuations, overall resulting in hysteresis effects. These relaxation contributions to the MNP magnetization can only be accurately characterized by combined Néel-Brownian rotation relaxation dynamics, which are described mathematically as follows: The internal magnetic moment of a particle, $\boldsymbol{m}_i$, rotates within the particle magnetic volume as described by the Landau-Lifshitz-Gilbert (LLG) equation [26,27]:

$$\frac{\mathrm{d}\boldsymbol{m}_i}{\mathrm{d}t} = \frac{\gamma_0}{1 + \alpha'^2} \cdot (\boldsymbol{H}_{\text{eff}} \times \boldsymbol{m}_i + \alpha' \cdot \boldsymbol{m}_i \times (\boldsymbol{H}_{\text{eff}} \times \boldsymbol{m}_i)), \tag{2.33}$$

with the electron gyromagnetic ratio γ_0, the (phenomenological) damping parameter $\alpha' \in [0, 1]$ and the effective field $\boldsymbol{H}_{\text{eff}}$ (defined below, eq. (2.35)). Similarly, the Brownian rotation dynamics can be described in terms of a generalized torque [28], $\boldsymbol{\Theta}$, acting on the easy axes of a particle, $\boldsymbol{n}_i$, and depending on the surrounding fluid viscosity, η, and the hydrodynamic volume, $V_H = \frac{\pi}{6} d_H^3$:

$$\frac{d\boldsymbol{n}_i}{dt} = \frac{\boldsymbol{\Theta}}{6\eta V_H} \times \boldsymbol{n}_i. \tag{2.34}$$

These two differential equations, eqs. (2.33) and (2.34), describe the combined Néel-Brownian rotation relaxation dynamics for the general case of non-zero fields and at non-equilibrium conditions. The physics governing the relaxation process is encoded in $\boldsymbol{H}_{\text{eff}}$ and $\boldsymbol{\Theta}$, which are determined using the Helmholtz free energy $F = U - T \cdot \Sigma$ of the system, with the internal energy U, temperature T and entropy Σ. When considering monodisperse MNP, entropy is negligible, $\Sigma \approx 0$, so that the effective field and generalized torque read:

$$\boldsymbol{H}_{\text{eff}} = \frac{1}{\mu_0} \frac{\partial F}{\partial \boldsymbol{m}} \approx \frac{1}{\mu_0} \frac{\partial U}{\partial \boldsymbol{m}} \tag{2.35}$$

and

$$\boldsymbol{\Theta} = \frac{\partial F}{\partial \boldsymbol{n}} \times \boldsymbol{n} \approx \frac{\partial U}{\partial \boldsymbol{n}} \times \boldsymbol{n}. \tag{2.36}$$

The internal energy, U, includes contributions arising from the applied field $\boldsymbol{H}$ (ε_{Zee}), particle interaction ($\varepsilon_{\text{pp-IA}}$) and magnetic anisotropy (ε_{ai}):

$$U = \varepsilon_{\text{Zee}} + \varepsilon_{\text{pp-IA}} + \varepsilon_{\text{ai}}, \tag{2.37}$$

with the Zeeman term[2]

$$\varepsilon_{Zee} = -\boldsymbol{m} \cdot \boldsymbol{H}, \tag{2.38}$$

the magnetic dipole-dipole interaction exerted on an individual particle, with magnetic moment $\boldsymbol{m}_0$ (cf. eq. (2.30)) located at an arbitrary point R_0, by all other particles that are at the distance r_i away from R_0 and have the magnetic moment $\boldsymbol{m}_i$

$$\varepsilon_{\text{pp-IA}} = \sum_i \frac{\mu_0}{4\pi r_i^3} \left(\frac{3(\boldsymbol{m}_0 \cdot \boldsymbol{r}_i) \cdot (\boldsymbol{m}_i \cdot \boldsymbol{r}_i)}{r_i^2} - \boldsymbol{m}_0 \cdot \boldsymbol{m}_i \right), \tag{2.39}$$

and the magnetic anisotropy energy, ε_{ai}, as described above in eqs. (2.22) or (2.23) for uniaxial or cubic anisotropy, respectively.

This set of coupled equations, eqs. (2.33) - (2.37), describes the combined rotational relaxation dynamics fully deterministically. However, in order to include thermal fluctuation, a stochiastic

---

[2]The Zeeman energy describes the applied (external) field energy inside a magnetized body of volume V. Its more general form reads $\varepsilon_{\text{Zee}} = -\mu_0 \int \boldsymbol{M} \cdot \boldsymbol{H} dV$, which was solved here already for spherical MNP [14].

term must be introduced to eqs. (2.35) and (2.36), giving [29]:

$$\boldsymbol{H}_{\text{eff}} = \frac{1}{\mu_0} \frac{\partial U}{\partial \boldsymbol{m}} + \boldsymbol{H}_{\text{th}}, \tag{2.40}$$

and

$$\boldsymbol{\Theta} = \frac{\partial U}{\partial \boldsymbol{n}} \times \boldsymbol{n} + \boldsymbol{\Theta}_{\text{th}}. \tag{2.41}$$

These thermally generated fields, $\boldsymbol{H}_{\text{th}}$, and torques, $\boldsymbol{\Theta}_{\text{th}}$, are expressed as Gaussian-distributed with an approximately flat frequency distribution of noise (equivalent to white noise) defined under the initial conditions

$$H_{\text{th}}^i(t = 0) = 0 \quad \text{and} \quad \Theta_{\text{th}}^i(t = 0) = 0, \tag{2.42}$$

and with zero mean

$$\langle H_{\text{th}}^i(t) \rangle = 0 \quad \text{and} \quad \langle \Theta_{\text{th}}^i(t) \rangle = 0. \tag{2.43}$$

The magnitude of the thermal fluctuations is encoded in the averaged variances reading:

$$\langle H_{\text{th}}^i(t) H_{\text{th}}^j(t') \rangle = \frac{2k_{\text{B}}T}{\gamma_0 |m|} \frac{1 + \alpha'^2}{\alpha'} \delta_{ij} \delta(t - t'), \tag{2.44}$$

$$\langle \Theta_{\text{th}}^i(t) \Theta_{\text{th}}^j(t') \rangle = 12 k_{\text{B}} T \eta V_H \delta_{ij} \delta(t - t'), \tag{2.45}$$

with $i, j \in x, y, z$, Cartesian spacial coordinates. The magnitude of fluctuations depends on MNP parameters, as well as the Boltzmann constant, k_{B}, and the temperature, T. Furthermore, the magnitude is unambiguously defined in 3D-space (Kronecker-Delta function δ_{ij}) and in time (Dirac-Delta function $\delta(t - t')$). The introduction of thermal fluctuations by inserting eqs. (2.40) and (2.41) in eqs. (2.33) and (2.34) makes this a set of coupled *stochastic differentiable equations* (SDE), whose solution requires stochastic calculus schemes and numerical integration, discussed in detail in the next Chapter 3; esp. Section 3.2.1.

In practice, calculating the relaxation dynamics of a single particle is meaningless for predicting magnetization loops, $M(H)$, for MNP solutions of typically concentrations of $\sim 10^{13}$ individual particles per mL. Therefore, an ensemble of at least 1000 particles is usually simulated and by calculating the ensemble average magnetization for different applied fields, one can predict $M(H)$ for any arbitrary set of particles core or hydrodynamic sizes, anisotropy constants, saturation magnetizations and thermal fluctuations. This kind of numerically solving stochastic problems in the limit of large populations in an ensemble is commonly called a *Monte Carlo* method. Throughout this thesis, the simulations based on magnetic particle relaxation will therefore be referred to as *Monte Carlo (MC-)simulations*, henceforth. The implementation and extraction of data from such MC-simulated $M(H)$-loops is presented in detail in Chapter 3, Section 3.2.

2.3. Physics of Magnetic Particle Imaging

The relaxation of superparamagnetic MNP in an AMF described above in Sections 2.1.3 and 2.2 can be exploited for direct (i. e. positive contrast[3]) imaging of particles in magnetic particle imaging (MPI) [30]. In MPI, the derivative of the MNP magnetization $M(t)$, $M'(t) = dM(t)/dH$ is measured by a receive coil via electromagnetic induction (s. below). The MNP are excited by an external sinusoidal AMF of typical field amplitude $H_0 \leq 20\,\mathrm{mT}/\mu_0$ and frequency $f \sim 25\,\mathrm{kHz}$. MPI relies on the superparamagnetic non-linear response of $M(t)$, described by the Langevin function $L(\xi(H))$ (cf. eq. (2.27)) [31]: When a sufficiently high field amplitude is applied, $M(t)$ saturates and $M'(t)$ is zero. Contrastingly, $M'(t)$ will only be non-zero, at sufficiently small fields that do not saturate $M(t)$.

MPI uses a send coil to apply a sinusoidal alternating magnetic field (AMF), cf. eq. (2.29), to change the magnetization of MNP, $M'(t)$. By using Faraday's law of induction, $M'(t)$ can be measured from the voltage, $V(t)$, induced in a receive coil by a temporary change in the magnetic induction, dB/dt, as will be outlined in the following: For simplification, considering the receive coil as a single conductor loop spanning the surface S_{RC}, then the voltage at the end points of the loop reads according to Faraday's law in integral form:

$$V(t) = \oint_{\partial S_{\mathrm{RC}}} \boldsymbol{E}(\boldsymbol{l}) \cdot \mathrm{d}\boldsymbol{l} = -\frac{\mathrm{d}}{\mathrm{d}t}\Phi, \tag{2.46}$$

with the electric field, $\boldsymbol{E}(\boldsymbol{l})$, induced along the conductor loop of length l, by the magnetic flux, Φ, defined by

$$\Phi = \oint_{S_{\mathrm{RC}}} \boldsymbol{B}(\boldsymbol{r}) \cdot \mathrm{d}\boldsymbol{A}, \tag{2.47}$$

through the surface of the receive coil S_{RC}, the magnetic induction $\boldsymbol{B}$ and the differential vector $\mathrm{d}\boldsymbol{A}$ perpendicular to S_{RC}, as outlined in Fig. 2.8. Inserting eq. (2.47) in eq. (2.46) and knowing that $\boldsymbol{B} = \mu_0(\boldsymbol{H} + \boldsymbol{M})$ from Section 2.1, eq. (2.2), one acquires:

$$V(t) = -\mu_0\frac{\mathrm{d}}{\mathrm{d}t}\oint_{S_{\mathrm{RC}}} \boldsymbol{H}(\boldsymbol{r},t) \cdot \mathrm{d}\boldsymbol{A} - \mu_0\frac{\mathrm{d}}{\mathrm{d}t}\oint_{S_{\mathrm{RC}}} \boldsymbol{M}(\boldsymbol{r},t) \cdot \mathrm{d}\boldsymbol{A} = V_{\mathrm{AMF}}(t) + V_{\mathrm{MNP}}(t). \tag{2.48}$$

Eq. (2.48) shows that the (time-varying) alternating magnetic field (AMF), $\boldsymbol{H}(\boldsymbol{r},t)$, is directly picked up by the receive coil as the voltage $V_{\mathrm{AMF}}(t)$ and superimposed with the voltage $V_{\mathrm{MNP}}(t)$ induced by the MNP's change in magnetiaztion. The latter can be expressed as an integration over the sample volume, V, containing MNP within the receive coil by using the law of reciprocity

---

[3]Positive contrast is defined by the fact that the imaging signal is proportional to the amount of material generating the signal.

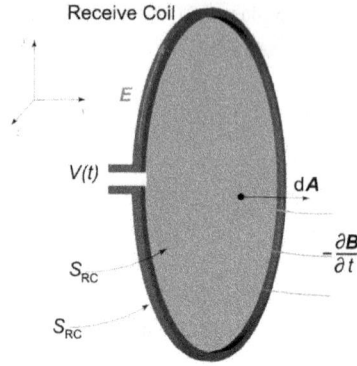

Fig. 2.8.: The voltage, $V(t)$, induced by a temporal change in the magnetic induction, $\boldsymbol{B}$, in a receive coil spanning the surface S_{RC}. The voltage is equal to the line integral of the electric field $\boldsymbol{E}$ along the receive coil loop. Adapted from [32].

[32], yielding:

$$V_{\mathsf{MNP}}(t) = -\mu_0 \int_V \boldsymbol{\zeta}(V) \cdot \frac{\mathrm{d}\boldsymbol{M}(V, t)}{\mathrm{d}t} \mathrm{d}V, \tag{2.49}$$

with the coil sensitivity $\boldsymbol{\zeta}$, summarizing the geometrical parameters of the coil. To discriminate the superimposed voltages one chooses a sinusoidal AMF for the MNP excitation, $H_{\mathsf{ac}}(t) = H_0 \cdot \cos(2\pi f \cdot t)$, cf. eq. (2.29): Then, the particle signal shows two distinct peaks, which are clearly distinguished from the excitation signal, as is shown in Fig. 2.9. Due to the sinusoidal AMF, the particle and excitation signal can be mathematically decoupled in the frequency domain by expanding into a discretized Fourier series [32]:

$$V(t) = \sum_{k=-\infty}^{\infty} \hat{V}_k \cdot \exp(2\pi \cdot ik \cdot f \cdot t). \tag{2.50}$$

Its spectrum consists of discrete lines at multiples of the frequency of the applied AMF, f:

$$f_k = k \cdot f, \qquad k \in \mathbb{Z}. \tag{2.51}$$

f_k will be denoted as *harmonics* throughout this thesis. Since $V(t)$ is real, the Fourier coefficients follow the relation:

$$\hat{V}_k = f \cdot \int_0^{1/f} V(t) \cdot \exp(-2\pi \cdot ik \cdot f \cdot t) \mathrm{d}t \tag{2.52}$$

$$= f \cdot \int_0^{1/f} (V(t) \cdot \exp(2\pi \cdot ik \cdot f \cdot t))^* \mathrm{d}t \tag{2.53}$$

$$= (\hat{V}_{-k})^*. \tag{2.54}$$

From eq. (2.54) it follows, that the negative frequencies do not carry any additional information on the MPI signal and are therefore neglected henceforth.

Since the excitation AMF is purely sinusoidal, the contribution of the excitation signal, V_{AMF},

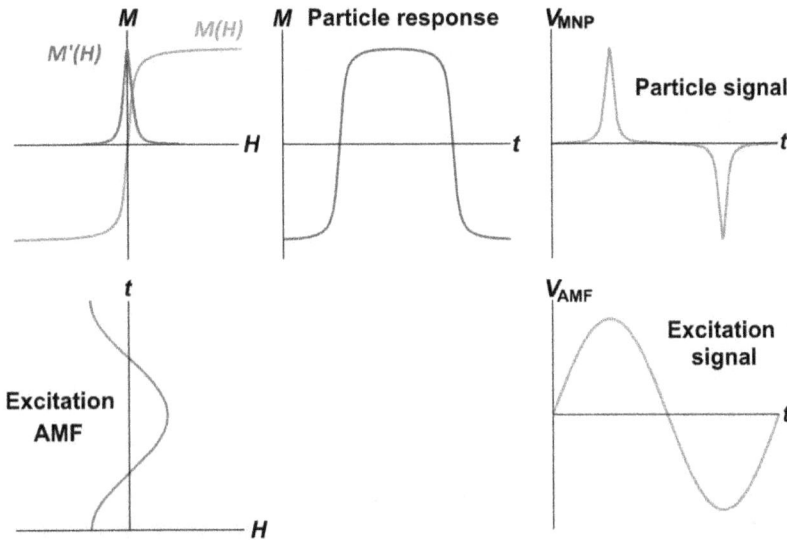

Fig. 2.9.: Schematic of the signal generation of MPI (reading from bottom left clockwise to bottom right): Applying a sinusoidal excitation field, H, drives the MNP magnetization M from negative saturation through zero to positive saturation according to the Langevin function $L(\xi)$. The derivative of MNP magnetization, $M'(H) = dM/dH$, is only non-zero for small field amplitudes. The MNP response to the applied field, $M(t)$, resembles a rectangular function. The voltage, V_{MNP}, induced in the receive coil by the MNP shows two distinct peaks as the particle signal, directly resulting from $M'(H)$, while the excitation signal resembles a sinusoidal function. Adapted from [32].

in frequency space is limited to a single peak at the frequency f. Contrastingly, the non-linear relationship $M(H)$ of the particles allows particle signal contributions not just for f but rather for all higher harmonics, f_k. This is illustrated in Fig. 2.10, where the excitation signal, the particles signal and their superposition is shown in time and frequency domain. From this one observes that the particle signal is hardly detected in the superposition signal in the time domain. This is due to the low amplitude of the particle signal compared to the excitation signal[4] [32]. In the frequency domain however, the particle signal can be easily seen in the superimposed signal spectrum for higher harmonics. Only the first harmonic, $f = f_{k=1}$, is covered by the excitation signal. By ignoring the first harmonic, one removes the influence of the excitation signal as well as any background signal. Therefore, all higher harmonics of the MPI signal carry only the undisturbed particle signal and can be used for imaging[5]. The fact that the even harmonics are missing in the signal spectrum a consequence of the non-linear MNP response (described by the Langevin function) to the sinusoidal excitation AMF, as derived in Appendix A.1.1.

To image a sample volume, MPI uses a field-free-point (FFP) to spatially encode the particle signal according to the non-linear magnetization particle response to the AMF. Since the

---

[4]For a typical MPI experiment the difference between V_{MNP} and V_{AFM} ranges from six to ten orders of magnitude, depending on the MNP concentration imaged, the coil sensitivities and the quality of the MNP in order to generate a good signal.

[5]The higher harmonics include all non-linear contributions to $M(H)$ in the scanned sample volume. This usually includes MNP only for an in vivo MPI scan, as e. g. bodily iron is present in the atomic form, therefore much smaller than the MNP and with a linear magnetization response for the considered AMF range [32].

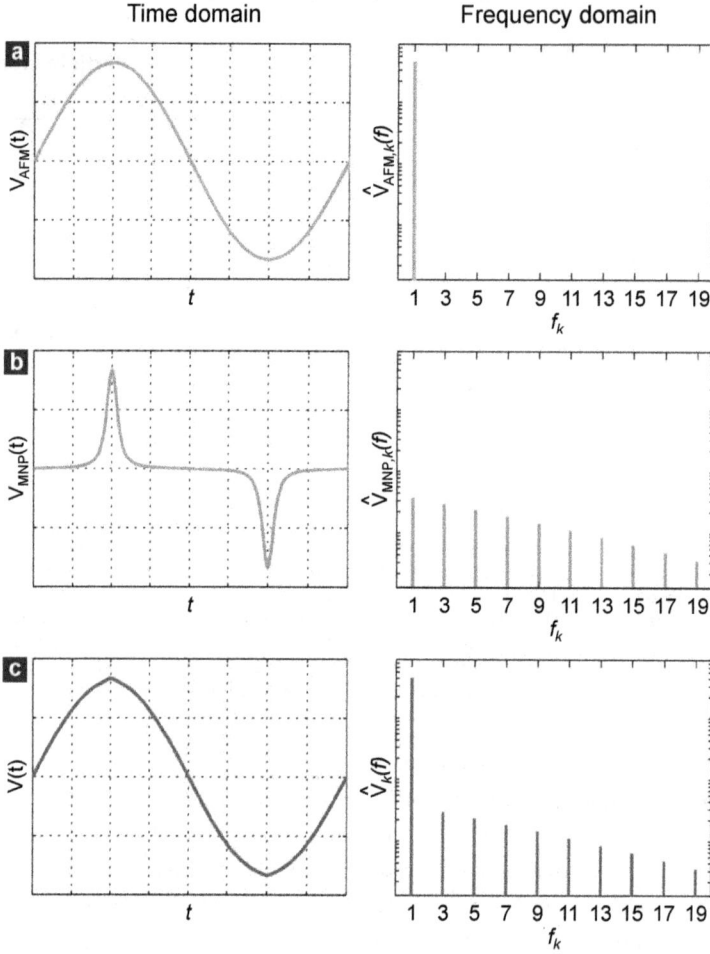

Fig. 2.10.: Comparison of the the signal induced in the receive coil for a sinusoidal AMF in time domain and frequency domain for (a) the excitation signal, (b) particle signal and (c) superposition of (a) and (b). Adapted from [32].

excitation AMF is known (cf. eq. (2.29)), eq. (2.49) can be expanded by $\frac{\mathrm{d}H}{\mathrm{d}H}$, reading:

$$V_{\mathrm{MNP}}(t) = -\mu_0 \int_V \boldsymbol{\zeta}(V) \cdot \frac{\mathrm{d}M(V,t)}{\mathrm{d}H(t)} \cdot \frac{\mathrm{d}H(t)}{\mathrm{d}t} \mathrm{d}V =: -\mu_0 \int_V \boldsymbol{\zeta}(V) \cdot M'(H(t)) \cdot H'(t) \mathrm{d}V. \quad (2.55)$$

Only considering contributions by the MNP, $V_{\mathrm{MNP}} \propto M'(H(t))$ holds. Since the magnetization of MNP is described by the Langevin function, $M(H(t)) \propto L(\xi(H(t)))$ (cf. eq. (2.27)), the particle signal is determined by the derivative of the Langevin function:

$$L'(\xi(H(t))) := \frac{\mathrm{d}L(\xi(H(t)))}{\mathrm{d}\xi} = (\xi^{-2} - \sinh^{-2}(\xi)), \quad (2.56)$$

with $\xi = \frac{\mu_0 V_M M_S H_0}{k_B T}$. $L'(\xi(H(t)))$ has a maximum at $H = 0 \Rightarrow \xi = 0$ with $L'(\xi = 0) = 1/3$ and therefore the particle response is maximal for $H = 0$. By scanning the FFP (or field-free line in more advanced systems) through a sample volume containing a MNP distribution, the signal induced in the receive coil is non-zero only where MNP are located at the FFP [33]. Around the FFP the MNP are saturated by a sufficiently high gradient field, and therefore do not generate a signal $V_{MNP} \propto M'(H(t))$. In this way, a MNP distribution can be conveniently mapped by gridding the particle signal to the location of the FFP (as depicted in Fig. 2.11).

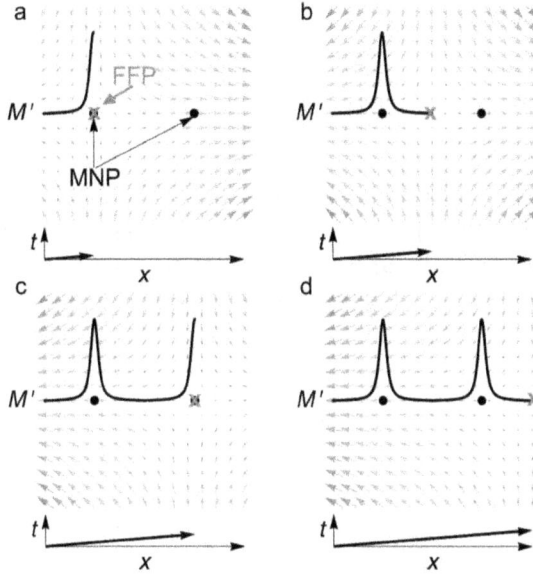

Fig. 2.11.: Characteristic response of MNP to a field-free point (FFP), generated by a combination of gradient fields and scanned along the x-axis at four different time points (a) through (d): The MNP response, $M'(H(t))$, is only non zero, when a distribution of MNP (symbolized as black point sources here) is crossed by the FFP. Note that $M'(H(t))$ is indicated as a black line here, showing the accumulated signal one line scanned by the FFP over time. Adapted from [34].

In practice, 2D- and 3D-MPI require complex trajectories of moving the FFP to record the particle signal efficiently [14, 32]. Furthermore, the raw MPI data must be mathematically post-processed to reconstruct an MPI image [35–38]. Both concepts go beyond the scope of this thesis, but are covered in more detail in the references cited above.

The suitablility of MNP as MPI tracers is determined by the magnitude of $M'(H = 0)$, $|M'(H = 0)|$ (i.e. the signal intensity or MPI signal-to-noise ratio (SNR), respectively) and the full width at half maximum (FWHM)[6] of $M'(H)$, ΔH_{FWHM}, (i.e. the spacial resolution) [34]. The signal intensity can be derived from $M(H) = M_S \cdot L(\xi(H))$ (s. eq. (2.26)) and the derivative of the

---

[6]The full width at half maximum (FWHM) is defined as the width at which a (peak) function drops to 50 % of its maximum value.

Langevin function (s. eq. (2.56)), yielding [32]:

$$|M'(H = 0)| = c \cdot \frac{|\boldsymbol{m}|^2 \cdot \mu_0}{3k_{\mathrm{B}}T}, \tag{2.57}$$

with the MNP concentration c, and particle magnetic moment $|\boldsymbol{m}| = \mu_0 \cdot M_{\mathrm{S}} \cdot V_{\mathrm{M}}$ (s. eq. (2.30)). The FWHM can be similarly approximated from the derivative of the Langevin function reading:

$$\Delta H_{\mathrm{FWHM}} = \frac{4.16 \cdot k_{\mathrm{B}}T}{\mu_0 \cdot |\boldsymbol{m}|}. \tag{2.58}$$

From eq. (2.57) one sees that the particle signal is directly proportional to the MNP concentration. Therefore, MPI is a quantitative, positive contrast imaging technique, as the signal is directly related to the amount of the (MNP) tracer.

If a MPI scanner is not accessible easily, the suitablity of MNP as MPI tracers described by the signal intensity (eq. (2.57)) and the spacial resolution (eq. (2.58)), can be assessed by magnetic particle spectoscopy (MPS) measurements. MPS is basically a zero dimensional MPI scan [23], where a sinusoidal field is applied, $H_{\mathrm{ac}}(t)$ cf. eq. (2.29), with typical frequencies in the range of $f = (3 - 25)\,\mathrm{kHz}$ [34]. Since the sample volume, V, and coil sensitivity, ζ, are accessible, the particle signal induced in the receive coil from eq. (2.55) simplifies to [39,40]:

$$V_{\mathrm{MNP}}(t) = -\mu_0 \cdot \zeta \cdot M'(H) \cdot H'(t), \tag{2.59}$$

with the MNP response $M'(H(t))$ and the derivative of the field $H'(t) = \mathrm{d}H(t)/\mathrm{d}t$. Rearranging eq. (2.59) for M' gives

$$M'(H(t)) = \frac{-1}{\mu_0 \cdot \zeta \cdot 2\pi f \cdot H_0} \cdot \frac{V_{\mathrm{MNP}}(t)}{\cos(2\pi f \cdot t)}. \tag{2.60}$$

Equivalently to an MPI scanner, the received signal, $V(t)$, is Fourier transformed $\mathcal{F}(V(t)) = \hat{V}_k(f_k)$ to derive the harmonic spectrum in the frequency domain [41]. Here, the first harmonic, $f = f_{k=1}$ is excluded usually as it carries the excitation signal (as discussed above) and to the remove background signal. Typically, the performance of MNP as MPI tracers is analyzed by comparing the MPS harmonic spectra: A higher absolute spectral magnitude indicates superior SNR of the MNP tracer (cf. eq. (2.57)) and a broader spectrum (i. e. smaller FWHM) indicates improved spacial resolution (2.58) [23,32].

2.4. Physics of Magnetic Particle Heating

As described in Section 2.2 above, an ensemble of MNP relaxes under the influence of an applied alternating magnetic field (AMF), $H_{\mathrm{ac}}(t)$, c. f. eq. (2.29). Both relaxation processes, Brownian and Néel, require magnetic energy to align the magnetic moment parallel to the direction of the applied field. During the field-driven relaxation processes, heat is generated equivalent to the magnetic energy consumed for this alignment. The heating rate is commonly referred to as the *specific loss power* (*SLP* [W/g]), defined as energy per unit time and per unit mass of nanoparticles [42]. It is defined in terms of the area of the hysteresis loop $M(H)$, denoted as A, the frequency f and the MNP materials' mass density ρ:

$$SLP = \frac{1}{\rho} \cdot A \cdot f \tag{2.61}$$

Two main theories are commonly used to describe the SLP on the basis of simple MNP magnetization models: The *Stoner-Wohlfarth Model Based Theory* (SWMBT), which approximates the hysteresis loop $M(H)$ for large particles in high fields (typically $d_M > 18\,\text{nm}$ and $H_0 > 20\,\text{mT}/\mu_0$ for magnetite) and the *Linear Response Theory* (LRT), based on the MNP linear relaxation approximation in low fields for small particles (typically $d_M < 10\,\text{nm}$ and $H_0 < 15\,\text{mT}/\mu_0$ for magnetite). In the following, SWMBT and LRT will be discussed briefly, building an understanding of the physics of magnetic particle heating.

Note that furthermore, a third way of simulating the hysteresis loop $M(H)$ within the framework of particle relaxation (s. previous Section 2.2) exists, which is generally not restricted to a certain field amplitude or particle size. These two theories will be compared to results from the third theory of simulating $M(H)$ in Chapter 3.

Stoner-Wohlfarth Model Based Theory of Magnetic Particle Heating

This theory is based on the Stoner-Wohlfarth model and describes the hysteresis losses for single-domain magnetic particles in the ferromagnetic regime with uniaxial anisotropy [43]. At $H_{ac} = 0$ and for fields below a certain amplitude (definded below), the magnetic moments of MNP are aligned along the easy axes due to anisotropy, which can be expressed in terms of an anisotropy field, H_K, eq. (2.15). The magnetic moments of the MNP remain aligned, until a sufficiently high field is applied, which can be either expressed in terms of the anisotropy field H_K or the coercivity H_C, (cf. Section 2.1): For $H_0 \geq H_C \approx \frac{H_K}{2}$, the magnetic moments align with the direction of the applied field $H_{ac}(t)$ and develop a hysteresis loop $M(H)$. In its original form, the Stoner-Wohlfarth model does not take thermal activation into account [44], thus is only valid for $T = 0$ or in the limit of infinite frequency $f \to \infty$.

Based on earlier works on the temperature dependency of the coercive field [45, 46], Carrey et al. [42] proposed an extension named Stoner-Wohlfarth model based theory for randomly oriented particles (SWMBT), calculating the hysteresis loop area, A, as:

$$A = 2\mu_0 \cdot M_S \cdot H_C(\kappa) \tag{2.62}$$

with the coercive field

$$H_C(\kappa) = 0.48 \cdot H_K \cdot (1 - \kappa(T, f, H_0, V_M)^{0.8}) \tag{2.63}$$

which includes the anisotropy field H_K, eq. (2.15) and the dimensionless parameter

$$\kappa(T, f, H_0, V_M) = \frac{k_B T}{K_u V_M} \cdot \ln\left(\frac{k_B T}{4 \cdot \mu_0 H_0 \cdot M_S V_M \cdot f \tau_0}\right), \tag{2.64}$$

where τ_0 is typically approximated as $\tau_0 \approx 10^{-10}$ s and a restriction of $\kappa \leq 0.7$ holds for eq. (2.63). Consequently, SWMBT is limited to high fields where $H_0 \geq H_C \approx \frac{H_K}{2}$ and generally to $\kappa < 0.7$. Experimental data on iron MNP support the validity of this concept for high field amplitudes of $H_0 = (15 - 60)$ kA/m [47], but do not consider the applicability at lower field amplitudes. One major drawback of SWMBT is the fact that it does not take particle relaxation processes into account and thereby neglects a fundamental effect of superparamagnetic particle relaxation on magnetic particle heating.

Linear Response Theory of Magnetic Particle Heating

This theory describes the magnetic particle heating in terms of a relaxation of the magnetic particle magnetization to equilibrium state within the material's linear response to the field, $M = \chi \cdot H$ (c.f. eq. (2.1)). Hence it is called linear response theory (LRT). It was derived under the assumption of an adiabatic system of monodisperse MNP by Rosensweig [48]. However, the linear response regime is limited to small fields and particle sizes, restricting the regime of validity of LRT. This limit is given by the ratio of magnetic versus thermal energy, in numbers $\xi \leq 1$, with

$$\xi = \frac{|m| \cdot |H|}{k_B T} = \mu_0 \cdot \frac{M_S V_M}{k_B T} \cdot H_0. \tag{2.65}$$

ξ is also called the reduced field parameter. For example, eq. (2.65) yields a limit of validity for LRT to a field amplitude $H_0 \leq 15.7$ kA/m for typical values of particle magnetic size of $d_M = 10$ nm and saturation magnetization $M_S = 400$ kA/m at temperature $T = 300$ K.

LRT can be derived in the framework of the Debye theory of dipolar fluids [49] under the assumption of spherical and non-interacting MNP: Let the MNP be subjected to an AMF; then the magnetic susceptibility of MNP becomes complex and shows both real (χ') and imaginary (χ'') parts, describing the in-phase and out-of-phase (or loss) component, respectively. This complex AC-susceptibility reads

$$\hat{\chi} = \chi' + i\chi'' = \chi_0 \cdot \left(\frac{1}{1 + (2\pi f \cdot \tau_R)^2} + i\frac{2\pi f \cdot \tau_R}{1 + (2\pi f \cdot \tau_R)^2}\right) = \frac{\chi_0}{1 + i2\pi f \cdot \tau_R}, \tag{2.66}$$

containing the initial DC susceptibility $\chi_0 = \frac{\mu_0 M_S^2 V_M}{3 k_B T}$ and the effective relaxation time τ_R that it takes for the system to relax back to equilibrium after a small step in the magnetic field (defined below, eq. (2.70)). $\hat{\chi}$ and $H_{ac}(t)$, from eqs. (2.66) and (2.29), respectively, are inserted in eq. (2.1), which yields for the magnetization:

$$M(t) = |\hat{\chi}| \cdot H_0 \cos(2\pi f \cdot t + \varphi), \tag{2.67}$$

with $|\hat{\chi}| = \dfrac{\chi_0}{\sqrt{1 + (2\pi f \cdot \tau_R)^2}}$,

where φ is the phase lag between magnetization and the alternating magnetic field. In this way, the heat is generated by the magnetization vector lagging behind the driving field vector in LRT [42]. Simple mathematical observation indicates that eq. (2.29) and eq. (2.67) correspond to the parametric equations for an ellipse, wherefore the area of the resulting hysteresis loop $M(H)$ is given by

$$A = \pi \chi'' \cdot H_0^2, \tag{2.68}$$

with $\chi'' = \chi_0 \cdot \dfrac{2\pi f \cdot \tau_R}{1 + (2\pi f \cdot \tau_R)^2}.$ \hfill (2.69)

The relaxation time τ_R can be approximated by adding both relaxation processes yielding the effective relaxation time:

$$\tau_R^{-1} = \tau_N^{-1} + \tau_B^{-1}, \tag{2.70}$$

where τ_B is the Brownian relaxation time and τ_N the Néel relaxation time, given by eq. (2.31) and eq. (2.32), respectively. Note that it is widely assumed that the faster relaxation process will dominate the relaxation time.

2.5. Magnetic Fluid Hyperthermia in Tumor Therapy

This Section describes various aspects of magnetic fluid hyperthermia (MFH), first describing the present-day status of applying MFH, including the main challenges and promises in Section 2.5.1. As magnetic nanoparticles (MNP) used in MFH will inevitably interact with cells inside the body, Section 2.5.2 deals with a general description of pathways of particles inside cells. The nanoscale heating effects of intracellular MNP in MFH and their direct effects on cells are presented in Section 2.5.3. Finally, a treatment approach with MFH is motivated and described for pancreatic ductal adenocarcinoma (PDAC) in Section 2.5.4.

This Section is partly based on an original publication by the writer [50] (s. Appendix C). Where entire sentences are cited directly from the publication, this is marked with [†].

2.5.1. Magnetic Fluid Hyperthermia in Tumor Therapy

The term *hyperthermia* derives from ancient Greek combining 'hyper' meaning *over* and 'thermia' meaning *heating*. Hyperthermia is applied to purposefully induce local heating in parts of the body with temperatures of $(42 - 46)\,^\circ$C. At these elevated temperatures, the function and structure of many enzymes and proteins alters within cells, which can lead to apoptosis (induced cell death) [51]. Thus, hyperthermia is an alternative to open surgery, X-ray irradiation and chemotherapy in modern tumor therapy [52]. Nevertheless, hyperthermia must be clearly distinguished from thermoablation, where temperatures up to 56°C are applied to destroy cells by necrosis and coagulation [53].

Hyperthermia can be applied using different techniques such as microwave [54] and laser irradiation [55], resistive heating via implanted electrode devices [56], and whole body treatment with water baths [57]. Additionally, *magnetic fluid hyperthermia* (MFH) offers a novel and promising approach [58, 59], which is also the main application focus of this thesis. In MFH, magnetic nanoparticles are accumulated locally by magnetic targeting or injection in target tumor tissue and subsequently heated by applying an external alternating field, as described in the previous Section 2.4. At the target tumor size, the MNP inevitably interact with the surrounding cells and tissue, leading to an uptake of MNP inside cells (s. Section 2.5.2 for details). Once internalized inside cells, the intracellular MNP can potentially deliver therapeutic heat from the inside of the cells, increasing the efficacy of MFH by so-called *nanoheating effects*, discussed in detail in Section 2.5.3.

The MNP material used for MFH should show colloidal stability, adjustable magnetic properties, low cytotoxicity, sufficient circulation time within the body during treatment and eventually appropriate biodegradability [60]. These requirements limit the MNP-type that can be used for MFH mainly to the magnetic iron oxides maghemite (γ-Fe_2O_3) and magnetite (Fe_3O_4), which are well tolerated in the human body [23, 61]. Within the field amplitudes, H_0, and frequencies, f, applied in hyperthermia ($H_0 < 25\,\text{mT}$, $f < 1\,000\,\text{kHz}$), these iron oxides display controllable heating characteristics. However, the field itself can stimulate non-selective and undesired heating within the body due to the formation of eddy currents [62]. The heat of such eddy currents Q_{eddy} depends on the field and the diameter of the induced current loop in the body, D^*, according to $Q_{\text{eddy}} \propto (D^* \cdot H_0 \cdot f)^2$ [62]. Medically, there is a limit to the AMF tolerated within the body, which experimentally has been determined to be $H_0 \cdot f < 4.9 \cdot 10^8\,\text{A}/(\text{m} \cdot \text{s})$ in 1984 [63]. The first successful clinical trials at the Charité Berlin, however, have shown slightly higher tolerances of $H_0 \cdot f < 1.4 \cdot 10^9\,\text{A}/(\text{m} \cdot \text{s})$ for fields applied to the head [64] and $H_0 \cdot f < 5.0 \cdot 10^8\,\text{A}/(\text{m} \cdot \text{s})$ for application of MFH at the lower body (rectum) [65]. Both tolerances agree well with the field-frequency limit of $H_0 \cdot f < 8.0 \cdot 10^8\,\text{A}/(\text{m} \cdot \text{s})$ suggested by the International Commission on Non-Ionizing Radiation Protection (ICNIRP) in their Guidelines for a public exposure limit to alternating magnetic fields [66, 67]. Depending on the desired medical application, either $1.4 \cdot 10^9\,\text{A}/(\text{m} \cdot \text{s})$ (head) or $0.5 \cdot 10^9\,\text{A}/(\text{m} \cdot \text{s})$ (body) will serve as the medical limit for the AMF throughout this thesis.

In hyperthermic treatment, both temperature T, and exposure time to elevated temperatures t_{HT}, have to be monitored closely, as their interaction with each other determines the cell's response in a complex manner: If cells are heated above $T > 43\,°\text{C}$ for a short time, the surviving cells prove more vulnerable to subsequent heating. This effect is known as step-down heating [68]. Contrastingly, if the exposure was $T < 43\,°\text{C}$, the surviving cells appear to be much more resistant to heating to even higher subsequent temperatures. Moreover, if cells are heated above $T > 43\,°\text{C}$ but afterwards incubated at $37\,°\text{C}$ for several days, they present themselves more resistant to subsequent heating as well [69]. This process has been coined as thermotolerance [70] and been shown to exist for both healthy and tumor tissue [71]. In order to benefit from the effects of step-down heating, while avoiding thermotolerance in therapy, the heating must be generally well controlled in both temperature and exposure time. It seems reasonable to devise therapies of applying temperatures $T \geq 43\,°C$ for extended periods of time, e. g. $t_{HT} > \frac{1}{2}$ hour [10, 72, 73]. However, the individual success of hyperthermia treatment depends on the superposition of challenges, such as (tumor) blood perfusion, targeting efficiency and accessibility of the tumor. To this end, personalized treatment for a given patient is necessary for successful MFH treatment, as proposed by the Berlin company *MagForce AG* [74–76]. In fact, based on these treatments, MFH has reached successful clinical trial stages I+II one decade ago [64, 65, 77, 78].

Most promise for MFH application lies with its enhancement of the efficacy of existing standard treatments. It has been shown that hyperthermia improves the efficacy of irradiation therapy [79–81]. Specifically, one third less radiation dose is needed to kill the same amount of cells, if they are heated to $T = 43\,°C$ either before or after irradiation [82]. Moreover, the efficacy of certain chemotherapeutic drugs increases significantly when administered together with local heating of the tumor tissue [83, 84]. New models even envision the combination of magnetic drug targeting (MDT) with MFH to form an entirely new, local cancer treatment technique [85, 86]. Here, MNP of thermosensitve coatings are loaded with chemotherapeutic drugs and injected into the body, where they are magnetically guided to and accumulated at the tumor site [87, 88]. There, the MNP are heated in an external AMF to trigger a local drug-release. Such a triggering has been demonstrated in vitro by the controlled release of fluorecent markers [89–91] and the anticancer drug doxorubicin [92] upon heating.

Interestingly, the potentially ground-breaking advantage of local heating restricted to tumor tissue in magnetically targeted MFH also includes its greatest challenge to enter clinical translation: According to Wilhelm et al., who analyzed 117 independent targeting studies, only $< 1\,\%$ of the administered MNP acutally reach the desired tumor [93]. Furthermore, Southern et al. recently cast doubt on whether MFH can actually be effective at low MNP concentrations in such a targeting scenario [94]. However, assuming an MNP concentration as chosen in the aforementioned clinical trials of $1\,M(Fe)^7$ [64, 81], a targeting efficiency of $1\,\%$ results in $10\,mM(Fe)$ of MNP at the tumor site. In Section 2.5.4 the targeting efficacy of MDT will be discussed in more detail for the example of a treatment approach of the pancreatic ductal adenocarcinoma (PDAC).

2.5.2. Pathways into Cells

Cells are the basic structural and functional unit of any living entity and the smallest units of life in general. Cells are enveloped by a membrane that separates the intracellular and extracellular

---

[7]This thesis differentiates between the MNP iron mass concentration, c [mg(Fe)/mL], and the MNP iron molar concentration, c_M [M $\hat{=}$ mol/L]. Typically the iron mass concentration, c, is given. Using the molar mass of iron, $m_{mol,Fe} = 55.85\,g/mol$, the two concentrations can be converted via $c[mg(Fe)/mL] = \frac{1}{55.85} \cdot c_M[M]$.

space [95]. The intracellular space is divided in the *nucleus*, containing mostly deoxyribonucleic acid (DNA) necessary for cell proliferation, and the *cytoplasma*, comprised of cellular organelles and metabolic end-products as well as foreign substances [96]. The cellular (plasma-)membrane consists mostly of a double layer of amphiphilic phospoholipids (so-called *phospholipid bilayer*), which is selectively permeable and regulates the exchange of substances between intra- and extracellular space, enables chemo-electrical communication to neighboring cells and provides mechanical adhesive forces to fix the cell in its surrounding [97].

Cells interact with their surrounding by exchanging substances (e. g. particles) via the processes of *endocytosis* (uptake inside the intracellular space) and *exocytosis* (secretion to the extracellular space) [98]. These pathways will be discussed in the following with special focus on the transport of nanoparticles. Generally, particles can enter cells by different endocytic pathways, depending on their size and surface ligands interacting with the cellular membrane. Large particles ($\sim 1\,\mu m$) are internalized via actin-driven phagocytosis, where the plasma membrane forms cup-shaped invaginations, which engulf the particle. Although quick and effective, phagocytosis is primarily used for the uptake of dead cells and debris, usually reserved to professional phagocytic cells (e. g. macrophages, dendritic cells and neurophils) [99]. For smaller particles ($\sim 100\,nm$), receptor-mediated, and more specifically clathrin- and caveolin-mediated, endocytosis occurs, which is the most common pathway into cells for viruses and nanoparticles [100, 101]. In clathrin-mediated endocytosis, clathrin-ligand binding causes the formation of invaginations in the cellular plasma membrane, which wraps the particle. Similarly, in the caveolin-mediated endocytic pathway, caveolin assembly at the cytoplasmatic side of the membrane triggers the formation of "flask-like" invaginations of $60 - 80\,nm$ in diameter (so-called "caveolae"), induced again by receptor-ligand binding [102]. Clathrin assembly at the membrane is also involved in the formation of vesicle necks necessary in the late wrapping stage in the pinch-off process for final internalization of particles [103]. Additional non-specific endocytic pathways are pinocytosis and macropinocytosis: In pinocytosis, biological fluid and small particles ($< 10\,nm$) are directly absorbed across the plasma membrane, whereas macropinocytosis is an actin-driven process of membrane engulfment of large amounts of extracellular fluid and substances, observed for the uptake of larger nanoparticles ($\sim 1\,\mu m$) [104]. Endocytosis pathways are summarized in Fig. 2.12. Note that the endocytosis pathways can be inhibited by passivating the cell, which can e. g. be achieved by cooling it down to temperatures of $T \approx (4 - 10)\,°C$ [98].

Upon internalization, the particles are wrapped in a membrane-enclosed intracellular vesicle and remain separated from the intracellular fluid. Inside the cell, these endocytic vesicles (called early endosomes) develop in late endosomes that increase in size by fusion with other endosomes [105], reaching final sizes of approximately $500\,nm$ in diameter [106]. Sometimes, endocytic vesicles are merged with their vesicle membranes still intakt, forming multivesicular bodies. Generally, late endosomes are of increasing acidity in preparation for finally fusing with lysosomes [107]. Lysosomes are responsible for the digestion of exogen (cell-foreign) material and therefore carry digestion enzymes (so-called hydrolase), working in an acidic environment with pH-value approximately 4 [96]. Sometimes, substances escape endosomes into the intracellular fluid, from where they are secreted quickly. Cargo, which does not reach late endosomal stage or cannot be digested by the lysosomes is secreted via exocytosis. Exocytosis is basically the reversal of the endocytosis process: Intracellular vesicles transmigrate to the plasma membrane and are secreted either as membrane-wrapped vesicles or expelled freely into the extracellular space [99]. The entire pathways of nanoparticles into, across and out of cells are summarized in Fig. 2.12.

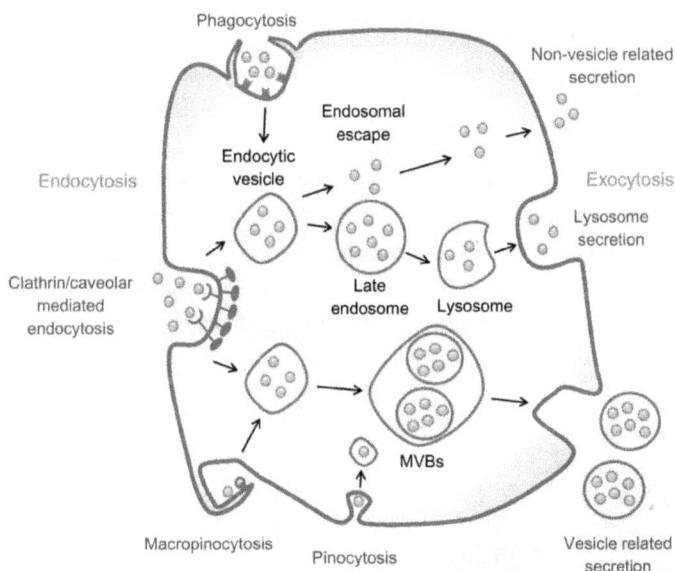

Fig. 2.12.: Pathways of nanoparticles into (endocytosis), through and out (exocytosis) of the cell. Endocytosis comprises of phagocytosis, clathrin- and caveolar-mediated endocytotis, macropinocytosis and pinocytosis. Inside the cell, endocytic vesicles develop either in late endosomes and lysosomes or multivesicular bodies (MVBs). In seldom cases, nanoparticles escape the endosome. Nanoparticles exit cells via lysosome secretion, vesicle related secretion or via non-vesicle related exocytosis. Adapted from [99].

2.5.3. Nanoheating Effects

As previously described in Section 2.5.1, hyperthermic cancer treatment usually aims at raising the temperature of tumor tissue to $(42 - 46)\,°\mathrm{C}$, leading to tumor cell death, while at the same time sparing healthy tissue. This was a long standing goal for MFH since the first in vitro studies conducted by Jordan et al. [108] in 1996: In that study, human adenocarcinoma cells subjected to a bulk temperature of approx. $43\,°\mathrm{C}$ reached via MFH showed a roughly equal inactivation of cell activtiy as the same cells heated to $43\,°\mathrm{C}$ externally with a water bath. However, evidence is mounting that MFH is dealing intracellular cytotoxic damage even without a perceptible rise in bulk temperature [109–112]. The effect was coined nanoscale thermal phenomena or short *nanoheating* [113] and describes a dramatic temperature increase in the direct (nanoscale) vicinity of MNP upon interaction with AMF. Evaluating the intensity of thermosensitive fluorescent dye Dylight549 attached to the surface of MNP, Huang et al.

demonstrated that the surface temperature of the MNP was raised by approx. $5\,°C$ after $45\,s$ of AMF application, limiting particle heating clearly to the nanoscale vicinity of MNP [114]. Similarly, Polo-Corrales et al. attached fluorescent dye bound in thermosensitive polymer poly(N-isopropylacrylamide)(pNIPAM) to iron-oxide MNP to indirectly measure MNP surface temperatures by monitoring the increase in fluorescent intenstiy upon particle heating [115]: They applied an AMF and found an immediate increase in fluorescence. Using a water bath in control experiments, the same intensity as for AMF application was matched by reaching a global temperature increase of $\Delta T \approx 15\,°C$ (initial temperature $20\,°C$). Therefore, the authors further concluded that nanoheating on the order of $\Delta T \approx 15\,°C$ must occur on the surface of MNP under particle heating, which however does not affect the global temperature. Dong et al. used $NaYF_4$:Yb^{3+}-Er^{3+} crystal nanothermometers inside silica nanoparticles ($d_C \approx 100\,nm$) loaded with MNP ($d_C \approx 20\,nm$) to optically measure the temperature dependent emission spectra, reporting an increase of up to $\Delta T \approx 30\,°C$ inside the silica particles [116]. Using temperature-sensitively bound Fluoresceineamine attached to the surface of iron oxide MNP Riedinger et al. were even able to approximate an immediate region of nanoheating of up to $100\,nm$ around the MNP, beyond which no substantial heating was measured [117]. Recent theoretical finite-

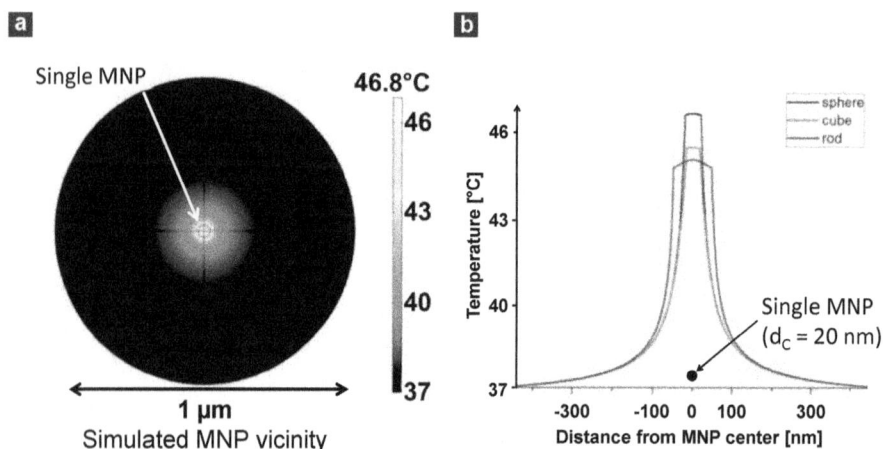

Fig. 2.13.: Simulated temperature distribution in the vicintiy of a single MNP with particle core diameter $d_C = 20\,nm$: (a) temperature distribution of the MNP located in the origin of a spherical compartment with diameter $d = 1\,\mu m$. (b) cross-sectional temperature distribution in the vicinity magnetic nanostructures of different shapes, all located at 0. Based on methods developed for a sperical MNP with $d_C = 20\,nm$, a nanocube and nanorod with the same nominal volume were simulated. Adapted from [118].

element simulation performed by Taloub et al. for a single iron oxide MNP with particle core size of $d_C = 20\,nm$ and assuming an SLP of $500\,W/g(Fe)$[8], confirms the above-mentioned experimental findings (Fig. 2.13) [118]: The MNP heats up by almost $10\,°C$ inside and at its surface (Fig. 2.13a), from where the temperature is exponentially decreasing, dropping to

---

[8]Actually, Taloub et al. assume a volumetric power density of $Q = 10^{16}\,W/m^3$ for their simulations. The authors base this assumption on [119], where SLP values of up to $500\,W/g(Fe)$ are reported, which also fits to the SLP values reported throughout this thesis.

approx. 15% of the maximum temperature at a distance of $100\,nm$ from the MNP surface (Fig. 2.13b). Approximately $500\,nm$ away from the MNP, there is no heating observed. Interestingly, nanostructures of different shapes but of the same nominal volume as the $20\,nm$-sized spherical MNP reached slightly lower maximum temperatures, but the temperature gradient in their outer vicinity ($100\,nm$ and larger) was independent of the shape (cf. 2.13b).

All of the studies mentioned above clearly indicate the existence of nanoheating in the vicinity of MNP during particle heating, which is not perceptible in bulk temperature [113]. It is assumed, that such nanoheating effects have the ability to damage cells, provided that MNP are uptaken inside lysosomes or at least attached to the cell membrane [12, 112]. In fact, the expression of reactive oxygen species (ROS) can be detected within $30\,min$ of AMF application [111], which is directly connected to lysosomal membrane permeabilization, leading to cell death [120].

In addition to nanoheating, a second mechanism of intracellular damage on the nanoscale can arise from (direct) mechanical rupture of the membrane due to the physical rotation of membrane-bound MNP with the applied magnetic field [121, 122]†. It has been demonstrated that this mechanism leads to apoptosis of INS-1 cells in low frequency AMF $f = (5...20)\,kHz$ [123]†. Further, U87 brain tumor cells were reliably killed by membrane rupture due to slow rotation of membrane-bound $2\,\mu m$ CoFeB/Pt microparticles [124]†.

In this way, intracellular nanoheating and mechanical rupture together could support the efficacy of MFH treatment especially for lower MNP concentrations at the tumor site†, for which therapeutic bulk temperatures of $T \geq 43\,°C$ cannot be reached.

2.5.4. Targeting Pancreatic Tumors

Cancer is one of the most challenging deseases to treat worldwide and approximately 25% of the total deaths in Germany were caused by cancer in 2015 [4]. Among the most aggressive types, the pancreatic ductal adenocarcinoma (PDAC)† has a 5-year-survival-rate of only less than 5% [125]. Moreover, PDAC is predicted to rank second in the total number of deaths caused by carcinoma in 2020 in the United States of America [126]†. At present, surgical removal (resection) is the only curative therapy among established treatment routines, as PDAC has proven to be highly resistant to chemo- and radiotherapy [127, 128]†. Unfortunately, resection is only possible in approximately 20% of the cases, as by the time the PDAC is diagnosed, the tumor has often metastasized already [129]†. Of these 20% resectable tumors, many are encasing the superior mesenteric artery by more than $180\,°$, making resection very risky [130, 131]†. Thus, there is desperate need for alternative therapies that are either stand-alone techniques or assist in partial regression of at least such 20% the tumor to make it accessible to resection eventually†.

Current treatment strategies are either focused on combining chemo- and radiotherpy (so-called neoadjuvant therapy) — such as the CONKO-007 study [132] — or are exploring entirely new and innovative therapies. Among those innovative cancer therapies, MFH becomes increasingly attractive due to the local and minimally-invasive delivery of heat with therapeutic temperatures to tumors (Section 2.5): For advanced therapy, biocompatible MNP are either directly injected into the tumor or administrated intravenously and accumulated at the tumor site via external magnetic fields (magnetic targeting)[†].

In clinical trials mentioned before (cf. Section 2.5), effective intratumoral temperatures up to approx. $47\,°C$ were reached during treatment [64][†]. These elevated temperatures could be achieved mainly due to a relatively high local concentration of MNP of up to approx. $128\,mg(Fe)/mL$ after a direct MNP intratumoral injection[†]. Nevertheless, an intratumoral injection is an invasive procedure with high risks of developing metastasis[†]. These risks can be omitted when magnetic targeting of MNP is intravenously applied, however, at the cost of reaching comparatively low MNP concentrations of approx. $150\,µg(Fe)/g(Tumor)$ $(3\,mM)$ [133] to $400\,µg(Fe)/g(Tumor)$ $(7\,mM)$ [134][†]. Such low concentrations were achieved for a mouse tumor model using permanent magnets[†]. Endoluminal tumor that can be reached endoscopically, such as PDAC, offer additional potential for magnetic targeting, as these allow the minimally-invasive insertion and accurate positioning of miniaturized coils or permanent magnets for magnetic targeting [18]. Using an endoscopic targeting setting and individualized magnets, it was recently demonstrated that the targeting efficiency could be enhanced by a factor of 40 [135][†]. This study is part of the same therapy approach as the MFH investigations in this thesis. Consequently, the local MNP concentration at disposal for MFH treatment would be much higher than the one for simple magnets mentioned above and the effective temperatures for treatment might be reached more easily[†]. In this way, an individualized and less stressful cancer therapy for each patient may be possible[†]. In particular, PDAC tumors could achieve regression and, in this way, be accessible for secondary resection[†].

This work is embedded in a treatment approach for PDAC and therefore the in vitro studies of the present thesis are focused on pancreatic tumor cells, adding to the full treatment strategy of combining endoscopic MNP magnetic targeting and MFH application. Using low MNP concentrations of $(3-7)\,mM(Fe)$ that are achieved for MNP targeting in animal models (s. above), this thesis explicitly studies the uptake of MNP inside pancreatic tumor cells in Chapter 6 and examines further the effects of nanoheating upon intracellular MFH application in Chapter 8.

2.6. Hydrogels and Rubber Elasticity Theory

Hydrogels are three-dimensional crosslinked polymer networks that swell, but do not dissolve, in water. Hydrogels have many applications, especially in biomedical technology, as they posses tissue-equivalent mechanical properties, while being easy to synthesize reproducibly. Hydrogels mixed with MNP are used in this thesis as a biocompatible model system to study the effect of gradual immobilization of MNP on magnetic particle heating in situ in Section 7.2. In the following, a general description of hydrogels is given, followed by a brief introduction in the rubber elasticity theory, which is used to describe the mechanical properties of hydrogels mathematically. The description of the specifications of agarose and poly(acrylamide) hydrogels, which are specifically used throughout this thesis, concludes this section.

General Description of Hydrogels

Hydrogels are nonfluid natural or synthetic three-dimensional polymer networks resulting from cross-linking of polymer strands. These networks maintain their structure when being subjected to water, but at the same time exhibit a remarkable ability to absorb a large fraction of water within their stucture (typically containing up to $> 90\%$ of water)[9] [137]. Consequently, they swell in water, combining solid-like properties (e. g. internal polymer network structure) and liquid-like properties (e. g. flexibility, thermal and electrical properties similar water), and are therefore denoted as *hydrogels*[10]. Hydrogels offer a versatile tool for many applications in biomedicine and medical technology due to their biocompatibility, tissue-equivalent characteristics and tunable mechanical properties [138,139]. Most importantly, hydrogels are used in tissue engineering and cell biology as 2D and 3D extracellular cell matrices (ECM), promoting tailored (stem) cell growth and differentiation [140,141]. But hydrogels also have major applications in e. g. macromolecular biology as a biocompatible electrophoresis embedding [142], and generally as tissue-equivalent phantoms in (pre-)clincial assessment of thermal dosiometry, ultrasound and magnetic resonance imaging [139,143,144]. Essentially, hydrogels can be differentiated in natural polymers - such as gelatine, matrigel, polysaccharides and the protein-based fibrin and collagen - and synthetic polymers, containing poly(acrylamide) (PAAm), polyethylenglycol (PEG), poly(oxazolines) (POZ), poly(vinyl alcohol) (PVA), and biodegradable poly(lactic-co-glycolic acid) (PLGA) and oligo(ethylene glycol) fumarates (OPF) [141]. All of these different hydrogels have unique properties useful for the applications mentioned above. The range of applications can even be expanded, however, by incorporating magnetic nanostructures inside hydrogels, forming so-called *ferrohydrogels* (FHG). As magnetically manipulable hybrid structures, FHG attract much interest in fundamental materials analysis and for biomedical applications: For example, $CoFe_2O_4$ MNP incorporated in PAAm hydrogels have been proposed as magneto-mechanical probes to non-invasively analyize the mechanic properties of the respective hydrogel network [145]. Furthermore, remote controlled cell and drug delivery systems were designed, which exploit the deformation of FHG in magnetic field gradients [146]. When investigated specifically for MFH applications, FHG enable active control of therapeutic heating temperatures for MFH [147] and thermosensitive drug delivery [148].

Rubber Elasticity Theory

The general term *elasticity* describes the ability of a material body to change its shape under the action of an external force and return to its initial shape when the external force is removed. *Rubber elasticity* is a special form of elasticity, which is thermodynamically based on a reversible

---

[9]There are also special hydrogels, classified as *superabsorbent hydrogels*, which can contain up to 99.9% water [136].

[10]In this thesis, water-swollen gels are used, therefore they are named *hydrogels*. In principle, any fluid could be used as an extender to swell the gel, however these gels are not the focus of this thesis and are thus neglected.

change in the entropy of the material rather than a change the internal energy [149]. The *rubber elasticity theory* (RET), describes the mechanical properties of ideal rubber-like materials by a combination of thermodynamics and polymer chain statistics. RET is derived under to following assumptions [150]:

(I) the (Helmholtz) free energy, $F = U - T \cdot \Sigma$, with the temperature, T, and entropy, Σ, is minimized in the resting state,

(II) the internal energy, U, does not change under deformation, i. e. $dU = 0$,

(III) the material is incompressible, i. e. the material volume does not change under deformation, and

(IV) the deformations of the material are microscopically and macroscopically the same (affine deformation assumption).

From combining assumption (I) and (II) it follows that for an isothermal process an increase in the free energy due to deformation of the material will result in an a decrease in the system entropy, $d\Sigma < 0$, according to [151]:

$$dF = -T \cdot d\Sigma. \tag{2.71}$$

Furthermore, RET holds for materials whose matrix consists of long (polymer) chains of identical building blocks, which must be highly flexible and joined by crosslinkers in a network structure [151]. These chains rotate about their joints randomly due to activation from thermal energy[11], so that the distance between their ends is governed purely by statistical considerations: Applying assumption (III) and (IV) within the so-called Gaussian network theory [152], the entropy of deformation of these chains is fully described by the number of chains per unit volume, ν_{el}, and the Boltzmann constant, k_B. For *pure* strain (i. e. strain that does not involve rotating the principal axes), with the extension ratios λ'_x, λ'_y, λ'_z — describing the deformation parallel to the principal axes x, y, z in comparison to the initial state — and assuming $\lambda'_x \lambda'_y \lambda'_z = 1$, the entropy of deformation reads [152]:

$$d\Sigma = -\frac{1}{2} \nu_{el} k_B \cdot ((\lambda'_x)^2 + (\lambda'_y)^2 + (\lambda'_z)^2 - 3). \tag{2.72}$$

For shear strain in the (x, y)-plane the extension ratios may be defined as $\lambda'_x = \lambda'$, $\lambda'_y = 1/\lambda'$, $\lambda'_z = 1$ and the corresponding shear strain is $\gamma_{el} = (\lambda' - 1/\lambda')$ [153]. Now eq. (2.72) simplifies to:

$$d\Sigma = -\frac{1}{2} \nu_{el} k_B \cdot \gamma_{el}^2. \tag{2.73}$$

Inserting eq. (2.73) in eq. (2.71) and applying the definition of shear stress, one obtains:

$$\sigma_{el} = \frac{dF}{d\gamma_{el}} = \nu_{el} k_B T \cdot \gamma_{el}. \tag{2.74}$$

---

[11]This implies that the molecular segments of the chains have sufficient thermal energy to rotate freely, independent of their neighbors, i. e. RET holds only for those polymers in which the intermolecular forces between chains are sufficiently weak to satisfy this condition [152].

It can be seen from comparing eq. (2.74) to the general definition of the shear modulus, $G := \frac{\sigma_{el}}{\gamma_{el}}$, that the shear modulus is equivalent to the following expression in RET:

$$G = \nu_{el} \cdot k_B \cdot T. \tag{2.75}$$

Consequently, the elastic properties (i.e. the stress-strain relations) of an ideal rubber-like material can be described by only a single physical parameter, the shear modulus G. Furthermore, the stress-strain relations have the same form for all rubber-like materials (assuming constant temperature), only subjected to a scale factor (G), which is determined by the number of chains per unit volume, ν_{el} [152]. Note that ν_{el} is also referred to as the number of active polymer strands throughout this thesis.

Specifics of Agarose and Poly(acrylamide) Hydrogels

For this thesis, the natural polymer hydrogel agarose and the synthetic polymer hydrogel poly(acrylamide) (PAAm) are used for immobilization studies of MNP in Section 7.2. The polymer network forms differently in these two hydrogels, as will be shortly sketched in the following description:

Agarose

As a linear polysaccharide, agarose consists of $(1 \rightarrow 3)$-β-D-galactopyranose-$(1 \rightarrow 4)$-3, 6-anhydro-α-L-galactopyranose and contains a few ionized sulfate groups [154]. In highly purified agarose powders the amount of sulfate groups is below 0.2% to better control its elastic properties. The gelling mechanism is temperature dependent, forming intramolecular hydrogen-bonds upon cooling, which assemble as aggregates of double helices resuliting from the entanglement of anhydro bridges [155]. The chemical formula of one such building blocks of agarose gels is depicted in Fig. 2.14a. Agarose hydrogels are prepared from powder dissolved in hot deionized water (DI-H_2O): Standard agarose fully dissolves at $T = (90-100)\,°C$ solidifying at $T = (35-39)\,°C$ [156], while low-melting (LM-) agarose only requires temperatures around $T = (60-70)\,°C$ for dissolving in DI-H_2O and solidifies at $T \approx 28\,°C$ [157]. By controling the mass fraction of agarose in solution, the elastic moduli, E, and the mean mesh size, d_{mesh}, can be easily tuned between $E \sim (0.1-1\,000)\,kPa$ [141] and $d_{mesh} \sim (10-1\,000)\,nm$ [158], respectively.

Poly(acrylamide)

Poly(acrylamide) (PAAm) hydrogels form when monomers from one polymer are chemically joined at junction points by crosslinking monomers of a second polymer (s. Fig. 2.14b for a typical building block of such a crosslinking junction). Typically, PAAm hydrogels are produced by radical polymerization of a solution containing acrylamide (AAm) monomers and bi-functional[12] N,N'-methylenebisacrylamide (BIS) crosslinkers. The gelling mechanism is initialized on-demand by adding ammonium persulfate (APS) and N,N,N',N'-tetramethylethylene-diamine (TEMED) to start the exothermal free radical polymerization [159]. The reaction can

---

[12]I. e. consisting of two functional groups; here monomers and crosslinkers.

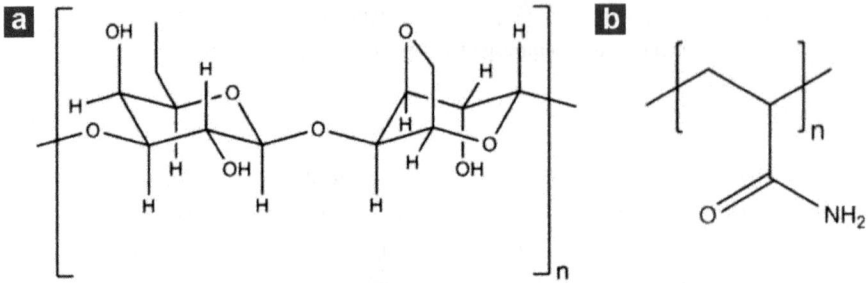

Fig. 2.14.: Chemical formula of one building block (multiplied block in brackets marked with n) of (a) agarose and (b) (poly)acrylamide hydrogels. Modified figures from [141].

be discribed by a chain polymerization mechanism [149]: After initiation, the radical reacts with one AAm monomer molecule, elongating the chain by one unit and moving the radical to the end of the chain. One bi-functional BIS crosslinker molecule can react with two separetly growing polymer chains, thereby forming crosslinked networks with "rubberlike" elasticity [160] (s. above). The elastic moduli, E, and the mean mesh size, d_{mesh}, of such a PAAm hydrogel can be tuned ranging between $E \sim (1 - 10\,000)\,\text{kPa}$ [141] and $d_{\text{mesh}} \sim (1 - 100)\,\text{nm}$ [158], respectively, by controlling the crosslinker fraction, i. e. the BIS mole fraction

$$\alpha = n_{\text{BIS}}/(n_{\text{AAm}} + n_{\text{BIS}}),\tag{2.76}$$

with the amounts of AAm, n_{AAm}, and BIS, n_{BIS}, respectively.

3. Simulation of Magnetic Particle Heating

This chapter deals with the prediction of magnetic particle heating on the basis of magnetic particle relaxation Monte Carlo (MC-)simulations. Section 3.1 summarizes the literature on relaxation simulations applied to predict magnetic particle heating. The implementation of the MC-simulations used here are described in Section 3.2, introducing stochastic calculus and the Stratonovich-Heun scheme. The simulated particle heating results as a function of particle size, effective anisotropy and applied AMF parameters are discussed in Section 3.3. The MC-simulation results are then compared to the predictions from SWMBT and LRT in Section 2.4. This chapter concludes with a summary of the magnetic particle heating trends predicted by each of the three methods (resp. theories) and a brief outlook on future research opportunities using MC-simulations to predict particle heating in Section 3.4.

The MC-simulations were performed in cooperation with the Krishnan Lab at the University of Washington (Seattle, WA, USA) using the Python language (Python Software Foundation, Wilmington, DE, USA). The implementation is presented courtesy of C. Shasha from the Krishnan Lab, who wrote the code.

3.1. State of Research

As has been demonstrated in Section 2.4, both common theories of magnetic particle heating — Stoner-Wohlfarth Model Based Theory (SWMBT) and Linear Response Theory (LRT) — are only valid for a specific range of field amplitudes H_0 and particle magnetic sizes d_M: SWMBT is limited to $H_0 > H_C$ and $\kappa < 0.7$, while LRT is restricted to $\xi(H_0, d_M) = \mu_0 \cdot \frac{M_S V_M}{k_B T} \cdot H_0 < 1$. Consequently, SWMBT is generally limited to larger field amplitudes and larger MNP with $d_M > 18\,\text{nm}$ and $H_0 > 20\,\text{mT}/\mu_0$, while LRT holds generally for smaller field amplitudes and small MNP: $H_0 < 15\,\text{mT}/\mu_0$ and $d_M < 10\,\text{nm}$. From this one sees that there is a significant gap for intermediate particles' magnetic sizes, $10\,\text{nm} < d_M < 18\,\text{nm}$, where neither theory holds. Therefore, the prediction of MNP heating from simulated hysteresis ($M(H)$-)loops has been receiving increasing attention over the past decade to close this gap for accurately predicting size-dependent particle heating. To simulate the $M(H)$-loop, it is common to assume uniaxial anisotropy with two energy minima (denoted A and B) (s. Section 2.1.3; cf. esp. Fig. 2.5) and perform simulations in the so-called 2-level-approximation. This assumes that particles can only reside in one of the two energy minima, separated by the anisotropy energy barrier $\Delta E = K_u V_M$. In an ensemble of MNP, the distribution of particles on these two minima is calculated from a probability term for finding a particle's magnetization pointing up (minimum A) or down (minimum B). This probability term depends on ΔE, temperature T, and the so-called attempt frequency ν_0, which comprises sample properties such as saturation magnetization M_S, damping constant α' and the effective anisotropy constant K_{eff}, as well as the external field parameters H_0 and f. By summarizing the contributions of these many properties and parameters in the one constant ν_0, the 2-level-approximation significantly simplifies the complex magnetic particle relaxation process, thus leading to some degree of inaccuracy. This is compensated by much faster computation times however, and the 2-level approximation has been successfully applied by Carrey et al. to reproduce the results from SWMBT and LRT by Carrey et al. [42]. Magnetic dipole-dipole interparticle interactions were included in the 2-level approximation by Tan et al. to predict the heating behavior of intracellularly clustered MNP [161]. This was expanded further to simulate the SLP of aggregated particles by Ovejero et al. [162]. A similar approach was taken by Ruta et al., who presented a unified model of hyperthermia to predict SLP values within and beyond the limitations of LRT, while including particle size distribution and magnetic dipole-dipole interparticle interactions [163]. This work is a good example of the potential for $M(H)$-loop simulations to go beyond the standard theories of SWMBT and LRT. Mamiya and Jeyadevan expanded the physics of the model by using the 2-level-approximation to simulate Néel relaxation of uniaxial MNP and further included a magnetic torque to simulate the MNP Brownian relaxation under the influence of thermal fluctuations. In this way, the authors achieved significant success in predicting ideal parameters for optimizing SLP values [25, 164, 165]. The approach of Mamiya and Jeyadevan comes closest to simulating particle relaxation in the framework of the Landau-Lifshitz-Gilbert (LLG) equation as proposed in this thesis (based on Section 2.2, esp. eq. (2.33)), but in direct comparison, their approach lacks the accuracy given by incorporating in the LLG equation.

3.2. Implementation

As seen in Section 2.2, including thermal effects in the mathematical description of magnetic particle relaxation (by inserting eqs. (2.40) and (2.41) in eqs. (2.33) and (2.34)) makes these equations a set of coupled stochastic differentiable equations (SDE) that require the use of stochastic calculus. Section 3.2.1 introduces the basics of stochastic calculus and derives a formulation for solving SDE by numerical integration. Section 3.2.2 then presents and explains the step-by-step implementation of numerical Monte-Carlo (MC-)simulation, which is used to simulate hysteresis loops for arbitrary ensembles of MNP. The use of hysteresis loops for extracting information about the particle heating will also be demonstrated. Section 3.4 concludes this chapter by summarizing the results of predicting particle heating via MC-simulations.

3.2.1. Stochastic Calculus

The thermal fluctuations introduced to magnetic particle relaxation modeling in Section 2.2 can be approximated as Gaussian distributed white noise. Imagine a white noise distribution $\lambda_\omega = \frac{1}{\sqrt{2\pi}} \int_{-\infty}^{\infty} e^{-i\omega t} \delta(t - t') dt' = \frac{e^{-i\omega t'}}{\sqrt{2\pi}}$ and uniform spectral density (*i.e.* flat) described by its magnitude $|\lambda_\omega| = \frac{1}{\sqrt{2\pi}}$ [166]. A differential equation containing λ_ω therefore has a non differentiable time trajectory, and stochastic calculus must be considered in order to solve this SDE. Imagine further a SDE for an arbitrary time trajectory with the differentiable function $\boldsymbol{f}(\boldsymbol{X}(t), t)$ and the non-differentiable function $\boldsymbol{g}(\boldsymbol{X}(t), t)$ of the form

$$\mathrm{d}\boldsymbol{X}(t) = \boldsymbol{f}(\boldsymbol{X}(t), t)\mathrm{d}t + \boldsymbol{g}(\boldsymbol{X}(t), t)\mathrm{d}\boldsymbol{W}(t). \qquad (3.1)$$

Where the independent Gaussian stochastic process is described in the Wiener integral

$$\boldsymbol{W}(t) = \int \lambda_\omega \mathrm{d}t, \tag{3.2}$$

with properties:

$$\boldsymbol{W}(t=0) = 0 \qquad \langle \boldsymbol{W}(t) \rangle = 0 \qquad \langle (\boldsymbol{W}(t)\boldsymbol{W}(t+\Delta t))^2 \rangle = \Delta t. \tag{3.3}$$

Eq. (3.1) can be solved by [167]

$$\boldsymbol{X}(t) = \boldsymbol{X}(t_0) + \int_{t_0}^{t} \boldsymbol{f}(\boldsymbol{X}(t'), t') \mathrm{d}t' + \int_{t_0}^{t} \boldsymbol{g}(\boldsymbol{X}(t'), t') \mathrm{d}\boldsymbol{W}(t'). \tag{3.4}$$

The crux of stochastic calculus arises from defining and solving the integral over the Wiener process [167]: Imagine the incremental partition $\{t_0, ..., t_N\}$ of the interval $[t_0, t_N]$, where τ_i is a value between t_{i-1} and t_i, $\tau_i = (1-\beta)t_{i-1} + \beta t_i$, solving the stochastic integral by discretizing in partial sums as follows [168]:

$$\int_{t_0}^{t} \boldsymbol{g}(\boldsymbol{X}(t'), t') \mathrm{d}\boldsymbol{W}(t') = \lim_{N \to 0} \sum_{i=1}^{N} \boldsymbol{g}(\boldsymbol{X}(\tau_i), \tau_i)[\boldsymbol{W}(t_i) - \boldsymbol{W}(t_{i-1})], \tag{3.5}$$

Depending on the choice of β, the limit in eq. (3.5) is different and thus the solution to the stochastic integral is different. Typically, the two commonly chosen interpretations are the Itō scheme with $\beta = 0$ and the Stratonovich scheme, where $\beta = \frac{1}{2}$ is chosen [169]. In other words, Itō's integral uses the minimum value of the discretized interval to evaluate the integral, while Stratonovich uses its midpoint. For additive noise, the Itō and Stratonovich integrals can be equivalently transferred into each other (mathematical proof of this can be found in e. g. [167,170]). Realistically, white noise is only an idealization of colored noise with an exponentially decaying correlation function. For thermal fluctuations expressed as a discretized Wiener process with timestep Δt, the autocorrelation function $\langle \boldsymbol{W}(t)\boldsymbol{W}(t+\Delta t) \rangle \propto e^{\Delta t/t_{\mathrm{corr}}}$ must yield an autocorrelation time, t_{corr}, which is much shorter than the effective relaxation of magnetic particles, i. e. $t_{\mathrm{corr}} \ll \tau_{\mathrm{R}}$ [24]. In this limit of zero correlation time of the Wiener process, the stochastic integration (eq. (3.5)) can be solved with the Stratonovich-Heun Integration using the Stratonovich scheme [171], as described next.

Consider a discretized Wiener process. From the variance given in eq. (3.3), the standard deviation of such a Wiener process scales as the square root of the chosen time step, $\sqrt{\Delta t}$, and the discretized integration step, $\Delta \boldsymbol{W}$, is distributed according to

$$\Delta \boldsymbol{W} = N(0, \sigma)\sqrt{\Delta t}, \tag{3.6}$$

where $N(0, \sigma)$ represents the Gaussian distribution

$$N(x, \mu, \sigma) = \frac{1}{\sqrt{2\pi\sigma^2}} \cdot e^{\frac{(x-\mu)^2}{2\sigma^2}} \tag{3.7}$$

with mean $\mu = 0$ and standard deviation σ. Using the Stratonovich scheme, eq. (3.4) can be solved by a predictor-corrector method, known as Heun method [172, 173]: Given the value x_i at time t_n of the discretization, the predictors at time $t_n + \Delta t$, $\overline{x}_i(t_n + \Delta t)$, are obtained via the Euler integration:

$$\overline{x}_i(t_n + \Delta t) = x_i(t_n) + f_i(x(t_n), t_n)\Delta t + g_i(x(t_n), t_n)\Delta W(t_n). \tag{3.8}$$

Then the actual value $x_i(t_n + \Delta t)$ is given in the Stratonovich-Heun integration by

$$
\begin{aligned}
x_i(t_n + \Delta t) = x_i(t_n) &+ \frac{f_i(x(t_n), t_n) + f_i(\overline{x}(t_n + \Delta t), t_n + \Delta t)}{2}\Delta t \\
&+ \frac{g_i(x(t_n), t_n) + g_i(\overline{x}(t_n + \Delta t), t_n + \Delta t)}{2}\Delta W(t_n).
\end{aligned}
\tag{3.9}
$$

Inserting Gaussian distributed, white noise thermal fluctuations (eqs. (2.40) and (2.41)) in the relaxation equations for the magnetic moment, m, and the direction of the easy axis, n (eqs. (2.33) and (2.34)), respectively, allows rearrangement in a SDE equivalent to eq. (3.1) with a differentiable, $f_{m,n}$, and a non-differentiable component $g_{m,n}$:

$$dm = f_m(m, n, t)dt + g_m(m, n, t)dW_m \tag{3.10}$$

and

$$dn = f_n(m, n, t)dt + g_n(m, n, t)dW_n. \tag{3.11}$$

Here, the white noise-approximated thermal fluctuations are included in the Wiener integrals W_m and W_n. Since the numerical Monte Carlo simulations are carried out in discretized time steps of length Δt, the Wiener processes W_m and W_n are also discretized and their magnitude can be calculated by combining eqs. (2.44), (2.45) (with $t' = t + \Delta t$) and (3.3), (3.6):

$$\Delta W_m = \sqrt{\frac{2k_B T}{\gamma_0 |m|} \frac{1 + \alpha'^2}{\alpha'}} \cdot \Delta t \cdot N(0, \sigma = 1) \tag{3.12}$$

$$\Delta W_n = \sqrt{12 k_B T \eta V_H \cdot \Delta t} \cdot N(0, \sigma = 1), \tag{3.13}$$

where the magnitude of the magnetic moment is set to $|m| = \mu_0 \cdot M_S \cdot V_M$. By setting $\sigma = 1$ and by multiplying with the square root of the magnitude of the variance, the implementation is simplified. (It would, however, also be correct to set σ equal to the variance in eq. 3.7). The length of one time step in the numerical Monte Carlo simulation is chosen as

$$\Delta t = \frac{1}{f \cdot N_t}, \tag{3.14}$$

where f is the frequency of the applied field (typically $f \sim 100\,\text{kHz}$ for MFH) and N_t is the number of time steps simulated for one full cycle of magnetization. In order to ensure a negligible correlation time *vs* MNP relaxation times so that $t_{\text{corr}} \ll \tau_R$ holds, the time steps must be chosen $N_t \geq 1000$. Under these conditions, the Stratonovich-Heun integration applies and the SDE, eqs. (3.10) and (3.11), can be solved according to eqs. (3.8) and (3.9):

$$\overline{m}(t + \Delta t) = m(t) + f_m(m, n, t)\Delta t + g_m(m, n, t)\Delta W_m \tag{3.15}$$

$$\overline{n}(t + \Delta t) = n(t) + f_n(m, n, t)\Delta t + g_n(m, n, t)\Delta W_n \tag{3.16}$$

from which the next step in relaxation of m and n under thermal fluctuations after the time step Δt evolves as

$$\begin{aligned} m(t + \Delta t) = m(t) &+ \frac{f_m(m, n, t) + f_m(\overline{m}, \overline{n}, t + \Delta t)}{2}\Delta t \\ &+ \frac{g_m(m, n, t) + g_m(\overline{m}, \overline{n}, t + \Delta t)}{2}\Delta W_m, \end{aligned} \tag{3.17}$$

$$\begin{aligned} n(t + \Delta t) = n(t) &+ \frac{f_n(m, n, t) + f_n(\overline{m}, \overline{n}, t + \Delta t)}{2}\Delta t \\ &+ \frac{g_n(m, n, t) + g_n(\overline{m}, \overline{n}, t + \Delta t)}{2}\Delta W_n. \end{aligned} \tag{3.18}$$

The initial conditions $m(t = 0)$ and $n(t = 0)$ are realized by individually placing the direction of the magnetization and easy axes of each particle in randomized directions when initializing the simulations. From this, the evolution of an entire $M(H)$-loop under the influence of an applied field and thermal fluctuations can be calculated with above equations using recursive iteration.

3.2.2. Step-by-Step Implementation of Magnetic Particle Relaxation Simulations

The result of the MC-simulation is a hysteresis loop M(H). As seen in the previous Section 3.2.1, the implementation of the magnetic moments' and easy axis' dynamics, eqs. (3.17) and (3.18), requires discretized time steps of step size of $\Delta t = \frac{1}{N_t \cdot f}$ (eq. 3.14), with the number of time steps, N_t, and the frequency of the applied field, f. Therefore, after $t = N_t \cdot \Delta t$ one full cycle of magnetization (i.e. on $M(H)$-loop) has been traced or in other words, the AMF has passed through one entire period of its oscillation, denoted as $T' = \frac{1}{f} = N_t \cdot \Delta t$. Inserting eq. (3.14) in the applied AC-field, eq. (2.29), consequently also discretizes the portion in which the field changes its magnitude, ΔH. In order to apply the Stratonovich scheme, $\Delta t \leq 10^{-8}\,\text{s}$ must hold, and for frequencies of the order of magnitude of $(10 - 1\,000)\,\text{kHz}$, $N_t \geq 10\,000$ holds for the implementation of MC-simulations. Furthermore, as discussed in Section 2.2, the relaxation dynamics must be simulated for a sufficiently large ensemble of particles to produce a meaningful result. A compromise between a large enough number of particles, P, to obtain statistical significance and the exponentially growing computation time with increasing P, is therefore being sought. For all results of MC-simulation presented in this thesis, the following parameters were chosen:

$$N_t = 10\,000,$$
$$n_{\text{cyc}} = 2,$$
$$P = 1\,000,$$
$$X = 20.$$

(3.19)

Here, n_{cyc} is the number of full magnetization cycles simulated, and X denotes the number of repetitions of each MC-simulation. In principle, the end result of each MC-simulation is a discretized list with N_t entries of the ensemble average magnetization in the direction of the applied field, here chosen as the z-direction, M_z, and furthermore averaged over X repetitions, related to the corresponding discretized field values ΔH:

$$\boldsymbol{M}_z = \frac{1}{X} \sum_{j=1}^{X} \left(\frac{1}{P} \sum_{i=1}^{P} \hat{m}_{z,i}(\Delta H) \right)_j$$

(3.20)

Note that for practical reasons during MC-simulations, all vectors of magnetization and the direction of easy axis are normalized and represented as $\hat{m}$ and $\hat{n}$, respectively. An implementation flow diagram is given in Fig. 3.1 and will be explained step-by-step in the following:

1. The input parameters, describing the MNP and the AMF as well as the implementation-relevant parameters, must be specified. They can generally be classified as *simulation input, particle properties* and *external parameters.*

   The simulation input is given by the number of particles simulated, P, the number of time steps simulated for a full oscillation of the field, N_t, the number of full magnetization cycles simulated, n_{cyc}, and the number of repetitions of simulations, X. Their values are used as specified in eq. (3.19).

   The class of particle properties includes all parameters characterizing the MNP, such as the (mean) magnetic diameter, d_M, with its log-normal distribution width, σ_{d_M} and the absolute thickness added to the particle diameter due to MNP coating, d_{coat} (together with the core diameter d_C, this constitutes the particle hydrodynamic size, $d_H = d_C + d_{\text{coat}}$). It also includes, the (mean) effective anisotropy constant, K_{eff}, which is assumed to be coupled to the particle size due to surface anisotropy effects (c.f. eqs. (2.22) and (2.23)) and therefore simulated as log-normal distributed with distribution width, $\sigma_{K_{\text{eff}}}$, the MNP saturation magnetization M_S and the phenomenological damping parameter α'.

   Finally, the external parameters describe the applied field with amplitude, H_0 and frequency, f, as well as the ensemble MNP concentration, C, the temperature, T, and the macroscopic viscosity of the carrier medium, η.

Input Parameters
P, N_t, X, n_{cyc}
$d_M, d_{coat}, \sigma_{d_M}, K_u, \sigma_K, M_S, \alpha'$
H_0, f, T, η, C

#1

#2

#4

Generate log-normal distributed sizes d_i with P entries:
$$p(d) = \frac{1}{\sigma d_i \sqrt{2\pi}} e^{\left(-\frac{(\ln(d_i)-\mu)^2}{2\sigma^2}\right)}$$
with $\mu = d_M$ and $\sigma = \sigma_{d_M}$

Generate log-normal distr. anisotropy constants K_i with P entries:
with $\mu = K_{eff}$ and $\sigma = \sigma_{K_{eff}}$
(uniaxial anisotropy)

#3

Generate magnetic and hydrodynamic volumes:
$$V_{M,i} = \frac{\pi d_i^3}{6}$$
$$V_{H,i} = \frac{\pi (d_i + d_{coat})^3}{6}$$

Randomly distribute P particles with $V_{M,i}, V_{H,i}, K_{eff,i}$ in box of length $L = \sqrt[3]{\frac{P}{C}}$ and average interparticle distance $r_{avg} = \sqrt[2]{3} \cdot \frac{1}{\sqrt[3]{C}}$.

#5

Assign randomly distributed directions of magnetization $\hat{m}_i$ and easy axes $\hat{n}_i$ for each particle.

#6

#10

Eqs. (1.32), (1.33), (1.39) & (1.40)

#8

Repeat X times

Thermalize system for $N_t/5$ time steps $\Delta t = \frac{1}{N_t \cdot f}$.
$$H_{eff,i} = H_{K,i} + H_{th} \left(+ H_{pp-IA,i}\right).$$

$N_t/5$

Eqs. (1.29), (1.32), (1.33), (1.39) & (1.40)

#9

Evolve system for N_t time steps under applied field
$$H_{eff,i} = H_{K,i} + H_0(\Delta t) + H_{th} \left(+ H_{pp-IA,i}\right).$$

N_t

Average over z-direction projection of all magnetic moments $M_z = \frac{1}{X} \sum_{j=1}^{X} \left(\frac{1}{P} \sum_{i=1}^{P} \hat{m}_i\right)_j$ for each time step (i.e. different value of applied field $\Delta H(\Delta t)$).

#11

Extract $M_z(\Delta H) = M(H)$.

#12

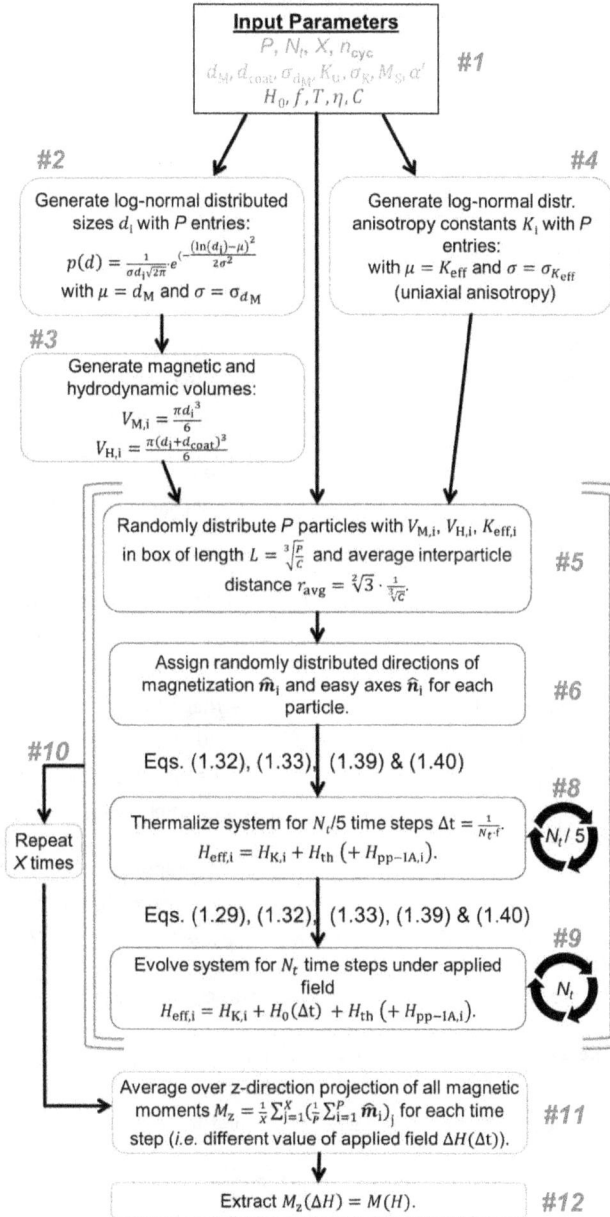

Fig. 3.1.: Flow diagram of Monte Carlo simulation implementation. The input parameters are classified by color as simulation input, particle properties and external parameters. The steps, explained in detail in the text, are marked at the appropriate position in the flow diagram with #x, x = (1...12). Note that step #7 is not shown.

2. From d_{M} and $\sigma_{d_{\mathsf{M}}}$, a one-dimensional array with P entries of log-normally distributed individual particle sizes, d_i, is generated.

3. From this d_i another array with P entries of magnetic volumes, $V_{\mathsf{M},i} = \frac{\pi d_i^3}{6}$, and a third array of the hydrodynamic volumes, $V_{\mathsf{H},i} = \frac{\pi (d_i + d_{\mathsf{coat}})^3}{6}$, is generated.

4. In the same way, a P-entry array of individual effective anisotropy values K_i is generated from K_{eff} and $\sigma_{K_{\mathsf{eff}}}$.

5. A cubic box of length $L = \sqrt[3]{\frac{P}{C}}$ is generated, in which the individual particles are randomly distributed with $V_{\mathsf{M},i}$, $V_{\mathsf{H},i}$, K_i and average interparticle distance, $r_{\mathsf{avg}} = \sqrt{3} \cdot \frac{1}{\sqrt[3]{C}}$.

6. These distributed particles are given randomly oriented axes of magnetization, $\hat{m}_i$, and directions of the easy axis, $\hat{n}_i$. This step completes the static arrangement of the MNP ensemble from which the relaxation dynamics are evolved in time in the next step.

7. The number of full magnetization cycles ($M(H)$-loops) simulated, n_{cyc}, is multiplied by the number of time steps, N_t, so that the new N_t' accounts for all time steps over n_{cyc} cycles: $N_t' = N_t \cdot n_{\mathsf{cyc}}$.

8. In order to account for any possible relaxation trends due to thermal activation, the system is first left to thermalize for $\frac{N_t'}{5}$ time steps with zero applied field, $H_{\mathsf{ac}} = 0$. For each time step, the magnetization and torque are calculated under thermal fluctuations from eqs. (2.33), (2.34), (2.40) and (2.41). The magnitude of the thermal fluctuation at each step is given by eqs. (3.12) and (3.13) and the effective field reads: $H_{\mathsf{eff},i} = H_{K_i} + H_{\mathsf{th}} \, (+H_{\mathsf{pp\text{-}IA},i})$, with the individual particle's anisotropy field, H_{K_i}, eq. (2.15), and, when used, the magnetic dipole-dipole interaction field, $H_{\mathsf{pp\text{-}IA},i}$.

9. After thermalizing, the external field, H_{ac}, eq. (2.29), is applied and the system is evolved for N_t time steps as described in step 7 above. In other words, a full cycle of the $M(H)$-loop is simulated and the effective field reads: $H_{\mathsf{eff},i} = H_{K_i} + H_{\mathsf{ac}} + H_{\mathsf{th}} (+H_{\mathsf{pp\text{-}IA},i})$.

10. Steps 5. - 9. are repeated X times to increase statistical precision.

11. The ensemble average magnetization over X repeated MC-simulation runs each with P particles is calculated for every discretized field step, $\Delta H(\Delta t)$, along the direction of applied field (in z-direction), $\Delta M_z(\Delta H)$, according to eq. (3.20).

12. From $\Delta M_z(\Delta H)$, the $M(H)$-loop is extracted.

Magnetic Particle Heating Predictions from Simulation

For further calculation of the specific loss power (SLP) values from MC-simulations, the area, A, of one $M(H)$-loops is calculated according to the composite trapezoidal rule. In this calculation, the area under a give curve, $g(H)$, is approximated as N_T trapezoids, so that the full length of the integration interval $[-H_0, H_0]$ is split in equidistant sub-intervals $\Delta h_i = h_i - h_{i-1}$ with

$-H_0 = h_0 < h_1 < ... < h_{N_{T_1}} < h_{N_T} = H_0$. Then the integration simplifies to

$$A = \int_{h_1}^{h_2} g(H)dH \approx \sum_{i=1}^{N_T} \frac{g(h_{i-1}) + g(h_i)}{2} \Delta h_i. \qquad (3.21)$$

Here, the length of one interval is chosen as $\Delta h = 0.1\,\text{mT}/\mu_0$ by the default, which is small enough to produce significant results for the applied field amplitude of $H_0 = 6...50\,\text{mT}$ used throughout this thesis. Using eq. (2.61), the SLP is derived from the area A, the applied field frequency f and the MNP material's mass density ρ.

Note that within the specific implementation chosen for the MC-simulations in this thesis, the resulting net magnetization ΔM_z is normalized to unity and the applied field H_{ac} is given in $[\text{mT}/\mu_0]$. Therefore, the integrated area must be multiplied by the saturation magnetization value, M_S, as well as the magnetic constant, μ_0, in order to arrive at the correct units for the extracted area, $A[\text{A}^2/\text{m}^2]$.

3.3. Results and Comparison to Theories

An exemplary $M(H)$-loop generated from MC-simulation is shown in Fig. 3.2. The input parameters are given in the caption. As described in the previous section, the SLP value can be

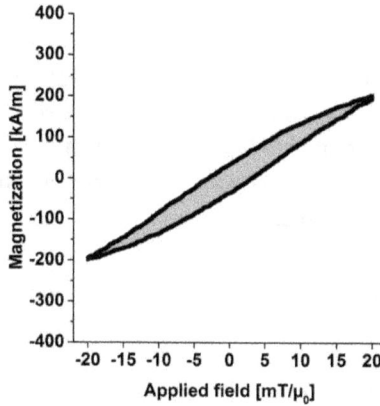

Fig. 3.2.: Exemplary $M(H)$-loop from MC-simulation with input parameters: $M_S = 400\,\text{kA/m}$, $K_{\text{eff}} = 11\,\text{kJ/m}^3$ (magnetite bulk value [20]), $d_M = 15\,\text{nm}$, $d_H = 75\,\text{nm}$, $H_0 = 20\,\text{mT}/\mu_0$ and $f = 176\,\text{kHz}$. The hysteresis area is marked in red.

predicted from the hysteresis area, A, derived from integrating the $M(H)$-loop via the composite trapezoidal rule, eq. (3.21). By simulating the $M(H)$-loops for various particle magnetic sizes, d_M, applied field amplitudes, H_0, and frequencies, f, the dependency of SLP(d_M), SLP(H_0) and SLP(f) can be assessed. Similarly, the dependency of SLP can be assessed for different anisotropy values, K_{eff}. Here, uniaxial anisotropy was used (by inserting eq. (2.13) in eq.

(2.37)). In this way, MC-simulation results are comparable to SWMBT and LRT, which both assume uniaxial anisotropy [42].

3.3.1. Comparison of Theories predicting Magnetic Particle Heating

Here, the SLP values predicted by MC-simulations are compared to SLP values from LRT and SWMBT for an evaluation and validation of the simulation performance of MC-simulations against established theories. As MC-simulations account for size and anisotropy distributions, LRT and SWMBT have also been extended to include a log-normal size (d_M) and anisotropy constant (K_{eff}) distributions for a more exact comparison to MC-simulations. The probability density function (PDF) of the log-normal distribution reads [174, 175]:

$$\text{PDF}(x, \mu, \sigma) = \frac{1}{\sigma x \sqrt{2\pi}} e^{-\frac{(\ln(x) - \mu)^2}{2\sigma^2}}, \tag{3.22}$$

with the parameters, μ, and σ. The mean, $\hat{\mu}$, and standard deviation, $\sigma_{\hat{\mu}}$, of the PDF are calculated from:

$$\hat{\mu} = e^{\mu + \frac{\sigma^2}{2}} \tag{3.23}$$

$$\sigma_{\hat{\mu}} = \sqrt{e^{(2\mu + \sigma^2)} \cdot (e^{\sigma^2} - 1)}. \tag{3.24}$$

Using the PDF (eq. (3.22)) and setting $\hat{\mu} = (d_M$ or $K_{eff})$ and $\sigma_{\hat{\mu}} = (\sigma_{d_M}$ and $\sigma_{K_{eff}})$, $N = 100\,000$ entries of log-normal distributed particle magnetic sizes $\{d_{M,1}, ..., d_{M,N}\}$ or effective anisotropy constants $\{K_{eff,1}, ..., K_{eff,N}\}$ were generated, respectively. For each entry $i = (1, ..., N)$, the particle heating $SLP_i(d_{M,i}, K_{eff,i})$ was calculated using either LRT or SWMBT (cf. Section 2.4). The overall SLP including size and anisotropy distribution was then calculated as the average over all entries according to:

$$SLP = \frac{1}{N} \sum_{i=1}^{N} SLP_i(d_{M,i}, K_{eff,i}).$$

The size-dependence of SLP(d_M) as well as the dependence on the field parameters, SLP(H_0) and SLP(f), was predicted for three effective anisotropy values, $K_{eff} = (5, 11, 17)\,\text{kJ/m}^3$. The middle value is chosen to match the anisotropy constant of bulk magnetite, $K_{Fe_3O_4} = 11\,\text{kJ/m}^3$ [20], while the others are used to assess the effect of anisotropy on SLP in general. For comparing SLP(d_M), the particle magnetic size was varied in the range of $10 \leq d_M \leq 30\,\text{nm}$ in increments of 0.5 nm. This size range was chosen for two reasons: First, for the field parameters used (s. below), this size range bridges the end points of both the regime of validity of LRT and that of SWMBT. This allows to evaluate the behavior of SLP(d_M) in the transition regime between LRT and SWMBT via predictions from MC-simulation. And second, it comprises the range of sizes that are usually of interest for medical applications such as MFH and MPI [23, 60].

For comparing the AMF-dependencies, SLP(H_0) and SLP(f), the field amplitude was varied between $1 < H_0 \leq 20 \, \text{mT}/\mu_0$ in increments of $1 \, \text{mT}/\mu_0$ (with fixed $f = 176 \, \text{kHz}$) and the field frequency was varied between $50 < f \leq 1\,000 \, \text{kHz}$ with increments of $50 \, \text{kHz}$ (with fixed $H_0 = (6$ or $20 \, \text{mT}/\mu_0$). These parameters were chosen to match the AMF of experimental setups (cf. Section 5.2), while at the same time complying with the medical safety limitation of AMF discussed in Section 2.5. The standard parameters used for the following comparison of SLP values between LRT, SWMBT and MC-simulation are summarized in Table 3.1.

Table 3.1.: Standard parameters used for the comparison of SLP values between LRT, SWMBT and MC-simulation divided into classes of simulation input, particle properties and external parameters.

Simulation input	Particle properties		External parameters
$P = 1\,000$	$d_\text{M} = 15$ or $25 \, \text{nm}$	$\tau_0 = 10^{-10} \, \text{s}$	$H_0 = 6$ or $20 \, \text{mT}/\mu_0$
$N_t = 10\,000$	$\sigma_{d_\text{M}} = 0.05$	$M_\text{S} = 400 \, \text{kA/m}$	$f = 176 \, \text{kHz}$
$X = 20$	$d_\text{H} = 75 \, \text{nm}$	$\alpha' = 1.0$	$T = 300 \, \text{K}$
$n_\text{cyc} = 2$	$K_\text{eff} = (5, 11, 17) \, \text{kJ/m}^3$	$\rho_{\text{Fe}_3\text{O}_4} = 5\,180 \, \text{kg/m}^3$	$\eta = 8.9 \cdot 10^{-4} \, \text{Pa} \cdot \text{s}$ (water at $\sim 300 \, \text{K}$)
	$\sigma_{K_\text{eff}} = 0.05$		

Size-dependent Magnetic Particle Heating Predictions

The results for the size-dependent SLP values are depicted in Fig. 3.3. Comparing MC-simulation and LRT, their trends agree qualitatively, as both predict a maximum SLP value for a specific particle size, d_M^*. Quantitative agreement between LRT and MC-simulation is observed in the regions, where LRT is valid (e.g. $d_\text{M} \leq 15 \, \text{nm}$ at $6 \, \text{mT}/\mu_0$, cf. Fig. 3.3(a),(c) and (e)). For increasing K_eff at constant field, d_M^* shifts towards smaller sizes at which the maximum SLP is achieved, however, the absolute SLP value decreases, as seen from comparing Fig. 3.3(a,c,e) or Fig. 3.3(b,d,f), respectively. This is due to the increase in the anisotropy barrier, $\Delta E = K_\text{eff} V_\text{M}$, which in turn increases the Néel relaxation time (cf. eq. (2.32)), leading to the partial inhibition of Néel relaxation, especially for larger particles. These particles consequently do not contribute efficiently to particle heating in MC-simulation and LRT, consequently lowering the overall SLP and decreasing d_M^*, as already observed by Ruta et al. [163].

The exact position of d_M^* shifts to larger sizes for MC-simulation compared to LRT as the field amplitude is increased from $H_0 = 6 \, \text{mT}/\mu_0$ to $H_0 = 20 \, \text{mT}/\mu_0$ (compare Fig. 3.3(a,c,e) with Fig. 3.3(b), (d) and (f)). Note that at this field amplitude, LRT is not valid any more ($\xi > 1$), however, LRT predicts very high SLP values outside its regime of validity compared to MC-simulation. These high SLP values result from the fact that LRT approximates the area of the $M(H)$-loop, A, as an ellipse to calculate the SLP values, as discussed in Section 2.4 (cf. eqs. (2.61) and (2.68)). This is only correct for minor hysteresis loops, i.e. $\xi < 1$, but significantly overestimates A for higher field amplitudes outside the regime of validity of LRT, as already pointed out in literature [42, 163, 176]. Therefore, it is expected that the MC-simulations generally predict the more realistic SLP values [163]. Nevertheless, for high values of K_eff, the trends predicted by MC-simulation and LRT qualitatively agree (cf. Fig. 3.3(e,f)).

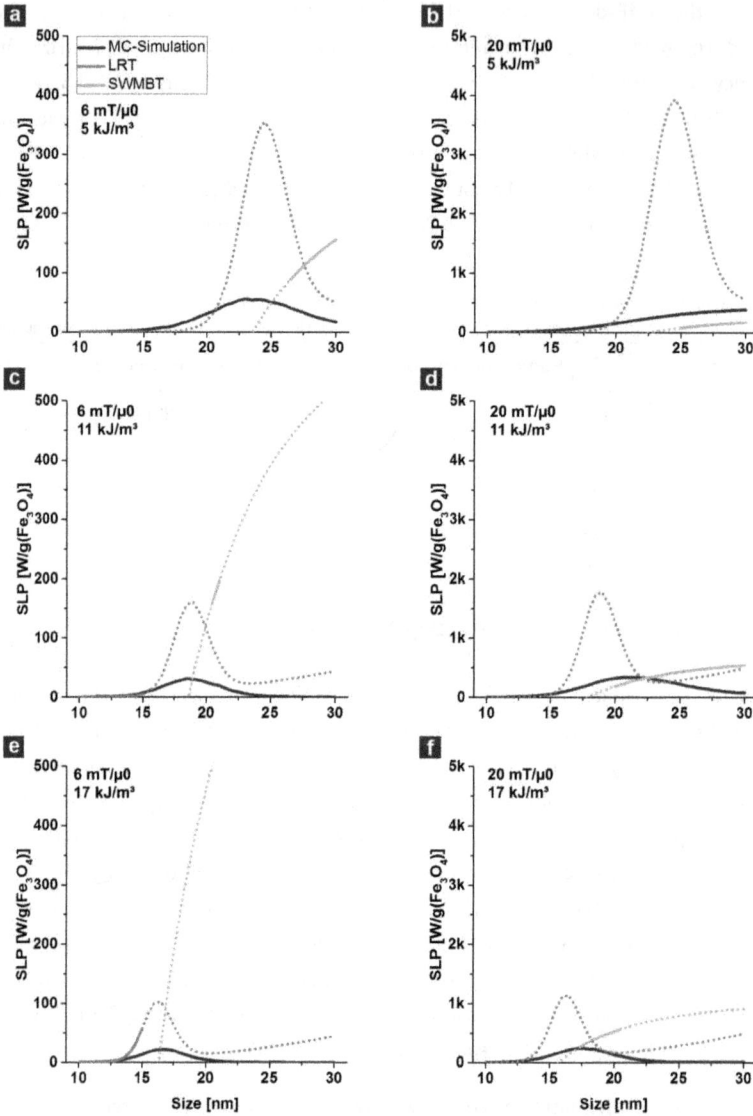

Fig. 3.3.: Comparison of MC-simulation, LRT and SWMBT for size-dependent SLP prediction under various field amplitudes, H_0, and anisotropy constants, K_{eff}: (a) $6\,\text{mT}/\mu_0$ and $5\,\text{kJ/m}^3$, (b) $20\,\text{mT}/\mu_0$ and $5\,\text{kJ/m}^3$, (c) $6\,\text{mT}/\mu_0$ and $11\,\text{kJ/m}^3$, (d) $20\,\text{mT}/\mu_0$ and $11\,\text{kJ/m}^3$, (e) $6\,\text{mT}/\mu_0$ and $17\,\text{kJ/m}^3$, (f) $20\,\text{mT}/\mu_0$ and $17\,\text{kJ/m}^3$. The field frequency is constant, $f = 176\,\text{kHz}$, for all plots. The dotted lines indicate SLP values outside the range of validity for LRT and SWMBT.

This is confirmed by literature, predicting agreement the SLP derived from LRT and dynamic relaxation simulations for high anisotropy constants $K_{eff} \geq 30\,\text{kJ/m}^3$ and small particles core sizes below $d_C \approx 10\,\text{nm}$ [42, 163].

In contrast, SWMBT predicts SLP(d_M) curves that are generally not consistent with the ones from MC-simulations. SWMBT's validity is highly restricted for the example of $H_0 = 6\,\text{mT}/\mu_0$, but its regime of validity expands to smaller particle sizes for increasing anisotropy K_{eff}. The trends observed for SLP(d_M) agree with MC-simulation only for low anisotropy, $K_{\text{eff}} = 5\,\text{kJ/m}^3$, and high field amplitude, $H_0 = 20\,\text{mT}/\mu_0$, cf. Fig. 3.3b. For higher anisotropy values and high field amplitudes, cf. Fig. 3.3d and (f), SWMBT overestimates SLP values compared to MC-simulation, as it does not take particle relaxation processes into account, instead assuming full alignment of the particle magnetic moments with the field.

Field-Dependent Particle Heating Predictions

Field amplitude-dependent SLP(H_0) values predicted by MC-simulation are compared to those SLP values calculated from LRT and SWMBT in Fig. 3.4.

For smaller particle sizes, e.g. $d_M = 15\,\text{nm}$, SWMBT is invalid in the range of field amplitudes investigated (cf. Fig. 3.4(a,c,e)). In fact, the SLP values predicted are negative and therefore not shown in the respective figures. In contrast, MC-simulation and LRT agree very well within and even beyond the regime of validity of LRT, $\xi < 1$ ($\hat{=} H_0 \leq 6\,\text{mT}/\mu_0$) for an anisotropy constant of $K_{\text{eff}} = 11\,\text{kJ/m}^3$ (Fig. 3.4c), resembling the value of bulk magnetite [20]. Under the assumption that MC-simulations predict the SLP(H_0) correctly, this could be an indicator that the regime of validity of LRT could possibly be expanded beyond $\xi > 1$ under certain conditions. For anisotropy values below bulk value, e.g. $K_{\text{eff}} = 5\,\text{kJ/m}^3$, LRT underestimates SLP(H_0) compared to MC-simulations (Fig. 3.4a). While for anisotropy values above bulk value, e.g. $K_{\text{eff}} = 17\,\text{kJ/m}^3$, the situation is opposite and LRT overestimates SLP(H_0) (again compared to MC-simulations; Fig. 3.4e). The trend can be summarized as SLP(H_0) $\propto H_0^2$ for both LRT and MC-simulation for $d_M = 15\,\text{nm}$.

For larger particle sizes, e.g. $d_M = 25\,\text{nm}$, Fig. 3.4(b,d,f), SWMBT is mostly invalid again, except for bulk anisotropy value, $K_{\text{eff}} = 11\,\text{kJ/m}^3$ and high field amplitudes, $H_0 \geq 16\,\text{mT}/\mu_0$ (Fig. 3.4d). However, SLP(H_0) predicted by SWMBT agrees with the SLP(H_0) of MC-simulations for small anisotropies and small field amplitudes quite well ($K_{\text{eff}} = 5\,\text{kJ/m}^3$, $H_0 < 6\,\text{mT}/\mu_0$), even though SWMBT is invalid in this situation (Fig. 3.4b). For higher-than-bulk-value anisotropy constants, e.g. $K_{\text{eff}} = 17\,\text{kJ/m}^3$, SWMBT largely overestimates SLP(H_0) values compared to MC-simulation (Fig. 3.4f). For $d_M = 25\,\text{nm}$, $\xi > 1$ for all situations simulated and therefore LRT is invalid here for all field amplitudes investigated. However, the trends for LRT and MC-simulation agree for high anisotropy constants, e.g. $K_{\text{eff}} = 11\,\text{kJ/m}^3$ and $K_{\text{eff}} = 17\,\text{kJ/m}^3$ (cf. Figs. 3.4(d,f), respectively).

Magnetite particles with $d_M \geq 18\,\text{nm}$ are assumed to be ferromagnetic at excitation frequencies of $f \sim 100\,\text{kHz}$ [23] and their particle heating is determined by hysteresis losses. These hysteresis losses are well described wihtin the Stoner-Wohlfarth model [43,177], which however is only valid for $T = 0$ (s. Section 2.4): At (relatively) high applied field amplitudes $H_0 \geq H_C \approx H_K/2$,

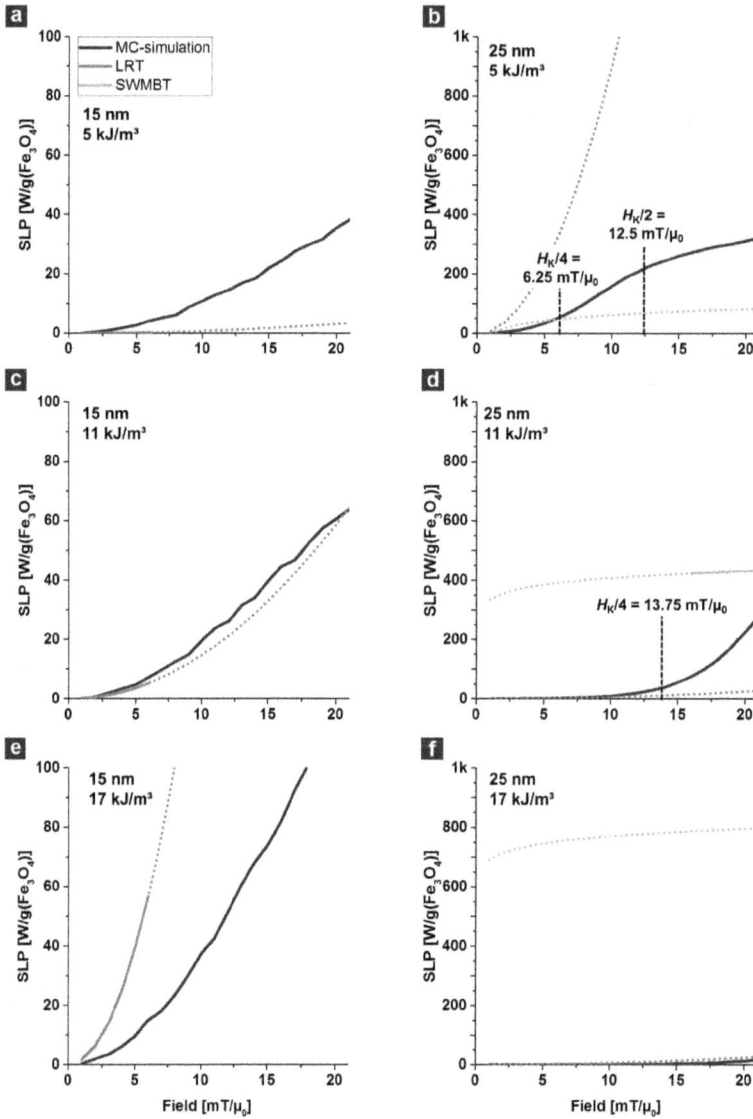

Fig. 3.4.: Comparison of MC-simulation, LRT and SWMBT for field amplitude-dependent SLP prediction for various magnetic sizes, d_M, and anisotropy constants, K_{eff}: (a) 15 nm and 5 kJ/m^3, (b) 25 nm and 5 kJ/m^3, (c) 15 nm and 11 kJ/m^3, (d) 25 nm and 11 kJ/m^3, (e) 15 nm and 17 kJ/m^3, (f) 25 nm and 17 kJ/m^3. The field frequency is constant, $f = 176$ kHz, for all plots. The dotted lines indicate SLP values outside the range of validity for LRT and SWMBT. For ferromagnetic particles ($d_M = 25$ nm), the values of $H_K/4$ and $H_K/2$ are marked in (b) and (d), where $H_K = 2K_{eff}/(\mu_0 M_S)$ (cf. eq. (2.15)).

the intrinsic anisotropy field, H_K (eq. (2.15)), is overcome by the particles' magnetic moment, m, and m is able to align with the applied field [14, 25, 42]. Thus, a steep increase in SLP is observed for $H_0 = H_K/2$, but SLP is zero for $H_0 < H_K/2$, where the particles' magnetic moments are oriented by anisotropy against alignment with the applied field. This can be reproduced by MC-simulations performed at $T = 0$, as shown by the dotted lines in Fig. 3.5 for $d_M = 20$ nm and various anisotropy values, K_{eff}. Under thermal activation, e.g. at $T = 300$ K,

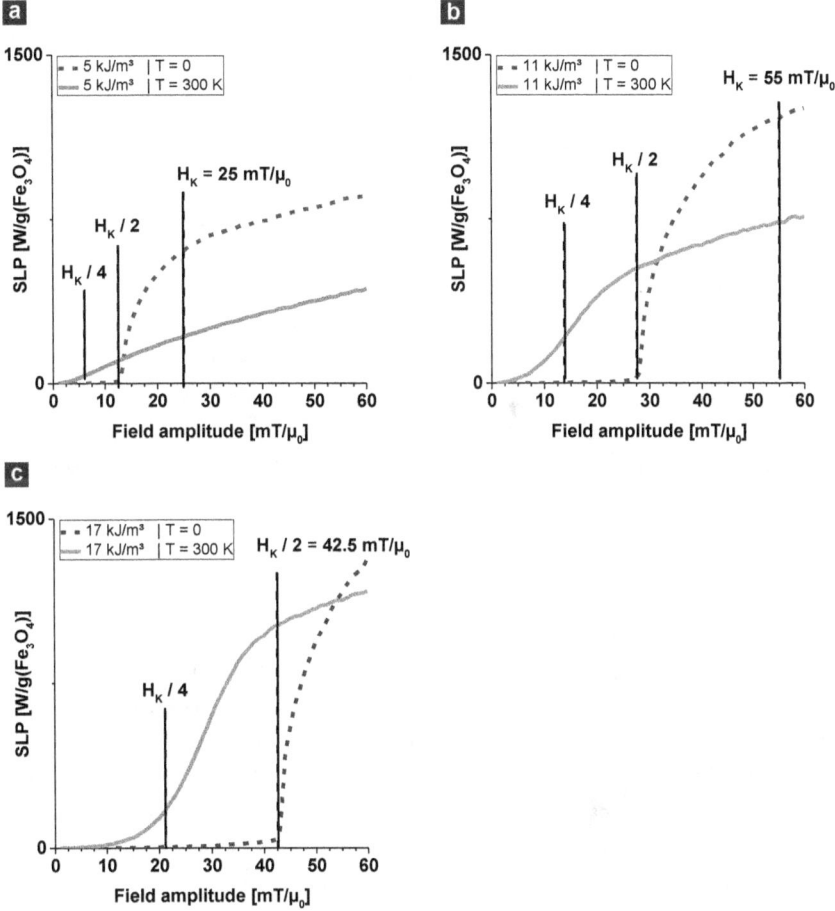

Fig. 3.5.: MC-simulation for field amplitude-dependent SLP for various anisotropy constants (a) $K_{eff} = 5$ kJ/m^3, (b) $K_{eff} = 11$ kJ/m^3, and (c) $K_{eff} = 17$ kJ/m^3. The particle magnetic size was $d_M = 20$ nm. For $T = 0$ the results are equivalent to Stoner-Wohlfarth model predictions, while for $T = 300$ K thermal fluctuations enable particle thermally-driven relaxations of the particles' magnetic moment. The values of $H_K/4$, $H_K/2$, and H_K are marked in each plot.

thermal energy dominates anisotropy energy, $\varepsilon_{therm} = k_B T > K_{eff} V_M$, allowing for thermally-driven Néel and Brownian particle relaxation, cf. eqs. (2.31) and (2.32). Such thermally-driven particle relaxation smoothes the sharp edge in SLP for $H_0 \approx H_K/2$ observed at $T = 0$, allowing a monotonous onset of particle heating already for $H_0 \leq H_K/2$. As MC-simulation is able to account for thermal activation by including thermal fluctuations, this onset of particle heating is observed for MC-simulations with $T = 300$ K in Fig. 3.4b and (d) and Fig. 3.5. From this, the trend of SLP(H_0) predicted by MC-simulation holds as follows: SLP(H_0) $\propto H_0^2$ for

$H_0 \leq H_K/4$, which is in line with the predictions from LRT (cf. eq. (2.68)). This is followed by a linear dependency, $\text{SLP}(H_0) \propto \alpha_1 H_0$, for $H_K/4 \leq H_0 \leq H_K/2$. For $H_K/2 < H_0$ a linear dependency is observed as well, however with a lower increase in particle heating, therefore the trend can be described by $\text{SLP}(H_0) \propto \alpha_2 H_0$ with $\alpha_2 < \alpha_1$.

The results for $\text{SLP}(f)$ are shown for constant $d_M = 15\,\text{nm}$ in Fig. 3.6 and $d_M = 25\,\text{nm}$ in Fig. 3.7, respectively. The frequency behavior of the SLP generally predicts a linear dependency, $\text{SLP}(f) \propto f$, within the frameworks of LRT and SWMBT, Figs. 3.6 and 3.7. For SLP predictions from MC-simulation, the frequency-dependence can be approximated with $\text{SLP}(f) \propto f$ as well, within the modeled frequency interval $f = [50, 1000]\,\text{kHz}$. However, the onset of the $\text{SLP}(f)$ curve shows slight deviations from linear dependencies from MC-simulations for different K_{eff}-values: For small fields, e.g. $H_0 = 6\,\text{mT}/\mu_0$, one observes a steeper-than-linear increase in SLP with f for $K_{\text{eff}} = 5\,\text{kJ/m}^3$ (Fig. 3.6(a)), a linear increase in SLP with f for $K_{\text{eff}} = 11\,\text{kJ/m}^3$ (Fig. 3.6(c)) and a less-than-linear increase in SLP with f for $K_{\text{eff}} = 17\,\text{kJ/m}^3$ (Fig. 3.6(e)). The same trends in $\text{SLP}(f)$ are observed at higher fields, e.g. $H_0 = 20\,\text{mT}/\mu_0$, when comparing Fig. 3.6(b) with Fig. 3.6(d) and Fig. 3.6(f). This reveals a more complex dependency of SLP on anisotropy constant, K_{eff} (as is further discussed in the next paragraph. Nevertheless, the frequency-dependent trend predicted by MC-simulations can be approximated as linear: $\text{SLP}(f) \sim f$.

Anisotropy-Dependent Heating Predictions

The dependence of $\text{SLP}(K_{\text{eff}})$ is arguably the most complex both in LRT and SWMBT as well as in MC-simulations, and so only trends for $\text{SLP}(K_{\text{eff}})$ are discussed in the following. As has already been pointed out in the previous discussion on size- and field-dependent prediction of SLP values, K_{eff} has a major impact on the SLP predictions, however, this impact is different for different models: For SWMBT, the trends for $\text{SLP}(K_{\text{eff}})$ show that an increase in K_{eff} generally increases the SLP value, as can be seen from eqs. (2.62), (2.63) and (2.64) as well as Figs. 3.3, 3.4, 3.6 and 3.7. The $\text{SLP}(K_{\text{eff}})$-dependency for LRT follows the same trend of increasing SLP with increasing K_{eff} within its regime of validity, as observed from the same figures.

For MC-simulations the trend in $\text{SLP}(K_{\text{eff}})$ is more complex, as it also depends on the particle magnetic size d_M and applied field amplitude H_0: From predictions of $\text{SLP}(d_M)$ in Fig. 3.3, one sees that the absolute SLP generally decreases with increasing K_{eff}. However, if specifically considering particles of a certain size, e.g. $d_M = 17.5\,\text{nm}$, one observes an increase in SLP when increasing anisotropy from $K_{\text{eff}} = 5\,\text{kJ/m}^3$ to $K_{\text{eff}} = 11\,\text{kJ/m}^3$ and $K_{\text{eff}} = 17\,\text{kJ/m}^3$, cf. Figs. 3.3(b,d,f), respectively. The same ambiguous trends can be observed for $\text{SLP}(H_0)$ in Fig. 3.5: Generally, the absolute SLP increases for increasing K_{eff}, however, if considering a specific situation again, e.g. $H_0 = 10\,\text{mT}/\mu_0$, the $\text{SLP}(K_{\text{eff}})$ actually shows no monotonous trend in Fig. 3.5 with $\text{SLP}(K_{\text{eff}} = 11\,\text{kJ/m}^3) > \text{SLP}(K_{\text{eff}} = 5\,\text{kJ/m}^3) > \text{SLP}(K_{\text{eff}} = 17\,\text{kJ/m}^3)$. The reason for this behavior lies in the dependence of $\text{SLP}(H_0)$ on the anisotropy field, as discussed in the paragraph on field-dependency of magnetic particle heating above: As SLP is generally

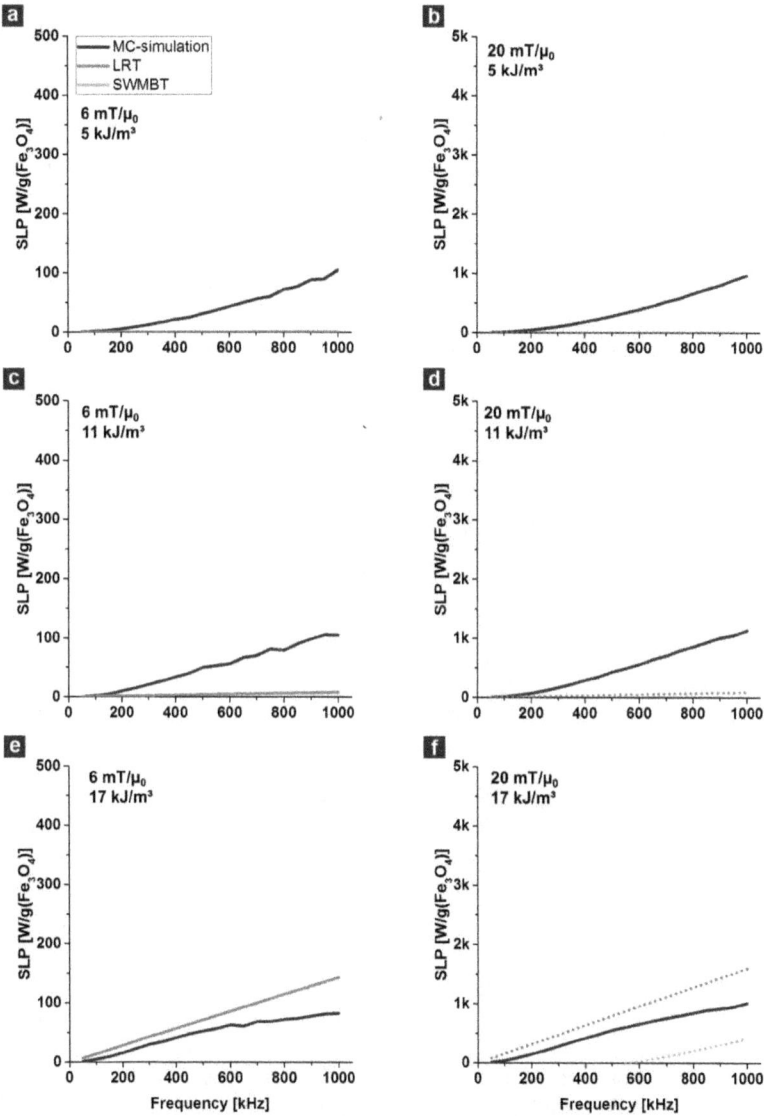

Fig. 3.6.: Comparison of MC-simulation, LRT and SWMBT for frequency-dependent SLP prediction for various field amplitudes, H_0, and uniaxial anisotropy constants, K^u: (a) $6\,\text{mT}/\mu_0$ and $5\,\text{kJ/m}^3$, (b) $20\,\text{mT}/\mu_0$ and $5\,\text{kJ/m}^3$, (c) $6\,\text{mT}/\mu_0$ and $11\,\text{kJ/m}^3$, (d) $20\,\text{mT}/\mu_0$ and $11\,\text{kJ/m}^3$, (e) $6\,\text{mT}/\mu_0$ and $17\,\text{kJ/m}^3$, (f) $20\,\text{mT}/\mu_0$ and $17\,\text{kJ/m}^3$. The magnetic size is constant, $d_M = 15\,\text{nm}$ for all plots. The dotted lines indicate SLP values outside the range of validity for LRT and SWMBT.

low for $H_0 < H_K/4$ (cf. Fig. 3.5) and $H_K \propto K_{\text{eff}}$ (cf. eq. (2.15)), at such low fields, lower anisotropies can show higher SLP values, as discussed above. Consequently, $\text{SLP}(K_{\text{eff}})$ from MC-simulation depends on a delicate balance of the effects of particle size and field amplitude, therefore actually reading $\text{SLP}(d_M, H_0, K_{\text{eff}})$ and a general trend cannot be given.

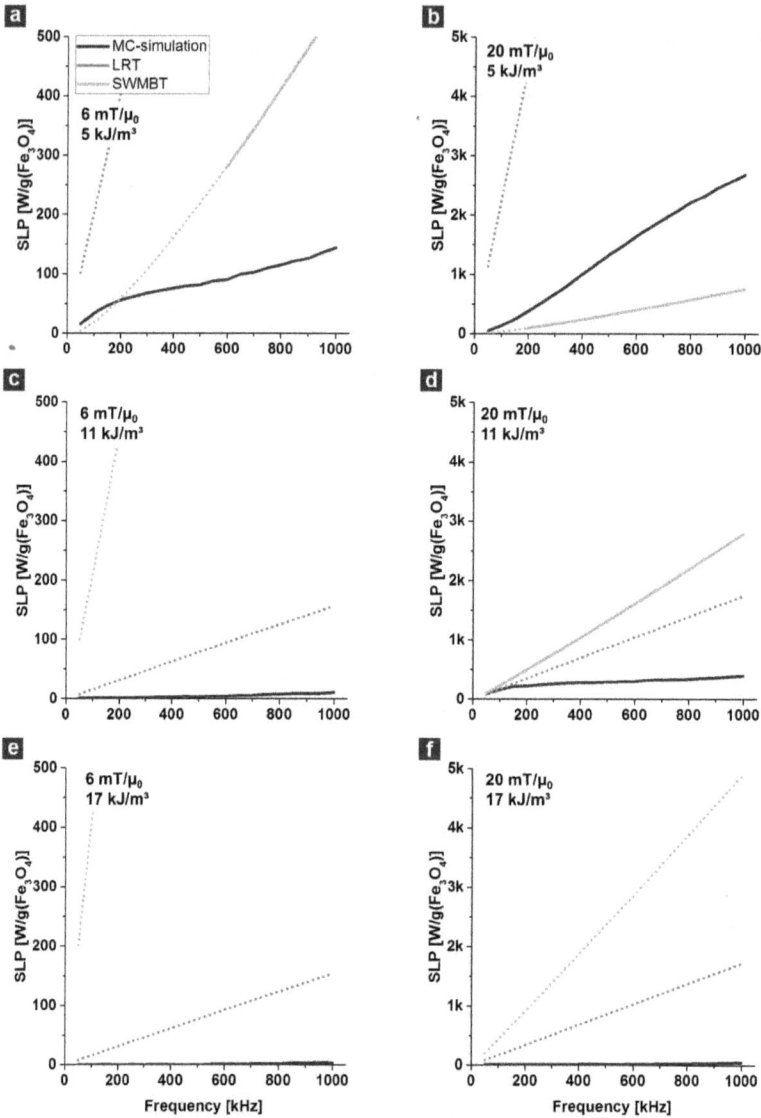

Fig. 3.7.: Comparison of MC-simulation, LRT and SWMBT for frequency-dependent SLP prediction for various field amplitudes, H_0, and uniaxial anisotropy constants, K^u: (a) $6\,\mathrm{mT}/\mu_0$ and $5\,\mathrm{kJ/m^3}$, (b) $20\,\mathrm{mT}/\mu_0$ and $5\,\mathrm{kJ/m^3}$, (c) $6\,\mathrm{mT}/\mu_0$ and $11\,\mathrm{kJ/m^3}$, (d) $20\,\mathrm{mT}/\mu_0$ and $11\,\mathrm{kJ/m^3}$, (e) $6\,\mathrm{mT}/\mu_0$ and $17\,\mathrm{kJ/m^3}$, (f) $20\,\mathrm{mT}/\mu_0$ and $17\,\mathrm{kJ/m^3}$. The magnetic size is constant, $d_M = 25\,\mathrm{nm}$ for all plots. The dotted lines indicate SLP values outside the range of validity for LRT and SWMBT.

3.4. Summary and Outlook

The trends of SLP predicted by LRT, SWMBT and MC-simulations are summarized in Table 3.2. In the case of LRT and SWMBT, the dependencies can be determined from eqs. (2.68) and (2.69) for LRT and from eqs. (2.62), (2.63) and (2.64) for SWMBT, respectively. Please

Table 3.2.: Comparison of SLP dependencies from theories (LRT and SWMBT) and MC-simulation as a function of field amplitude, H_0, and frequency, f, as well as particle size, d_M. A generalized trend for the dependency on the anisotropy constant, SLP(K_{eff}), could not be specified for MC-simulation (s. text). Details on the regimes of validity are found in Section 2.4, eqs. (2.65) (LRT) and (2.64) (SWMBT).

	SLP(H_0)	SLP(f)	SLP(d_M)	SLP(K_u)	Valid for:
LRT	$\propto H_0^2$	$\propto f$	peak at $d_{M,LRT}^*$	$\downarrow$ for $K_{eff} \uparrow$	$\xi < 1$
SWMBT	$\propto -\ln(H_0^{-1})$	$\propto f$	$\propto -\ln(d_M^{-1})$	$\uparrow$ for $K_{eff} \uparrow$	$H_0 \geq H_K/2$; $\kappa < 0.7$
MC-simulation	$\propto H_0^2$ $(H_0 \leq H_K/4)$ $\propto a_1 H_0$ $(H_0 \leq H_K/2)$ $\propto a_2 H_0$ $(H_0 > H_K/2; a_2 < a_1)$	$\sim f$	peak at $d_{M,MC}^*$	—	no general restriction

note that due to the complex behavior of SLP(H_0, d_M, K_{eff}) in MC-simulations, it is difficult to predict *ad hoc* SLP(K_{eff}) behavior. Therefore, no generalization of the trend for SLP(K_{eff}) has been given in Table 3.2.

Several potential improvements for the MC-simulation of magnetic particle heating are briefly discussed here, to conclude the chapter with an outlook of future research opportunities.

In principle, both uniaxial and cubic anisotropy can be assessed with the MC-simulations, as no approximations are made in the MC-simulation that restrict a general description of anisotropy for e. g. three easy axis as used for cubic anisotropy, described by the three directional cosines (cf. eq. (2.14)). Nevertheless at the current stage of the program code used in this thesis, only uniaxial anisotropy could be realized using eq. (2.22) in eq. (2.37). Implementing cubic anisotropy in MC-simulations would therefore naturally be of great interest for future magnetic particle heating prediction, as uniaxial anisotropy is only an approximation for magnetite particles with actual cubic anisotropy [14, 45].

Additionally, magnetic dipole-dipole particle interactions have been neglected for the MC-simulations in the present chapter. This is due to the fact that the MNP show negligible interaction energy compared to thermal activation at the MNP concentrations used throughout this thesis. In anticipation of the results of Chapters 4 and 5, a respective estimation can be found in Appendix A.3.2. For comparison, exemplary SLP values predicted from MC-simulations with and without including magnetic dipole-dipole particle interaction have been compared for a representative MNP concentration of $c = 1\,\text{mg(Fe)}/\text{mL}$. Results for $K_{eff} = (5; 11; 17)\,\text{kJ/m}^3$ at $f = 176\,\text{kHz}$ and $H_0 = 6$ and $20\,\text{mT}/\mu_0$ are shown in Fig. 3.8. Due to the highly increased computation time, the simulation parameters were set to $N_t = 1\,000$, $n_{cyc} = 2$, $P = 100$, $X = 20$ for the MC-simulations. The lower N_t and P increases the statistical inaccuracy of the simulated SLP values, causing the fluctuations observed in Fig. 3.8. However, even within these fluctuations, the comparison of the SLP values simulated for interacting and non-interacting MNP clearly demonstrates that there is no effect of magnetic dipole-dipole interaction at such concentrations. Consequently, for simulating the MNP systems used throughout this thesis,

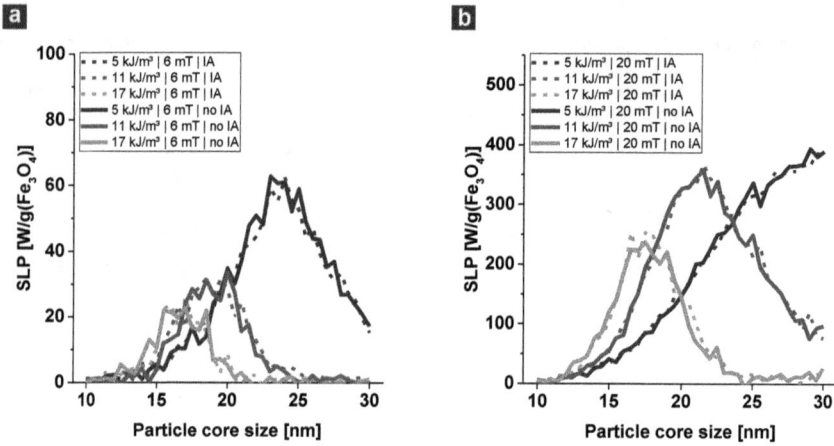

Fig. 3.8.: MC-simulations with and without magnetic dipole-dipole particle interactions (IA) for an MNP concentration of $c = 1\,\mathrm{mg(Fe)/mL}$, various effective anisotropy constants $K = (5; 11; 17)\,\mathrm{kJ/m^3}$, frequency $f = 176\,\mathrm{kHz}$ and (a) $H_0 = 6\,\mathrm{mT}/\mu_0$ and (b) $H_0 = 20\,\mathrm{mT}/\mu_0$.

magnetic dipole-dipole particle interactions can be neglected.

As will be demonstrated later throughout this thesis, the in vivo application of MFH changes the arrangement of particles, strongly increasing the particle interaction due to intracellular particle clustering (cf. esp. Chapters 6 and 8). Furthermore, the general arrangement of particles in clusters and the particular arrangement of particles in chains greatly influences the magnetic particle heating (cf. Chapter 7). Therefore, implementing particle clustering and chaining in the MC-simulations would be of great interest to predict magnetic particle heating for clinically-relevant situations. The actual implementation could be achieved simply by predefining various particle arrangements (cubes, crosses, linear chains, etc.) and using the implementation presented in Fig. 3.1. For an accurate simulation result, magnetic dipole-dipole particle interactions would be required for these MC-simulations, which strongly increases the computational time.

4. Magnetic Particle Characterization

This chapter provides an overview over the materials' characterization techniques and their results for the MNP samples used for later application in MPS and MFH. Section 4.1 explains the basics of MNP synthesis and introduces the specific MNP systems uses throughout this thesis. The subsequent Sections 4.2 to 4.7 describe the experimental techniques employed for MNP characterization (s. Table 4.1) and presents the characterization specific results for each MNP systems. The chapter concludes by summarizing the main MNP properties determined from characterization in Section 4.8.

Though generally reproducible, each synthesis of MNP is carried out under unique conditions. Hence, it is a mandatory requirement to characterize the properties of each MNP batch synthesized individually to validate the success of the synthesis. In particular, as seen in the background Sections 2.2 and 2.4, MNP properties such as particle core and hydrodynamic sizes and the saturation magnetization contribute majorly to magnetic particle relaxation and heating. Thus, a thorough characterization of these MNP properties is essential and the preliminary step before MNP relaxation and heating can be quantified and applied in subsequent chapters. To this end, state of the art materials characterization techniques were used to investigate the MNP properties, as summarized in Table 4.1.

Table 4.1.: List of MNP properties and the respective characterization techniques.

MNP properties	Characterization technique	Note
MNP material, particle crystalline size d_{XRD}	X-ray diffractometry (XRD; Section 4.2)	FFLA, ML and FFAM analyzed with XRD
Particle core size d_C and size distribution	Transmission electron microscopy (TEM; Section 4.3)	
Particle hydrodynamic size d_H and size distribution	Dynamic light scattering (DLS; Section 4.4)	
Sample iron concentration c	Photometric analysis (PA; Section 4.5) Inductively coupled plasma optical emission spectroscopy (ICP-OES; Section 4.5)	FFLA, ML and FFAM analyzed with PA; SEA particles analyzed with ICP-OES
Saturation magnetization M_S, Initial (mass) susceptibility χ_0, Particle magnetic size M_M, and size distribution, estimate of effective anisotropy constant K_{eff}[1]	Superconducting quantum interference device magnetometry (SQUID; Section 4.6) and vibrating sample magnetometry (VSM; Section 4.7)	FFLA, ML and FFAM analyzed with SQUID; SEA particles analyzed with VSM

4.1. Magnetic Particle Systems

Throughout this work, a variety of different MNP systems (also denoted as *ferrofluids* throughout this thesis) will be studied, each characterized by an unique set of physical and magnetic properties: E. g. phospholipid-coated MNP are employed for in vitro studies (s. Chapters 6 and 8), while polymer-coated MNP are used for size-dependent heating studies (s. Section 5.3). In the following, a brief overview of the synthesis routes employed in this thesis will be given, which is concluded by a summary of the different MNP systems used in this thesis in Table 4.2.

Overview on MNP Synthesis

Many labs have studied the synthesis of nanoparticles over the past decades extensively and among many synthesis routines *(alkaine) co-precipitation* and *thermal decomposition* are the most investigated and reliable at present [178]. Both synthesis were used in this thesis and the following brief description of each synthesis is based on [179].

In co-precipitation, magnetite is synthesized by mixing a solution of Fe(II) and Fe(III) salts with a base (e.g. NH_4OH) [180]. After first nucleation of small MNP, rapid crystal growth promotes the precipitation of black Fe_3O_4, where larger particles are favored in growth above smaller ones (so-called Ostwald ripening [181]). The size, size distribution, shape and homogeneity of MNP, depends highly on the ingredients (the type of salts and the Fe(II):Fe(III) ratio), as well as the ambient conditions (reaction temperature, pH value and ionic strength of the medium) [182]. Once these parameters are fixed, the synthesis is fully reproducible and yields a large amount of MNP directly prepared in the aqueous phase. Nevertheless, size control lies only within a range of roughly 2 nm to 20 nm and the size distribution is generally broad. Co-precipitation MNP are sometimes denoted as polydisperse, i. e. the MNP are non-uniform (inconsistent) in size, shape and/or mass distribution. Contrastingly, monodisperse MNP are uniform and consistent in their size, shape and mass distributions.

For thermal decomposition synthesis a metallic precursor (nucleation seed) is pyrolysed with fatty acids (surfactant) in organic solution at temperatures $T > 300°$ C [183]. The resulting MNP are quite monodisperse compared to synthesis by co-precipitation, allowing accurate size control in the range of 5 nm to 30 nm, where the exact size depends on the precursor, solvents and stabilizing agent (and their ratios), as well as the reaction temperature and time. However, the resulting MNP are hydrophobic in organic solution (e. g. chloroform) and need careful phase transfer to the aqueous phase for further biomedical application.

Throughout this thesis, iron oxides, especially magnetite (Fe_3O_4) and maghemite (γ-Fe_2O_3), will be used due to their biocompatibility [184]: In 2009, commercial iron oxide MNP, *Ferumoxytol* (or *Feraheme*) [185], were approved by the U.S. Food and Drug Administration (FDA) for the treatment of iron deficiency for kidney disease [186]. While the same MNP received authorization in Europe in 2012 under the brand name *Rienso* [187]. In vivo clinical safety of these MNP has been evaluated in a one-year retrospecitve observation study in the U.S. involving more than 8 600 patients, reporting only very minor severe adverse effects, such as hypotension (0.12 %), hypersensitivity (0.06 %) or dyspnoea (0.05 %) [188]. In a comparable study observing gastrointernal effects of these MNP conducted in Europe on 1 562 patients, $\sim$ 7.9 % of the patients reported similar adverse effects [187].

On a last remark, magnetite is not stable and thermodynamic processes promote oxidation over time. When exposed to oxygen (e.g. in air), over time the following reaction occurs, turning magnetite into maghemite, a process called aging henceforth:

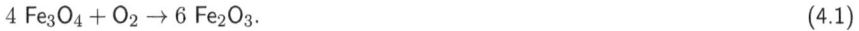

$$4\,Fe_3O_4 + O_2 \rightarrow 6\,Fe_2O_3. \tag{4.1}$$

Homogeneous coating of particles with protective shells prevents aging and promotes biocompatibility: In general, the magnetic particle core is synthesized as described above and coated with different biological (e.g. phospholipid) or synthetic (e.g. polymer) shells. Both types of coatings were applied in the present thesis. These coated particles have a larger size (diameter) compared to the core size, d_C, denoted as the *hydrodynamic size* d_H. An example of a phospholipid-coated particle is shown in Fig. 4.1, whose synthesis is described below.

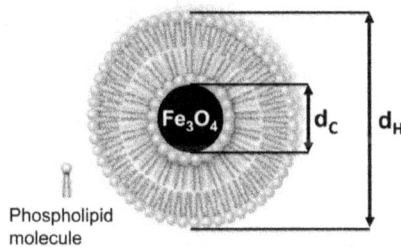

Fig. 4.1.: Schematic drawing of a magnetoliposome. The particle core consists of magnetite (Fe_3O_4) and has the size d_C. The phospholipid coating is considered in the hydrodynamic size d_H. Modified image taken from [189].

Throughout this thesis, the term *size* will always describe the *diameter* of a spherical object, usually that of MNP. In detail, three different particle sizes will be differentiated, denoted as:

- the particle core size, d_C, describing the physical size of the iron oxide core of MNP[2];

- the particle magnetic size, d_M, describing the magnetic size of the iron oxide core of MNP; d_M could be smaller than d_C if for example a magnetically dead layer exists around a MNP [14];

- the particle hydrodynamic size, d_H, describing the particle size including the particle coating and ion layer dissolved in liquid solution.

In the following, a short description of the MNP systems used in this thesis is given, divided according to the respective synthesis method and summarized in table 4.2.

---

[2]As will be shown throughout this chapter, d_C can either be determined with transmission electron microscopy (TEM), yielding d_{TEM}, or with X-ray diffractometry (XRD), yielding d_{XRD}. Throughout this thesis, $d_C \hat{=} d_{TEM}$ holds.

Synthesis of Co-Precipitation Particles

The co-precipitation particles were prepared according to the method of Khalafalla et al. [190]. Details on the exact route of synthesis are found in [18]. Due to their coating with lauric acid, these MP samples will be referred to as *FFLA*, FerroFluid stabilized with Lauric Acid, henceforth. The FFLA samples are kept in deionized water as a carrier liquid. The FFLA used in this thesis are from the same batch.

In order to increase biocompatibility of the MNP for in-vitro experiments (s. Chapters 6 and 8), MNP were coated with a different shell of a phospholipid bilayer as introduced by De Cuyper et al. [191]. The phospholipid consists of the two phospholipid headgroups DMPC[3] and DMPG[4] mixed in the ratio 9:1 (DMPC to DMPG). ML are prepared from FFLA particles and their lauric acid coating is exchanged for the phospholipid coating via ligand exchange. Further details on the synthesis are found in [18]. The resulting particles are called *magnetoliposomes* according to [192] and were kept in TES buffer[5] as the carrier solution. Samples will be referred to as *ML* throughout this thesis. Whenever ML are used for in vitro experiments, they were passed through a $0.22\,\mu$m syringe filter for sterilization. Explicitly, three batches, henceforth referred to as ML1, ML2 and ML3 were used in this thesis. Please note that ML3 was only characterized for its magnetic properties (cf. Section 4.6).

A third type of co-precipitation particles, kindly provided by the PHENIX Laboratories from Pierre and Marie Curie University (UPMC, Paris 6, France), was used for hyperthermia studies in hydrogels (s. Chapter 7). These particles were synthesized according to the procedure of Massart [193] and coated with sodium citrate, making these particles highly anionic. Details on the synthesis can be found in [194, 195]. These particles will be referred to as *FFAM* - FerroFluide Anionique Magnétique - henceforth and as with FFLA, all FFAM used in this thesis are from the same batch.

Synthesis of Thermal Decomposition Particles

Thermal decomposition particle synthesis followed the procedure described in [196], which is based on previous work [197, 198]. The coating, a PMAO[6] and PEG[7] polymer, was synthesized as described in [199]. PEG was chosen as it is biocompatible and nontoxic [200], therefore promoting suitablitiy for in vivo application. Deionized water (DI-H_2O) was used as the carrier liquid according to [196, 201]. Four different samples with different ratios of iron olate to oleic acid, reaction temperature and times were prepared in order to generate particles of different sizes. They are referred to as *SEA1* through *SEA4* in this thesis and used for size-dependent hy-

---

[3] dimyristoylphosphatidylcholine
[4] dimyristoylphosphatedylglycerol
[5] 6.88 mg TES (2-[[1,3-dihydroxy-2-(hydroxymethyl)propan-2-yl]amino]ethanesulfonic acid) buffer, dissolved in 900 mL NaOH (14 M) and 6 L deionized H_2O
[6] poly(maleic anhydride-alt-1-cotadecene)
[7] amine-terminated poly(ethylene glycol)

perthermia studies in Section 5.3. SEA particles were provided by the Krishnan Labs, University of Washington (Seattle, WA, USA).

Overview of the Particles Used in this Thesis

The following Table 4.2, lists the MNP system used and specifies their applications throughout this thesis.

Table 4.2.: Summary of the MNP samples used in this thesis.

Sample	Synthesis method	Coating	Note
FFLA	Co-Precipitation	Lauric Acid	Used for particle agglomeration studies (Section 7.4).
ML1	Co-Precipitation	Phospholipides	Used for MNP-Cell Interaction studies (Chapter 6).
ML2	Co-Precipitation	Phospholipides	Used for in vitro MFH studies (Chapter 8).
ML3	Co-Precipitation	Phospholipides	Used for intracellular magnetization studies (Section 6.4)
FFAM	Co-Precipitation	Sodium Citrate	Used for particle agglomeration and immobilization studies (Sections 7.2 & 7.4).
SEA1	Thermal Decomposition	PMAO-PEG polymer	Used for size-dependent magnetic particle heating studies (Section 5.3)
SEA2	Thermal Decomposition	PMAO-PEG polymer	s. above (Section 5.3)
SEA3	Thermal Decomposition	PMAO-PEG polymer	s. above (Section 5.3)
SEA4	Thermal Decomposition	PMAO-PEG polymer	s. above (Section 5.3)

4.2. X-Ray Diffractometry

X-ray diffractometry (XRD) is applied to determine the symmetry and crystal structure of a polycrystalline sample. The mean particle crystalline size, d_{XRD}, of the magnetic core of the MNP is also determined from the diffraction pattern that XRD yields.

General Principle

X-rays are generated in a radiation tube by accelerating electrons in a high energy electric field towards a metal anode, where the incoming electrons eject anode electrons from their binding orbitals, leaving vacancies in the electron shell of the anode material's atoms. These vacancies are filled by electrons from higher orbitals (or free electrons), causing characteristic high energy X-ray photon radiation. These X-ray photons are focused by a collimator on the sample from where they are diffracted at the atom lattice according to Bragg's law [202]. Parallel lattice planes fulfill the Bragg condition and yield diffraction reflexes of constructive interference at characteristic angles, θ, unique to the sample's crystalline structure. The general principle is shown in Fig. 4.2. By comparison to reference diffraction patterns of known materials, the

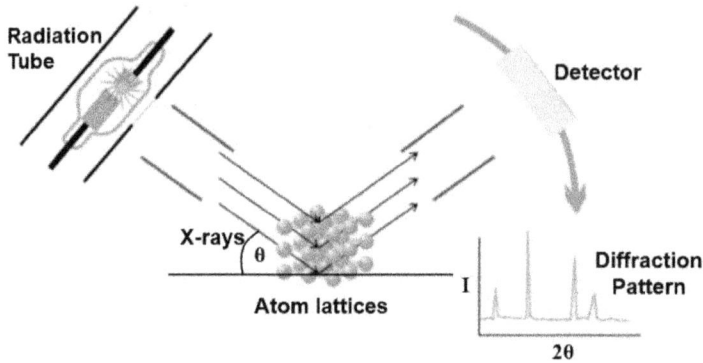

Fig. 4.2.: XRD general principle: X-rays are emitted by the radiation tube and diffracted at the atom lattices at angles θ characteristic to the sample material. A detector measures the angle-dependent intensity *I* of X-ray photons as it rotates; note that typically the angle 2θ is recorded. At certain angles of diffraction, θ', where the Bragg condition is met, constructive interference leads to material-specific peaks in the intensity diffraction pattern. Adapted from [203].

sample material can be identified (see below). Additionally, the average diameter d_{XRD} of the polycrystallites (denoted as particle crystalline size in this thesis) can be estimated from the peak width at full-width-half-maximum (FWHM), w, and the Bragg-angle, θ', by the modified Scherrer equation [204]:

$$d_{XRD} = \frac{K \cdot \lambda_{k_\alpha}}{w \cdot \cos(\theta')},$$

(4.2)

with the shape factor $K = 0.9$ for spherical shapes and the wavelength of X-ray light λ_{k_α}.

Sample Preparation

A glass substrate ($20 \times 20 \times 0,5\,\mathrm{mm}^3$; Menzel Glasbearbeitungswerk GmbH & Co. KG, Braunschweig, Germany) was cleaned in an acetone, followed by an isopropanol ultrasound bath ($49\,\mathrm{kHz}$; Emmi-5, EMAG AG, Mörfelden-Walldorf, Germany) for 15 minutes each and subsequently brushed and dried in an air stream. $200\,\mathrm{\mu L}$ ferrofluid was drop-cast on the clean substrate and dried the fume hood overnight.

MNP samples FFLA, ML1, ML2 and FFAM were prepared for XRD analysis. For FFLA, however, the glass substrate was immersed in $0.1\,\%$ poly-L-lysine (Sigma Aldrich, St. Louis, MO, USA) for $30\,\mathrm{min}$ and subsequently dried for $90\,\mathrm{min}$ before applying the ferrofluid. This step created a positively charged film on the substrate that promoted a homogeneous distribution of the negatively charged FFLA while drying. This was necessary as the sample preparation failed without applying poly-L-lysine to the sample substrate for FFLA, as the particles agglomerated heavily during drying. It is assumed that the additionally charged substrate surface after poly-L-lysine application stabilized the FFLA against clustering caused otherwise by increased dipole-dipole interaction between single MNP, when the lauric acid shell deteriorates during drying.

Experimental Procedure

An X-ray diffractometer of the X'pert series (PANalytical B.V., Almelo, The Netherlands) equipped with a cooper anode was used for the XRD measurements. The system had a wavelength, $\lambda_{k_\alpha} = 1.54\,\text{Å}$ and an acceleration current of $I_{acc} = 40\,\text{kA}$. The signal was recorded by a Xenon gas detector (proportional counter). The sample substrate was mounted on the sample holder and positioned in and aligned parallel to the X-ray beam by adjusting the three rotational angles of the sample plate (ψ, ω, φ, cf. Fig. 4.3).

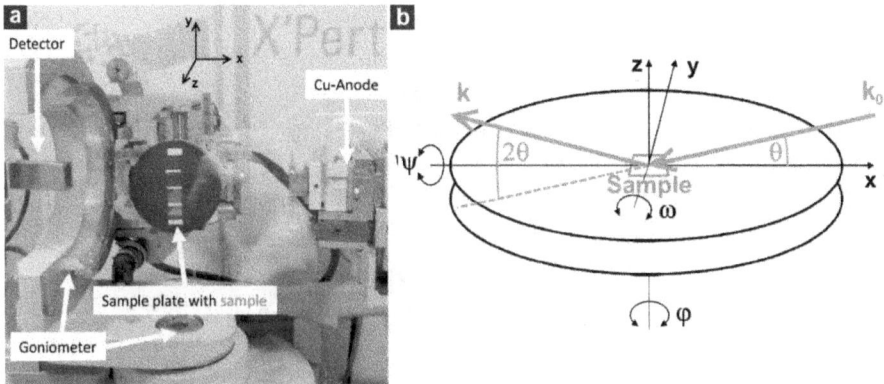

Fig. 4.3.: Working principle of the XRD-Setup: (a) depicts the X'pert XRD experimental setup with key components pointed out by white arrows. (b) describes the rotation angles of the sample plate and axes with respect to the position of the sample plate. It shows the incoming (k_0) and diffracted X-ray beam (k) as well as the diffraction angle θ. (b) adapted from [18].

Since polycrystalline samples (as used throughout this thesis) do not show long range structural order and in order to minimize artifact of the underlying substrate structure, the *grazing incident* measurement scheme has been chosen and the incident angle ω was fixed to $\omega = 0.7°$. This allowed to scan a larger sample surface area, so that more sample atoms contributed to the signal. The range of angles under which the diffracted signal was measured was $2\theta = (20° - 70°)$ and the angle increment was $0.02°$. At each increment, the intensity count was measured for $4\,\text{s}$. Data collection and sample positioning was PC-controlled via the software X'pert DataCollector (PANalytica B.V.).

Experimental Results and Analysis

Exemplary XRD spectra of the analyzed co-precipitation samples are shown in Fig. 4.4, plotted by convention over 2θ. Six reference peak positions for either magnetite or maghemite are plotted with the XRD sample spectra for comparison, obtained from the International Centre for Diffraction Data (ICDD) with reference numbers *PDF #19-0629* for magnetite [205] and *PDF #39-1346* for maghemite [206] (s. Appendix A.2.1). Five distinct peaks can be identified for each sample spectrum. Following common practice, these are labeled with the Miller indices (*hkl*) in the following. The sixth reflex at (422) is too weak to be unambiguously assigned.

Qualitative comparison of XRD spectra and reference peak positions allows to identify iron oxide as the sample material in all cases, however, an explicitly unique differentiation whether the samples are magnetite or maghemite is not possible. Indeed, an unambigous determination of the exact iron oxide phase is generally not possible by means of XRD only, but must be complemented with e. g. Raman spectroscopy, electron energy loss spectroscopy (EELS) and/or Mössbauer spectroscopy [207].

For FFLA (blue curve, Fig. 4.4), there are also distinct peaks between $2\theta = 20°$ and $2\theta = 25°$, as well as around $2\theta \approx 40°$ and $2\theta \approx 50°$, which are identified as lauric acid crystals, precipitating from the FFLA coating during sample preparation (details found in the Appendix A.2.1, Fig. A.1f). Note that the background signal slightly decreases between $2\theta = 20°$ and $2\theta = 40°$, which reflects the diffraction pattern of the glass substrate (s. Appendix A.2.1, Fig. A.1e).

In order to determine the Bragg angles of each sample's XRD spectrum, the five distinct peaks

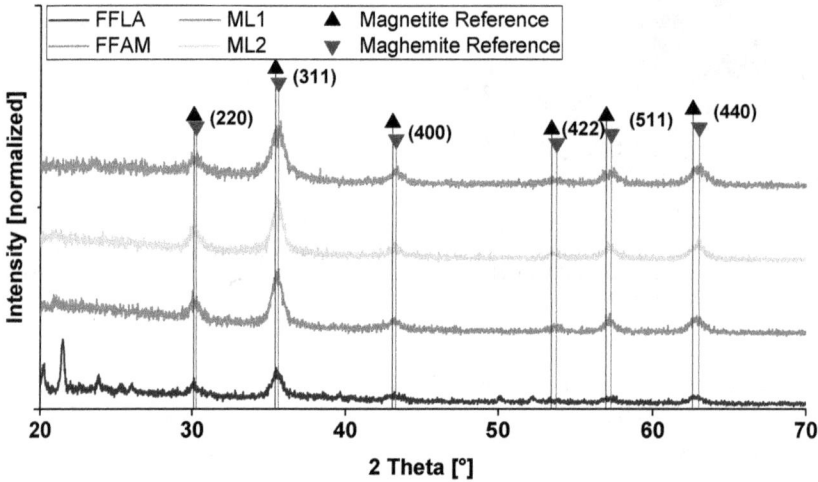

Fig. 4.4.: XRD diffraction patterns for the co-precipitation samples used. Reference angles (labeled with the receptive Miller indices) from the Powder Diffractometry Files are shown in black (magnetite) and brown (maghemite) for comparison and determination of the sample material. Note the additional peaks for FFLA for $2\theta \approx ((20 - 25), 40$ and $50)°$, caused by lauric acid crystallizing from the FFLA shell upon sample preparation.

were fitted with Pseudo-Voigt function [208]:

$$y = y_0 + A\left[m_u\frac{2}{\pi}\frac{w}{4(x - x_c)^2 + w^2} + (1 - m_u)\frac{\sqrt{4\ln(2)}}{\sqrt{\pi}w} \cdot \exp\left(-\frac{4\ln(2)}{w^2}(x - x_c)^2\right)\right] \quad (4.3)$$

where y_0 denotes the offset, A the amplitude, x_c the center position and w the width of the peak, while m_u is the profile shape factor. The peak center position is associated with the Bragg angle $\theta = \theta' = \frac{x_c}{2}$. The fitting parameters are listed in Table 4.3 and the peak fits are found in the Appendix A.2.1.

However, only the (311) peak gives a sufficiently good fit-quality (with $R^2_{adj} > 0.9$). Therefore, only this peak is used to determine the particle crystalline size d_{XRD} with eq. (4.2). Table 4.3 lists the results. The particle crystalline size, d_{XRD}, attained here describes the size of dried

Table 4.3.: Pseudo-Voigt peak fit parameters and particle crystalline size d_{XRD} calculated from these fit parameters using eq. (4.2) with the X-ray wavelength $\lambda_{k_\alpha} = 1,5418367\,\text{Å}$ and the (spherical) shape factor $K = 0.9$. Only peaks with a fit-quality of $R^2_{adj} > 0.9$ are considered for calculating d_{XRD}.

	(hkl)	θ' [°]	$\sigma_{\theta'}$ [°]	w [rad]	σ_w [rad]	R^2_{adj}	d_{XRD} [nm]	$\sigma_{d_{XRD}}$ [nm]
FFLA	(220)	15.058	0.011	0.0087	0.0017	0.607	10.71	0.41
	(311)	17.785	0.004	0.0136	0.0005	0.924		
	(400)	22.187	0.054	0.0457	0.0034	0.474		
	(511)	28.596	0.018	0.0158	0.0018	0.507		
	(440)	31.338	0.045	0.0211	0.0054	0.585		
ML1	(220)	15.089	0.012	0.0112	0.0009	0.618	10.05	0.22
	(311)	17.787	0.003	0.0144	0.0003	0.962		
	(400)	21.601	0.013	0.0122	0.0017	0.596		
	(511)	28.602	0.010	0.0142	0.0012	0.752		
	(440)	31.406	0.010	0.0178	0.0012	0.782		
ML2	(220)	15.092	0.009	0.0135	0.0011	0.757	10.55	0.26
	(311)	17.792	0.003	0.0138	0.0003	0.958		
	(400)	21.644	0.011	0.0054	0.0006	0.615		
	(511)	28.617	0.010	0.0142	0.0010	0.756		
	(440)	31.433	0.008	0.0164	0.0016	0.845		
FFAM	(220)	15.112	0.015	0.0130	0.0012	0.528	7.97	0.24
	(311)	17.792	0.005	0.0183	0.0006	0.928		
	(400)	21.655	0.012	0.0148	0.0014	0.693		
	(511)	28.626	0.014	0.0203	0.0017	0.682		
	(440)	31.456	0.009	0.0209	0.0038	0.821		

particles and is an estimate for the particle crystalline size. It will be compared to results from electron microscopy and magnetometry in Section 4.8.

The MNP systems SEA1 through SEA4 were characterized by XRD, EELS and Mössbauer spectroscopy elsewhere [196, 199, 207], confirming the material to be pure magnetite after synthesis. As the MNP were used for this thesis immediately after synthesis (within 1 month of preparation), their material is assumed to be magnetite.

4.3. Transmission Electron Microscopy

Transmission electron microscopy (TEM) is a versatile imaging technique, mapping sample structures on the micro- and nanoscale with high resolution. In this thesis TEM is applied to determine the particle core size, d_C, as described in this Section, and also to image MNP after internalization inside cells (s. Chapter 6, esp. Section 2.5.2).

TEM measurements on co-precipitation and cells (s. Chapters 6 and 8) were carried out in cooperation with the Electron Microscopy Facility at the Institute of Pathology from the RWTH Aachen University Hospital (Aachen, Germany).

TEM measurements on SEA particles were carried out in cooperation with the Krishnan Lab from the University of Washington (Seattle, WA, USA).

General Principle

In TEM, an electron source (e. g. a thermionic cathode) emits electrons, which are accelerated by a high-voltage field (e. g. 100 kV). By focusing these electrons through an arrangement of electromagnetic (condensing) lens(es) and apertures, a narrow and collimated electron beam is formed. This electron beam is incident via an objective lense and aperture on a thin sample, which must be transparent for electrons (therefore having a maximum thickness $\sim 0.1\,\mu m$) [209]. The electrons interact with the sample by scattering, refraction and transmission. The transmitted electrons, carrying the TEM image information, are collected and further magnified by an array of diffracting and projecting lenses and apertures and focused on a fluorescent screen or CCD camera. The fluorescent screen may be imaged with a binocular microscope, while the image captured by the CCD camera is projected on a computer screen for evaluation. The principle is sketched in Fig. 4.5. Note that the entire process is run under vacuum to ensure an undisturbed passage of electrons.

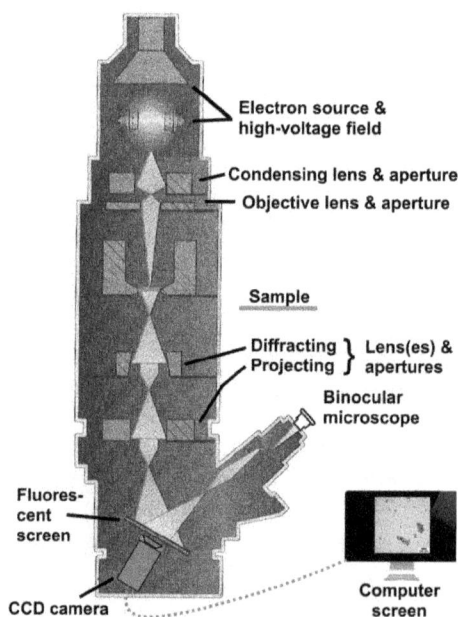

Fig. 4.5.: Principle of transmission electron microscopy (TEM) imaging: Electrons are emitted by a electron source and shaped by condensing lens(es) and aperture(s) to form an electron beam, which is focused on the sample (marked in red) by an objective lens and its aperture. The electrons transmitted through the sample are focused by diffracting and projecting lenses and their apertures either onto a fluorescent screen that can be monitored through a binocular microscope or onto a CCD camera that is connected to a computer screen. Adapted from [210].

For the application of TEM in this thesis (i.e. imaging iron oxide MNP), focus lies on TEM image contrast depicting the massdensity of the sample material, as denser materials cause more electron scattering and refraction and therefore appear darker in the image[8]. Thus, in the final image structures of different massdensity can be distinguished. At a typical acceleration voltage of 100 kV, the incident electrons acquire a speed of $1.64 \cdot 10^8$ m/s, approaching the speed of light [14]. If these electrons are considered as waves, they possess a de Broglie wavelength λ_{dB}, which — including relativistic effects — reads [14]:

$$\lambda_{dB} = \frac{h}{\sqrt{2m_e \cdot e \cdot V_e \left(1 + \frac{e \cdot V_e}{2m_e \cdot c_0^2}\right)}}, \tag{4.4}$$

with h Planck's constant, m_e the electron (rest) mass, c_0 the speed of light, e the elementary charge and V_e the acceleration voltage. The example above with $V_e = 100$ kV yields an electron wavelength of $\lambda_{dB} = 3.8$ pm, giving a theoretical limit for the possible resolution. However, aberrations in the electromagnetic lenses limit the resolution actually achieved in aberration-corrected TEM to 45 pm [211], while standard TEM achieve resolutions of ~ 100 pm [210].

Sample Preparation

MNP samples FFLA, ML1, ML2 and FFAM were diluted to $c = 300$ µg(Fe)/mL with its respective carrier liquid and carefully vortexed after one hour of rest. Subsequently, 1 µL sample was pipetted on formvar-carbon-coated nickel grids (200 mesh) (Electron Microscopy Sciences, Hatfield, PA, USA). The fluid was dried under ambient conditions for one hour prior to TEM measurement.

Details on the sample preparation of SEA particles are found in [196, 207].

Experimental Procedure

TEM measurements on FFLA, MLs, and FFAM were conducted with a Zeiss LEO 906 device (Carl Zeiss GmbH, Oberkochen, Germany) operated at 60 kV and equipped with a slow-scan 2k × 2k *Sharpeye* CCD camera (Tröndle TRS, Moorenweis, Germany).

SEA particles were imaged with a FEI Tenaci F-20 TEM (ThermoFisher Scientific, Hillsboro, Oregon, USA) at 200 kV equipped with a 2k × 2k *US1000* CCD camera (Gatan Inc., Pleasanton, CA, USA).

Experimental Results and Analysis

From the TEM images taken for each sample, the particle core size (i.e. physical diameter), d_C was measured for at least 500 individual particles. For FFLA, ML1, ML2, and FFAM particles, d_C was determined manually on the screen, using the software Paint.NET [212]. To this end,

---

[8]TEM image contrast is also sensitive to the crystallographic orientation and electronic composition of the sample material, especially at higher magnifications, where the TEM electron beam interacts with the materials' orbital electrons, modulating the image contrast intensity.

only isolated particles were considered, whose size was measured along the maximum concentric length a and along the length b perpendicular to a. a and b were averaged via $(a+b)/2$ to yield the particle core size d_M (cf. Section 6.2.3, Fig. 6.6 for an example of the procedure). The size of the SEA MNP was determined using the software ImageJ [213], where a threshold algorithm was applied to identify individual particles, counting and measuring several thousand particles [214]. TEM images and data on these particles are provided courtesy of Eric Teeman, from the Krishnan Labs, who conducted the actual measurements and analysis.

The experimentally particle core size values, x, were sorted according to their sequential sizes, accumulated and normalized to yield the cumulative percentage of the particle core size distribution. As the particle core size is assumed to follow the log-normal distribution [215], their accumulated distribution was fitted to the cumulative distribution function (CDF) of the log-normal distribution [174]:

$$\text{CDF}(x, \mu, \sigma) = \frac{1}{2} \cdot \left(1 + \text{erf}\left(\frac{\ln(x) - \mu}{\sqrt{2\sigma^2}}\right)\right), \tag{4.5}$$

with the fitting parameters, μ and σ, equivalent to the parameters of the log-normal distribution PDF, defined in eqs. (3.22)-(3.24). An exemplary TEM image and the corresponding CDF fit for FFLA are shown in Fig. 4.6. TEM images and CDF fits of all other MNP systems used in this thesis can be found in the Appendix A.2.2, along with the CDF fitting parameters.

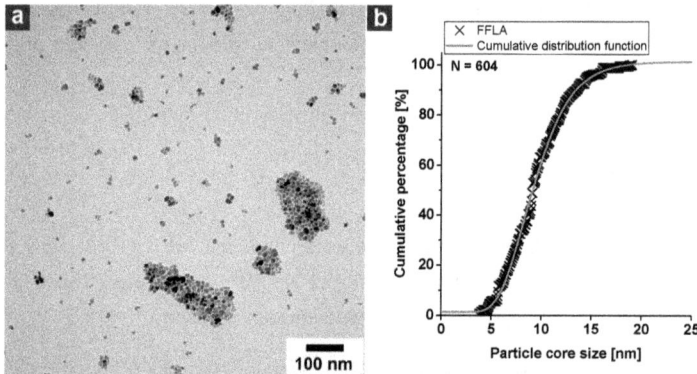

Fig. 4.6.: (a) TEM image of FFLA particles with (b) the cumulative size distribution of N particles with the cumulative distribution function fit.

From the CDF fit parameters the mean particle core size, d_C, and the respective standard deviation, σ_{d_C}, were calculated according to eqs. (3.23,3.24) and listed in Table A.2. The particle core size log-normal distribution function of all MNP systems is plotted for comparison in Fig. 4.7.

It is obvious that the particle core size distribution is rather broad for FFLA, ML1, and ML2, while it narrows down for FFAM. However, these MNP systems are all in the same mean size range of $d_C \approx (9 - 11)$ nm. Contrastingly, the MNP systems SEA1 through SEA4 are larger in

Table 4.4.: Mean particle core size calculated from CDF fits using eqs. (3.23) and (3.24).

Sample	d_C [nm]	Sample	d_C [nm]
FFLA	9.68 ± 3.10	SEA1	21.92 ± 0.88
ML1	11.03 ± 3.67	SEA2	23.13 ± 1.16
ML2	10.63 ± 3.35	SEA3	25.38 ± 2.03
FFAM	8.93 ± 1.31	SEA4	27.77 ± 1.95

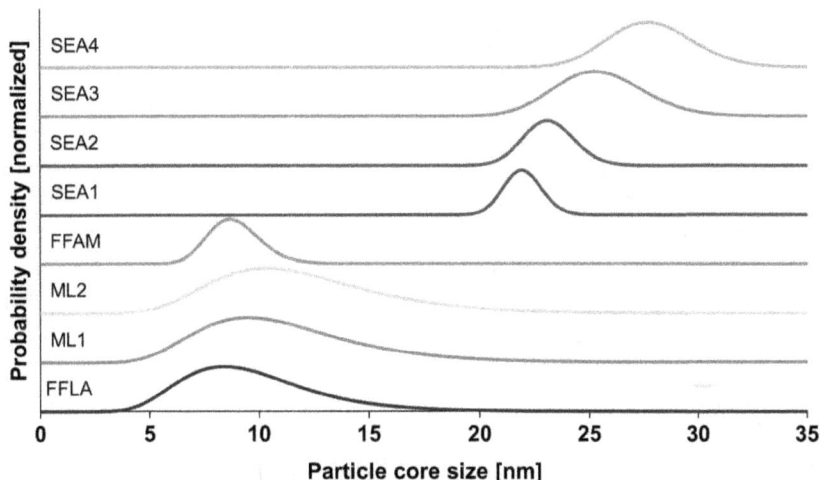

Fig. 4.7.: Probability density function (PDF) of the particle core size log-normal distribution plotted for parameters from Table A.2 with eq. (3.22). Note that the probability density was normalized to unity for all samples.

size and show a narrower size distribution. Taken together, all four SEA particles covering a size range of $d_C \approx (21 - 28)$ nm; cf. Table A.2.

The particle core size derived from TEM analysis will be compared to results from XRD and magnetic measurements in Section 4.8.

4.4. Dynamic Light Scattering

Dynamic light scattering (DLS), is employed to determine the average hydrodynamic size (diameter), d_H, of MNP in liquid suspension and allows an estimation of the polydispersity and tendency of particle agglomeration in a MNP sample. The description of the principle follows the explanations given in [210, 216].

General Principle

In DLS, MNP in liquid suspension are illuminated by a monochromatic laser beam and the intensity fluctuations of the scattered light (usually at a fixed angle) are evaluated over time to determine the average particle hydrodynamic size, d_H. These fluctuations are caused by Brownian (diffusive) motion of the particles in solution, where particle diffusive speed due to Brownian motion is given by the Stokes-Einstein equation:

$$d_H = \frac{k_B \cdot T}{3\pi \cdot \eta \cdot D_t}, \tag{4.6}$$

with d_H the particle hydrodynamic size, D_t the translational diffusion coefficient, k_B the Boltzmann constant, T the absolute temperature and η the viscosity of the liquid. Brownian motion causes the distance between scatterers to constantly change over time and therefore the scattered light undergoes constructive and destructive phase addition, resulting in an intensity fluctuations of the scattered light: As smaller particles move faster than larger particles, their intensity fluctuates faster in time and particle size-dependent information is encoded within the time-dependent intensity fluctuation (s. Fig. 4.8a). These fluctuations are evaluated by

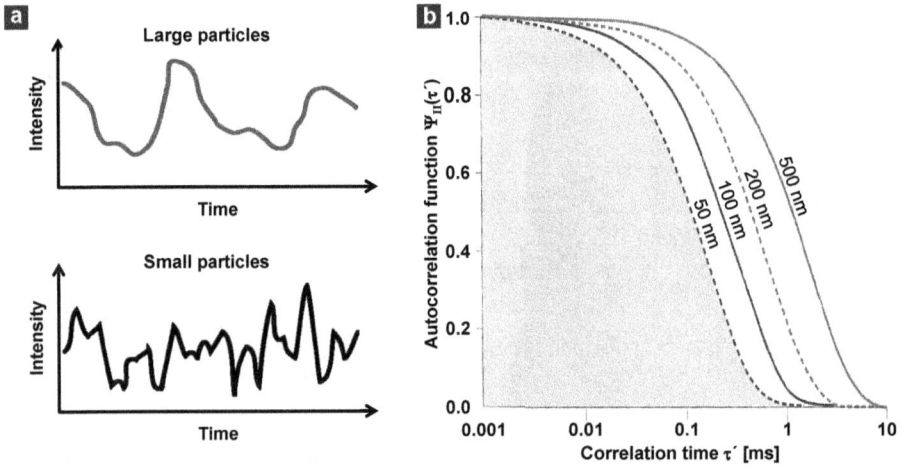

Fig. 4.8.: (a) the scattered light intensity fluctuates faster in time for smaller particles than for larger particles. The resulting autocorrelation function $\Psi_{\parallel}(\tau')$ therefore decays faster for smaller particles, i.e. the correlation is lost for shorter correlation times τ'. Adapted from [210].

calculating the autocorrelation function, $\Psi_{\parallel}$[9], from the time-dependent intensity fluctuations by comparing the intensity of scattered light at time t, $I(t)$, with the intensity at time $t + \tau'$, $I(t + \tau')$, and averaging its product, $I(t) \cdot I(t + \tau')$, over a long period T [210]:

$$\Psi_{\parallel}(\tau') = \lim_{T \to \infty} \frac{1}{2T} \int_{-T}^{+T} I(t) \cdot I(t + \tau') dt. \tag{4.7}$$

Denoting τ' as the correlation time, the autocorrelation is determined from calculating $\Psi_{\parallel}(\tau')$ for many correlation times, typically from the interval of $\tau' = [0, 10]$ ms. The correlation is lost (i.e. $\Psi_{\parallel} \to 0$) quicker for comparing intensities that are fluctuating faster in time, therefore,

---

[9]Actually, the second order autocorrelation function, Ψ_2, is measured by DLS and must be converted to the first order autocorrelation function Ψ_1, using the Siegert equation: $\Psi_2 = 1 + \beta \cdot \Psi_1^2$, where the parameter β accounts for the setup geometry and alignment of the laser beam. Details are found in [217]. For the following description, the autocorrelation function is assumed to include corrections for the setup geometry.

$\Psi_{||}$ decays faster for smaller particles (s. Fig. 4.8b). For a sample of monodisperse MNP, the autocorrelation function can be approximated as a single exponential decay with [217]:

$$\Psi_{||}(q, \tau') = \exp(-\Omega(q) \cdot \tau'), \tag{4.8}$$

with the decay rate $\Omega = q^2 \cdot D_t$, including the wave vector $q = \frac{4\pi n_0}{\lambda} \cdot \sin\left(\frac{\theta_{scat}}{2}\right)$, where λ denotes the DLS laser wavelength, n_0 denotes the sample refractive index and θ_{scat} denotes the angle at which the scattered light intensity is measured. As typically λ, n_0 and θ_{scat} are known for a DLS measurement, the translational diffusive coefficient, D_t, follows directly from fitting the autocorrelation function with eq. (4.8). Using D_t in eq. (4.6) finally yields an estimate for the particle hydrodynamic size d_H.

Sample Preparation

The MNP samples were diluted to a MNP concentration of approximately $c = 0.01\,\text{mg(Fe)}/\text{mL}$ dispersed thoroughly in $500\,\mu\text{L}$ of $DI\text{-}H_2O$ in dedicated sample cuvetts. Before the measurement, the cuvetts were brushed in a nitrogen air stream and the $DI\text{-}H_2O$ was filtered through a $0.22\,\mu\text{m}$ syringe filter, in order to avoid contamination of the samples with dust or other small objects, disturbing the DLS measurement.

Experimental Procedure

The samples were measured with a Zetasizer Nano S DLS device (Malvern Instruments Ltd., Worcestershire, UK). The device employed a $50\,\text{mW}$ He-Ne Laser at a wavelength $\lambda = 633\,\text{nm}$, measured the backscattered light at a fixed angle of $\theta_{scat} = 173\,°$ to the incident light and was operated at a constant temperature of $T = 20\,°\text{C}$. For FFLA, MLs and FFAM samples, each sample was measured 10 times individually and averaged. This measurement was repeated 3 times and again averaged. SEA samples were each measured 25 times individually and then averaged.

Experimental Results and Analysis

From the autocorrelation function, the DLS software derives an intensity size distribution, plotting the relative intensity of the light scattered by the particles, I_{Int}, distributed in various size intervals. As the length of each size interval increases with the size of the particles whose scattered light is measured, the respective intensity data of each interval was divided by the interval's length, $L_{Int}(n)$, for every interval n: $I_{Int}(n)[\text{normalized}] = I_{Int}(n)/L_{Int}(n)$. This accounts for a correctly weighted intensity size distribution.

As the wavelength employed by the DLS device ($\lambda = 633\,\text{nm}$) is much larger than the particle hydrodynamic size, i.e. $\lambda \gg d_H$, the light is scattered by the particles according to the Rayleigh scattering relation, $I_{Int} \propto d_H^6$ [218]. Consequently, the intensity size distribution of a polydisperse MNP sample will be significantly dominated by the intensity of the light scattered by the larger particles. The intensity size distribution of such a sample will show a shoulder for larger particles, which can also be used to detect agglomeration in a samples [217]. For a more accurate estimation of the particle hydrodynamic size, the intensity size distribution can be transformed in volume-weighted intensities with $I_{Vol} \propto d_H^3$ [216]. Such a volume-weighted intensity profile will attach less weight to larger particles, giving a more realistic size distribution of polydisperse MNP samples. Therefore, the volume-weighted intensity interpretation will be used for the remainder of this thesis, except if explicitly assessing MNP samples for agglomeration.

The results from DLS measurements are summarized in Fig. 4.9. Assuming a log-normal

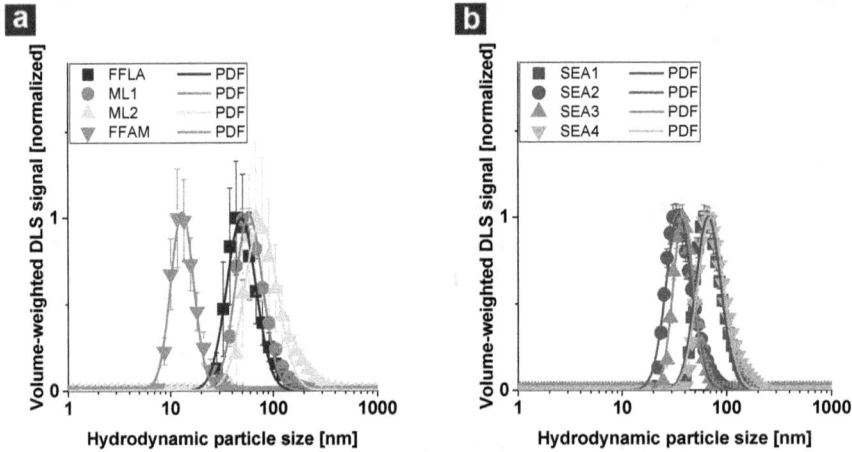

Fig. 4.9.: Particle hydrodynamic size distributions: The plots show the normalized DLS volume-weighted intensity vs. particle hydrodynamic size, (a) for co-precipitation samples FFLA, ML and FFAM and (b) for the thermal decomposition samples SEA1 through SEA4. The data was fitted with the probability density function (PDF) of the log-normal distribution, eq. (3.22), for which the fitting parameters are listed in the Appendix in Table A.3.

distribution of the particle hydrodynamic sizes [215], the experimental data was fitted with the PDF, eq. (3.22). All samples show good agreement for with the PDF with a fitting quality parameter $R_{adj}^2 > 0.95$. From the fitting parameters, μ and σ (listed in Appendix A.2.3, Table A.3), the mean particle hydrodynamic size, d_H, and its standard variation, σ_{d_H}, were calculated from eqs. (3.23) and (3.24). Results are listed in Table 4.5, along with the commonly reported values z_{avg} and the polydispersity index (PDI) for comparison, which are the values for the mean particle hydrodynamic size (z_{avg}) and the distribution width (PDI), calculated in accordance with ISO 22412:2017 [219] by the Zetasizer Software (Vers. 7.11, Malvern Instrumental Inc.)[10]. The PDI provides information about the mono-(or poly-)dispersity of the samples: Experimentally, values of $\sigma_{d_H} < 0.2$ are describing mostly monodisperse samples (i. e. with narrow size distribution) and $0.2 < \sigma_{d_H} < 0.4$ moderately polydisperse ones, while

---

[10]The z_{avg} denotes an overall mean hydrodynamic size, calculated from a forced fit of the single exponential decay (cf. eq. 4.8) to the initial part of the average autocorrelation function $\langle \Psi_{II} \rangle$ (down to 10 % of the zero intercept, i. e. down to $\langle \Psi_{II} \rangle = 0.1$). The PDI refers to the hydrodynamic size distribution width assuming a Gaussian size distribution of particles, calculated from the second order fitting term of the exponential decay [220].

Table 4.5.: DLS measurement results. The table lists the mean particle hydrodynamic size, d_H, calculated from eqs. (3.23) and (3.24), along with the z_{avg} and PDI (=*polydispersity index*), which are the respective average particle hydrodynamic size and distribution width, provided by the dedicated Zetasizer software for comparison.

Sample	z_{avg} [nm]	PDI	d_H [nm]
FFLA	82.43	0.194	57.67 ± 19.51
ML1	97.71	0.248	67.31 ± 21.13
ML2	121.5	0.214	93.00 ± 35.19
FFAM	20.64	0.189	14.35 ± 3.69
SEA1	94.09	0.130	75.17 ± 21.33
SEA2	55.40	0.127	39.39 ± 11.26
SEA3	48.57	0.104	41.50 ± 9.19
SEA4	96.96	0.121	80.04 ± 23.47

$\sigma_{d_H} > 0.4$ is describes samples with a broad distribtion of sizes, consisting of polydisperse MNP [221].

The samples from co-precipitation synthesis (Fig. 4.9a) all show a moderate PDI and are rather polydisperse. ML samples display the highest PDI (> 0.2) combined with larger hydrodynmic sizes of $d_H \approx 68$ nm (ML1) and $d_H \approx 93$ nm (ML2), when compared to FFLA with $d_H \approx 58$ nm. These size differences might be attributed to the different coatings: lauric acid for FFLA and liposome vesicles for MLs,. The results are in agreement with literature reporting on the same MNP systems [222]. The particle hydrodynamic size of the sodium citrate coated FFAM, yielding $d_H \approx 14$ nm, is in the same range as reported in literature [195].

The thermal decomposition samples (Fig. 4.9b) have hydrodynamic sizes between $d_H = 42$ nm (SEA3) and $d_H = 80$ nm (SEA4) with a rather monodisperse particle distribution PDI≤ 0.13. Although, the same amount of PMAO-PEG polymer (20 kDa) was used for coating the SEA particles, their individual d_H varies remarkably. This could be attributed to variations of the coating efficiency during each individual phase transfer, which depend on the sonification process required to disperse the MNP in aqueous solution [214]. As sonification also induces a substantial amount of heat to the particles and that heat can in turn influence the magnetic properties of the MNP [223], a balance between appropriate magnetic properties and d_H is required. In this case, the four SEA samples were sonicated for an equivalent amount of time to be consistent in regards to magnetic properties and total heat applied during sonification, while leaving d_H as a variable. Nevertheless, the low PDI (≤ 0.13) indicates a narrow and homogeneous size distribution for each individual SEA MNP sample.

4.5. Iron Concentration Measurements

The concentration of MNP in suspension, c is measured as the mass of iron per sample volume and determined from photo-spectroscopic analysis. Two methods for determining c will be presented in this section: First, photometric analysis (PA) of Tiron-chelated MNP, as performed for co-precipitation MNP, and second, inductively coupled plasma-optical emission spectroscopy (ICP-OES), used for SEA MNP.

General Principle

For measuring the iron concentration with PA, the MNP coating must first be removed and the iron (oxide) core must be dissolved in aqueous solution by adding a strong acid (e. g. hydrochloric acid, HCl) in order to isolate the iron ions in water. After adding an indicator (e. g. Tiron (s. Sample Preparation for details)), the aqueous iron form a chelating complex, which absorbs light at fixed wavelength (e. g. $\lambda = 400\,\text{nm}$) [224]. As the extinction value, E_{ex}, is proportional to the iron content, c, measuring $E_{\text{ex}}(c)$ via PA yields the sample iron concentration.

In ICP-OES, the liquid sample is heated to plasma temperatures of $(5\,000 - 10\,000)\,^\circ\text{C}$, which evaporates the liquid, breaks all ionic bonds and excites the valence electrons of the isolated sample atoms to higher energy levels [225]. When these valence electrons relax back to their initial energy level, photons are emitted with energies characteristic to the specific element of the respective atoms. The intensity of the emission is proportional to the quantity of the element present in the analyzed sample.

For both PA and ICP-OES, a calibration curve relating the iron concentration to the extinction value (PA) or the emission intensity (ICP-OES) must be determined from specimen of known iron concentration.

Sample Preparation and Experimental Procedure

Note that the experimental procedures start with the preparation of MNP samples of unknown iron concentration for illustrative purposes, while the preparation of reference samples of known iron concentration for the determination of the calibration curves follows afterwards. In reality, the calibration curve must be determined first in order to interpret the results of MNP samples.

For PA, $100\,\mu\text{L}$ diluted liquid MNP sample (typically, $5\,\mu\text{L}$ ferrofluid $+\ 95\,\mu\text{L}$ DI-H_2O, resulting in a dilution factor of $\Xi = 20$) was dissolved in $240\,\mu\text{L}$ of $37\,\%$ HCl (hydrochloric acid, Carl Roth GmbH $+$ Co. KG, Karlsruhe, Germany), resulting in the following reaction for magnetite ($Fe_2O_3 \cdot FeO$):

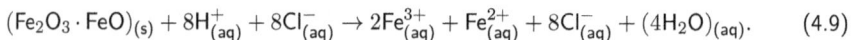

$$(Fe_2O_3 \cdot FeO)_{(s)} + 8H^+_{(aq)} + 8Cl^-_{(aq)} \rightarrow 2Fe^{3+}_{(aq)} + Fe^{2+}_{(aq)} + 8Cl^-_{(aq)} + (4H_2O)_{(aq)}. \qquad (4.9)$$

As the indicator used here (Tiron, s. below) only binds to Fe^{3+} ions, the transformation of Fe^{2+} into Fe^{3+} was accelerated by adding $40\,\mu\text{L}$ of $65\,\%$ HNO_3 (nitric acid, Merck KGaA, Darmstadt, Germany) and the mixture reacted for $15\,\text{min}$ at ambient conditions. Afterwards, $1620\,\mu\text{L}$ deionized water (DI-H_2O) was addad. From this mixture, $500\,\mu\text{L}$ were mixed with $4\,\text{M}$ KOH (potassium hydroxide solution; neoLab Migge GmbH, Heidelberg, Germany), to bind protons (H) and shift the pH of the mixture to ≈ 7. Now, $0.25\,\text{M}$ 4,5-Dihydroxy-1,3-benzenedisulfonic acid disodium salt monohydrate (Tiron, Carl Roth GmbH $+$ Co. KG) was added, which forms a chelating complex with the aqueous Fe^{3+} ions and turns the sample mixture to a reddish color. The mixture is finally stabilized at a pH ≈ 6 (where Tiron shows the highest sensitivity for photometric analysis [226]) by adding $1\,000\,\mu L$ phosphate buffer with pH of 5.5.

In the same way, reference samples of known iron concentration were prepared from Iron Standard for ICP ($c = 1\,mg(Fe)/mL$, Fluka Analytical, Seelze, Germany). Starting with the initial concentration $c = 1\,mg(Fe)/mL$, further dilutions were prepared from the previous solution at a ratio of $50 : 50$ iron solution and DI-H_2O, ending with the lowest iron concentration of $0.03125\,mg(Fe)/mL$. From the reference measurements, a calibration curve was extracted to determine the iron content of the sample of interest (s. Fig. 6.7).

Samples were prepared in triplicate, transferred in disposable micro cuvetts (polystyrene; SARSTEDT AG & Co. KG, Nümbrecht, Germany) and measured with an Ultrospect 2100 pro (Biochrom Ltd., Cambridge, UK) at a fixed wavelength of $\lambda = 480\,nm$.

For ICP-OES, $10\,\mu L$ liquid MNP sample was dissolved in $100\,\mu L$ of $37\,\%$ HCl (Sigma-Aldrich, St. Louis, MO, USA) and let react for approx. $15\,min$. Subsequently, $4\,890\,\mu L$ DI-H_2O was added, resulting in a dilution factor of $\Xi = 500$. Reference samples were prepared in the same way using Iron Standard for ICP (Fluka) with iron concentrations of $(0.0, 0.000\,1, 0.001, 0.01$ and $0.05)\ mg(Fe)/mL$. Reference and ferrofluid samples were analyzed with a fully automatized ICP-OES Optima 8300 (PerkinElmer, Inc., Waltham, MA, USA), which calculates the MNP sample concentration automatically from the calibration curve that is measured from the reference samples individually for each run.

Experimental Results and Analysis

The extinction value, $E_{ex}(c')$, measured in PA for the an exemplary set of reference samples is shown in Fig. 6.7. Fitting the data with the linear function $E_{ex}(c') = \alpha \cdot c' + E_{ex,0}$, with slope α and residual extinction value $E_{ex,0}$, allows to rearrange for $c' = (E_{ex}(c) - E_{ex,0})/\alpha$, to determine the iron concentration for any measured extinction value $E_{ex}(c')$. Note that the concentration c' determined for a MNP sample from the calibration curve must be multiplied by the respective dilution factor Ξ, yielding the final MNP sample iron concentration $c = \Xi \cdot c'$.

For ICP-OES, the calibration curve is recorded with every measurement automatically and internally processed by the measurements software Lab32 (PerkinElmer, Inc.), directly yielding the iron concentration, c', of the diluted MNP sample examined. As described above for PA, the final iron concentration of the initial MNP sample is calculated using the dilution factor from $c = \Xi \cdot c'$.

The iron concentrations measured after synthesis for the MNP systems used in this thesis are listed in Table 4.6. It can generally be seen, that the yield of particles was highest for co-precipitation samples FFLA and FFAM. Contrastingly, the yield for thermal decomposition particles, SEA1 through SEA4, was significantly lower.

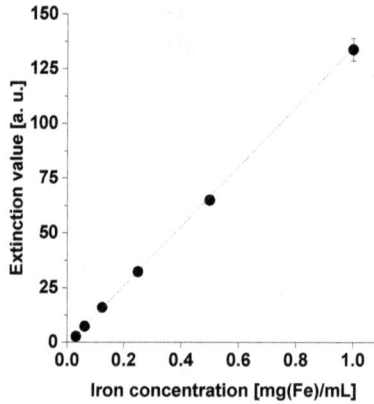

Fig. 4.10.: Extinction value versus iron concentration, c', for reference iron samples measured from photometric analysis. The data is fitted with a linear function $E_{ex}(c') = \alpha \cdot c' + E_{ex,0}$, as shown in red. Here, the parameters are $\alpha = (135.9 \pm 1.6)\,\text{mL/mg(Fe)}$, $E_{ex,0} = -1.4 \pm 0.2$, $R_{adj}^2 = 0.998$.

Table 4.6.: Iron concentration of each MNP sample measured with the respective method.

Sample	Method	Iron concentration [mg(Fe)/mL]
FFLA	PA	11.023 ± 0.083
ML1	PA	1.184 ± 0.027
ML2	PA	2.771 ± 0.020
FFAM	PA	4.444 ± 0.037
SEA1	ICP-OES	1.498 ± 0.012
SEA2	ICP-OES	1.315 ± 0.010
SEA3	ICP-OES	1.513 ± 0.009
SEA4	ICP-OES	1.499 ± 0.012

4.6. Superconducting Quantum Interference Device Magnetometry

Superconducting Quantum Interference Devices (SQUID) are among the most sensitive devices to measure magnetic properties, allowing to measure magnetic moments as low as $\leq 10^{-11}\,\text{A} \cdot \text{m}^2$ [179]. It is used to measure magnetization ($M(H)$-) and zero-field cooled field cooled (ZFC-FC) curves: From $M(H)$ measurements the sample's saturation magnetization, M_S, the initial susceptibility, χ_0 and the particle magnetic size, d_M are determined. ZFC-FC measurements allow to estimate the sample's blocking temperature, T_B, and effective anisotropy constant, K_{eff}.

General Principle

The SQUID sensing unit consists of a superconducting loop with two Josephson junctions in parallel (s. Fig. 4.11). These Josephson junctions are made of thin ($\sim 10\,\text{nm}$) insulating material, separating the two superconducting regions. When applying an external magnetic field, the superconducting loop is threaded by a magnetic flux, Φ[11], which is quantized in integer multiples of the magnetic flux quantum $\Phi_0 = 2.0678 \cdot 10^{-15}$ J/A. In the presense of the external magnetic field, a screening current, I_s, is induced in the superconducting loop due to the Meissner-Ochsenfeld-effect [44], compensating for this external magnetic flux. I_s is interrupted

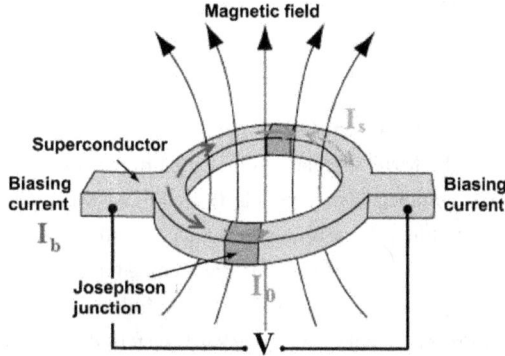

Fig. 4.11.: The magnetic sensor unit of a SQUID: two identical Josephson junctions (dark grey) are build in a superconductor loop. Across the junctions, a weak tunneling current, I_0, flows (red). If a magnetic field is applied, a screening current I_s flows (orange), which changes periodically with multiples of the magnetic flux quanta, Φ_0 (see text). Applying a biasing current, I_b, to the loop (blue) puts the device in resistive mode and allows to detecte the changes in Φ by measureing the voltage V across the device. Adapted from [227].

at the Josephson junctions and thereby experiences a phase shift that results in a Φ-dependent interference for I_s: For $\Phi = \frac{1}{2} + n\Phi_0$ with odd integers n, there is constructive interference and for $\Phi = \frac{1}{2} + n\Phi_0$ with even integers n, there is destructive interference (leading to zero current) (s. Fig. 4.12a). Therefore, the screening current I_s varies only for the difference between one

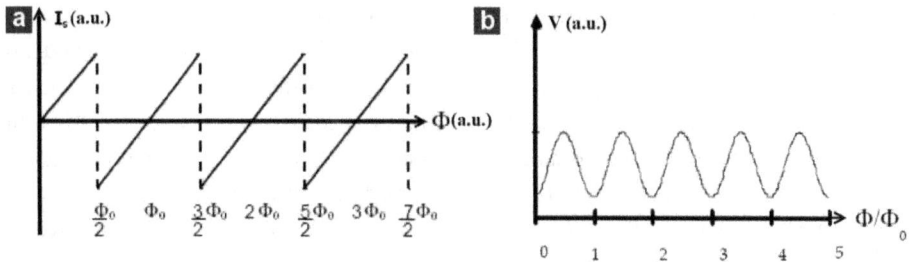

Fig. 4.12.: (a) Perodic dependence of the increasing screening current, I_s, on the magnetic flux Φ: The absolute maximum values of I_s occur from constructive interference of I_s at the Joseqphson junction for $\Phi = \frac{1}{2} + n\Phi_0$ with odd integers n and $I_s = 0$ due to destructive interference for $\Phi = \frac{1}{2} + n\Phi_0$ with even integers n. The dashed lines indicate the reversal in direction of I_s with every increase of Φ by one flux quantum Φ_0. (b) depicts the voltage measured across the Josephson junction, V, resulting from the periodically reversal of I_s with the magnetic flux. A resolution of down to one flux quanta Φ_0 is achieved. Adapted from [228].

admitted flux quantum Φ_0 and external flux just over $\Phi_0/2$ and reverses its direction for every

---

[11] The magnetic flux is generally defined by $\Phi = B \cdot A_\Phi$, with the magnetic induction, B (cf. eq. (2.2)) and the area, A_Φ, that B flows through.

$\Phi = \frac{1}{2} + n\Phi_0$ with odd integers n. This results in a periodically changing screening current $I_s(\Phi)$ as shown Fig. 4.12a. By applying a constant bias current, I_b, to keep the device in resistive mode, these changes in $I_s(\Phi)$ can be measured as a periodical oscillation in the voltage across the sensing unit, $V(\Phi)$, in dependence of the external magnetic flux (s. Fig. 4.12b). In this way, a SQUID is a magnetic-flux-to-voltage transducer, able to resolve even one flux quantum Φ_0. Further details are found in [44, 229].

Sample Preparation

The SQUID operates at high vacuum ($< 1\,\mu$bar); as such condition do not allow to measure liquid MNP suspension, the MNP samples require to be freeze-dried for SQUID measurements. For this, $30\,\mu$L of each co-precipitation sample (i. e., FFLA, ML1, ML2, ML3 and FFAM) of initial iron content as listed in Table 4.6, was mixed with $30\,\mu$L of $\mu = 13.0\,\%$ (weight fraction) mannitol aqueous solution[12]. Samples were directly prepared in dedicated $100\,\mu$L polycarbonate capsules (Quantum Design Inc., San Diego, CA, USA). Subsequently, the samples were frozen in liquid nitrogen. Prefrozen samples were quickly transferred in an Alpha 2-4 LDplus freeze dryer (Christ Gefriertrocknungsanlagen GmbH, Osterode am Harz, Germany) set to $T = -86\,°$C and freeze-dried under vacuum ($< 1\,\mu$bar) over night (for at least 12 h). Upon freeze-drying, mannitol acts as a spacer between MNP, thus ensuring a homogeneous particle distribution. Freeze-dried samples were stored at $4\,°$C until measured by SQUID.

Experimental Procedure

For recording the M vs. H (short: $M(H)$-)curves a dedicated Magnetic Properties Measurement System (MPMS) 5XL EverCool (Quantum Design Inc.) was utilized. Each measurement was performed in two steps, first measuring ZFC-FC curves, followed by measuring the initial magnetization and the hysteresis curves: For ZFC, the sample was cooled from $T = 295\,$K to $T = 5\,$K with $H = 0$. Subsequently, $H = 0.79\,$kA/m ($1\,$mT/μ_0) was applied and the temperature was slowly increased up to $T = 295\,K$ again, while the sample magnetization was recorded. For the following FC experiment, the sample was cooled back down to $T = 5\,$K with the field still switched on. In the second step, the initial magnetization curve (virgin curve) going from $H = 0$ to $H = 3979\,$kA/m ($5\,$T/μ_0) was recorded at $T = 295\,$K, followed by recording a hysteresis ($M(H)$-)curve from $H = -3979\,$kA/m to $H = 3979\,$kA/m, also measured at $T = 295\,$K. Before each measurement, the signal of an unloaded sample capsule was measured and used for a background subtraction.

The ZFC-FC measurement for ML3 was performed in the temperature range of $T = (5-400)\,$K.

Experimental Results and Analysis

The magnetic moment [emu] data obtained from SQUID measurements after background subtraction was normalized to the magnetite content per sample (using the MNP concentration,

---

[12] $1.5\,$mg mannitol dissolved in $10\,$mg of DI-H$_2$O and vortexed.

cf. Table 4.6) and expressed in [kA/m]. Furthermore, a linear function was subtracted from the resulting $M(H)$-curves to account for a diamagnetic portion of the sample. Magnetization curves for samples FFLA, ML1, ML2 and FFAM are shown in Fig. 4.13. The magnetization curve for ML3 is shown in Fig. 4.14. To evaluate the magnetic properties of the MNP samples,

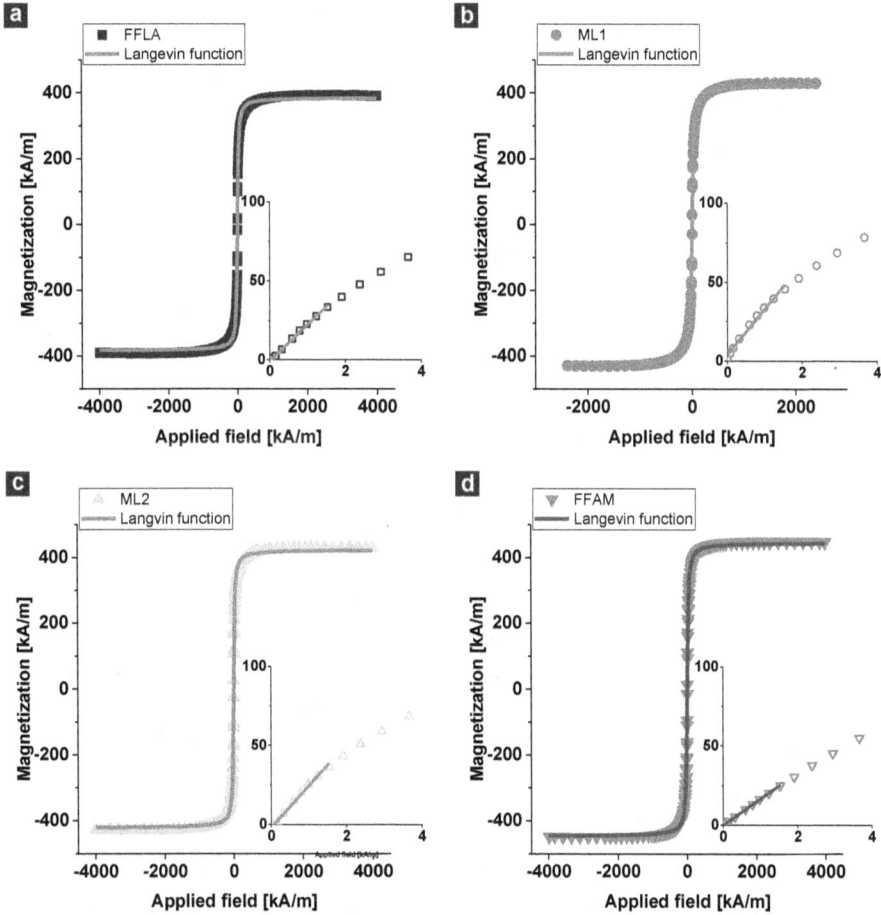

Fig. 4.13.: Magnetization curves of co-precipitation samples fitted with the Langevin function: (a) FFLA, (b) ML1, (c) ML2, (d) FFAM. Insets show virgin curves for low applied fields, with a linear fit between $H = 0$ and $H = 1.6\,\text{kA/m}$, to determine the initial susceptibility, χ_0 (s. Table 4.7).

the experimental data was fitted with the Langevin function:

$$M(H) = M_{\text{S}} \cdot L(\xi) = M_{\text{S}} \cdot (\coth(\xi) - 1/\xi), \tag{4.10}$$

with $\xi = \frac{m \cdot H}{k_{\text{B}} T} = \frac{\mu_0 V_{\text{M}} M_{\text{S}} H_0}{k_{\text{B}} T}$, as defined in eqs. (2.27) and (2.28). The Langevin function fitting allowed for accurate determination of the sample saturation magnetization M_{S}. For $H \to 0$, eq. (4.10) simplifies to $M \approx \frac{\mu_0 \cdot M_S^2 \cdot V_{\text{M}}}{3 \cdot k_B \cdot T} \cdot H = \chi_0 \cdot H$, with the initial magnetic (mass)

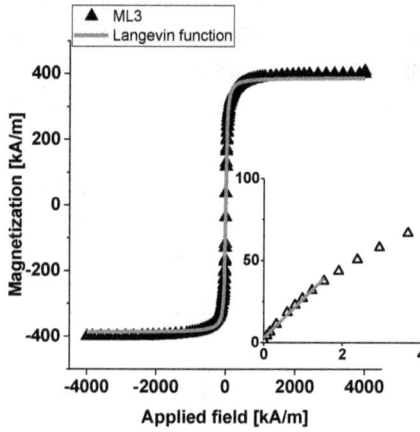

Fig. 4.14.: Magnetization curve of ML3 fitted with the Langevin function. Inset shows virgin curve for low applied fields, with a linear fit between $H = 0$ and $H = 1.6\,\text{kA/m}$, to determine the initial susceptibility, χ_0 (s. Table 4.7).

susceptibility, χ_0 (s. Section 2.1, cf. eq. (2.1)). χ_0 was derived by linear fitting to the virgin curves at the origin ($H = 0...1.6\,\text{kA/m}$), as depicted in the insets in Fig. 4.13.

Under the assumption that the particle magnetic sizes follow the PDF of a log-normal distribution, $\text{PDF}(d_\text{M}, \mu, \sigma)$ (cf. eq. (3.22)), and in the absence of particle-particle interation between particles, the particle magnetic size, d_M, can be derived from Chantrell fitting [230, 231]: The magnetization of MNP with log-normally distributed sizes is described with a Langevin integral, which weights the contribution of each particle size to the overall magnetization according to:

$$M(d_\text{M}) = \int_0^\infty L(\xi(d_\text{M}))\text{PDF}(d_\text{M}, \mu, \sigma)dd_\text{M} \tag{4.11}$$

with the Langevin function $L(\xi)$ (cf. eq.(2.27)) and the $\text{PDF}(d_\text{M}, \mu, \sigma)$ defined in eq. (3.22). For the limits of small ($\xi \ll 1$) or large fields ($\xi \gg 1$), $L(\xi)$ can be expressed as follows:

$$L(\xi \ll 1) \rightarrow \frac{\xi}{3} \tag{4.12}$$

$$L(\xi \gg 1) \rightarrow 1 - \frac{1}{\xi} \tag{4.13}$$

By inserting eq. (4.12) in eq. (4.11), an expression for the initial susceptiblity $\chi_0 = \frac{dM}{dH}\big|_{H \rightarrow 0}$ can be derived. In the same way, inserting eq. (4.13) in eq. (4.11) yields an expression where $M \propto \frac{1}{H_0}$, from which $\frac{1}{H_0}$ can be derived from linear fitting for high fields $H \rightarrow \infty$. From M_S, χ_0 and $1/H_0$, the median of the log-normal distributed particle magnetic sizes, μ^*, follows according to [230]:

$$\mu^* = \sqrt[3]{\frac{18k_B T}{\pi\mu_0 M_S}} \cdot \sqrt{\frac{\chi_0}{3M_S}\frac{1}{H_0}},$$

(4.14)

with the distribution width σ of the PDF(d_M, μ, σ) of the log-normal distribution function:

$$\sigma = \frac{1}{3} \cdot \sqrt{\ln\left(\frac{3\chi_0}{1/H_0}\frac{1}{M_S}\right)}.$$

(4.15)

Since the median μ^* is related to the PDF parameter μ with $\mu^* = \exp(\mu)$ [175], the magnetic particle size, d_M, and its standard deviation, σ_{d_M}, follow from eqs. (3.23) and (3.24) by using eqs. (4.14) and (4.15), with $\mu = \ln(\mu^*)$.

Results for the saturation magnetization M_S, initial susceptibility χ_0 and particle magnetic size, d_M are found in Table 4.7. Details on the Langevin function and Chantrell fitting parameters are listed in the Appendix A.2.4. For FFLA and ML3, the magnetization is approximately

Table 4.7.: Magnetic particle properties determined from SQUID measurements via Langevin function and Chantrell fitting: magnetic (mass) susceptibility, χ_0, saturation magnetization, M_S and particle magnetic size, d_M. $\Delta = \sigma_{d_M}/d_M$ gives the percentage error on d_M. For details on the fitting parameters see Appendix A.2.4.

Sample	χ_0	M_S [kA/m]	d_M [nm]	Δ[%]
FFLA	23.32 ± 0.55	385.18 ± 1.76	11.83 ± 5.42	45.8
ML1	28.97 ± 1.12	423.35 ± 1.60	12.01 ± 5.42	45.1
ML2	26.51 ± 0.85	422.13 ± 1.43	11.62 ± 5.32	45.8
ML3	23.56 ± 0.78	386.07 ± 1.39	11.75 ± 5.52	46.7
FFAM	16.51 ± 0.12	445.86 ± 1.08	10.43 ± 3.48	33.4

80 % of the bulk value for magnetite, which is 476 kA/m [232]. ML1 and ML2 reach up to approximately 89 %, and FFAM achieve even approximately 93 % of the bulk value. This is somehow contradicting to the experimental observations from literature for magnetite MNP having approximately 80 % of bulk magnetization [233] or even less than 50 % [234]. This lower-than-bulk value for the M_S of MNP is explained by a magnetic-dead layer at the surface of the MNP [235]: Here, the spins at the surface are canted and thus do not contribute to the overall MNP magnetization. Nevertheless, it was recently demonstrated that these canted surface spins can be aligned in very high magnet fields up to 4,000 kA/m (approx. 5 T/μ_0) [236]. As the fields applied in SQUID magnetometry are of the same order of magnitude this could explain the very high values for M_S here.

Another explanation was given by the synthesis method, as it has just recently been hypothized that the reduced MNP saturation magnetization originates from an antiferromagnetic (wustite, FeO) shell forming around MNP during synthesis, which does not contribute to the overall magnetization [237]. The same has been observed for core(FeO)-shell(Fe_3O_4) nanostructures [238,239]. The generation of wustite result from an incomplete oxidation during synthesis [240].

Therefore, this hypothesis concerns mostly particles produced by thermal decomposition that includes a separate oxidation step during synthesis. Contrastingly, co-precipitation MNP are less predisposed to wustite formation and, consequently, a reduction in M_S presumably does not occur for these MNP [179].

The magnetization curves of ZFC and FC experiments were normalized to the sample saturation magnetization M_S obtained from SQUID measurements (s. above; Table 4.7) and are plotted in Figs. 4.15 and 4.16. Additionally, the ZFC peak temperature, T_{max} and the branching temperature, T_{bra}, are estimated from the plots and listed in Table 4.8. The saturation temperature, T_{sat}, could however not be estimated unambiguously from the plots in Figs. 4.15 and 4.16.

The fact that the ZFC-magnetization drops below zero for samples FFLA and ML2 at low

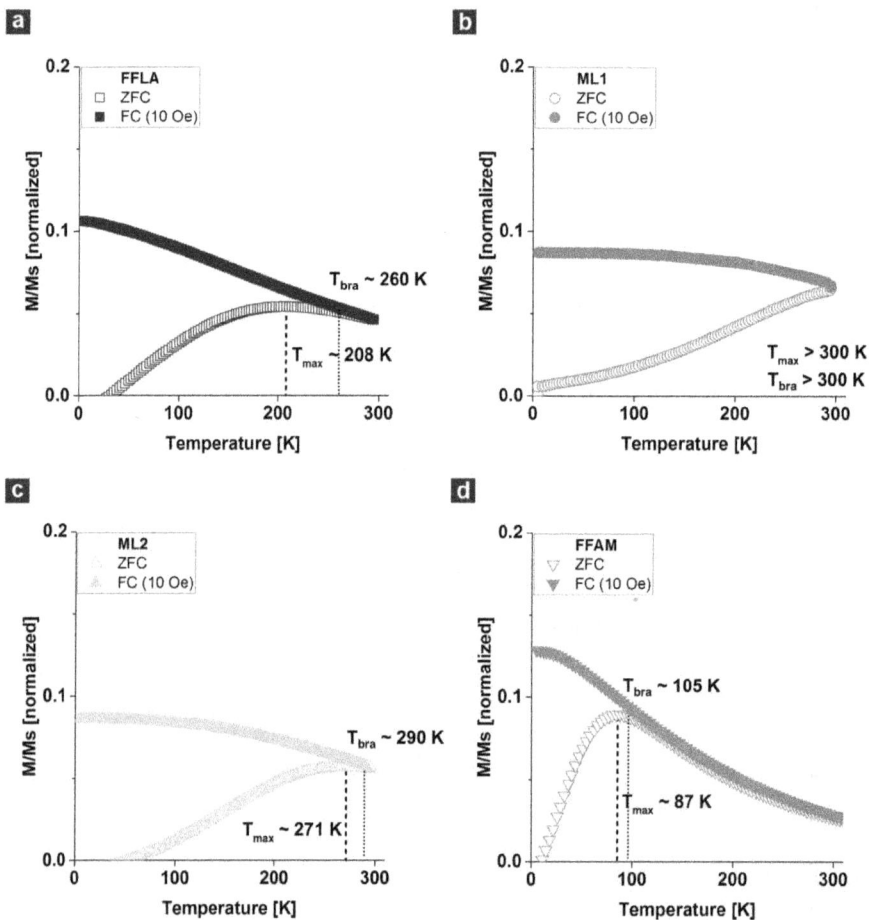

Fig. 4.15.: ZFC and FC magnetization curves normalized to the saturation magnetization, M_S (s. Table 4.7), for (a) FFLA, (b) ML1, (c) ML2, (d) FFAM. The ZFC peak temperature, T_{max} and the branching temperature, T_{bra} are marked.

fields presumably occurs due to a small shift in the position of the MNP sample resulting in

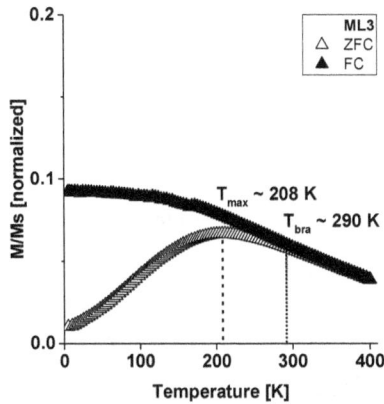

Fig. 4.16.: ZFC and FC magnetization curves normalized to the saturation magnetization, M_S (s. Table 4.7), for ML3. The ZFC peak temperature, T_{max} and the branching temperature, T_{bra} are marked.

acquiring the magnetic moment of the sample as a negative voltage. As this artifact does not influence the general shape of the curves though, it is neglected for the data analysis of the ZFC peak temperature, T_{max}, and the branching temperature, T_{bra}. The temperature peaking the ZFC curves, T_{max}, is smallest for FFAM particles, followed by FFLA, ML3 and ML2, as listed in Table 4.8. T_{max} is actually higher than the maximum measurement temperature for ML1, $T_{max} > 300$ K. Equally, the branching temperature, T_{bra}, increases in the same order. Generally, the width of the ZFC-curve peak increases with increasing the size distribution of the particle magnetic size or equally with the strength of the magnetic dipole-dipole particle interaction, resulting in a distribution of blocking temperatures, $T_B(d_M, K_{eff})$, across the temperatures $T_{sat} < T_B < T_{bra}$ [241, 242]. Therefore, the observations mentioned above are in line with expecting higher T_{max} and T_{sat} for MNP systems with larger particle magnetic sizes and corresponding size distribution widths (cf. TEM particle core size data, Appendix A.2.2, Table A.2). Or equivalently, if the particle magnetic sizes of two MNP samples are equal, but their peak temperatures in ZFC-curves differ significantly, $T_{max,1} < T_{max,2}$, this indicates an increased magnetic dipole-dipole particle interaction for the second sample.

From the peak temperature, T_{max}, one can roughly approximate the mean energy barrier and the mean blocking temperature via [243]:

$$\Delta E = K_{eff} \cdot V_M \approx 25 \cdot k_B \cdot T_B \approx 23.6 \cdot k_B \cdot T_{max}, \qquad (4.16)$$

assuming non-interacting MNP and a typical measurement time of $t_{ext} = \tau = 100$ s in eq. (2.25). It is obvious from eq. (4.16), that from knowledge of T_{max} and the particle magnetic volume, $V_M = \pi/6 \cdot d_M^3$, the effective anisotropy constant, K_{eff} can be estimated. However, Tournus and Tamion explicitly warn to use eq. (4.16) [241, 243], as it will be quite erroneous on K_{eff} due to the fact that T_{max} does not reflect the implicitly underlying size-dependent dispersion of the effective anisotropy (possibly also including magnetic dipole-dipole particle interactions): $T_B(K_{eff}(d_M, \sigma_{d_M}))$. At the same time, the same authors show that if the size-dependence of the effecitve anisotropy constant is implicitly dominated by the overall size distribution effects (i. e.: $T_B(K_{eff}(d_M), d_M) \rightarrow T_B(d_M)$), eq. (4.16) can be used to estimate T_B and K_{eff}. In numbers, particle size distribution effects dominate if for the relative size dispersion, i. e. the percentage of error on d_M, $\Delta = \sigma_{d_M}/d_M \geq 45\%$ holds [244]. As for samples FFLA, ML1, ML2 and ML3 $\Delta \geq 45\%$ holds, estimates for T_B and K_{eff} are derived with eq. (4.16). FFAM falls short of the limit with $\Delta \approx 33\%$, but is calculated anyways but must be treated with care. The results are listed in Table 4.8.

Table 4.8.: ZFC-FC peak temperature, T_{max}, and branching temperature, T_{bra} derived from Fig. 4.15. Furthermore, the mean blocking temperature, T_B, and effective anisotropy constant, K_{eff}, are estimated from eq. 4.16, using the magnetic core size data derived from SQUID data (cf. Tables 4.7, 4.9).

Sample	T_{max} [K]	T_{bra} [K]	T_B [K]	K_{eff} [kJ/m³]
FFLA	208 ± 10	260 ± 15	196	≈ 78
ML1	≥ 300	≥ 300	≥ 283	≥ 108
ML2	271 ± 10	290 ± 15	256	≈ 108
ML3	208 ± 10	290 ± 10	196	≈ 80
FFAM	87 ± 05	105 ± 05	82	≈ 48

The overall uncertainty of the method is reflected in the approximations given for K_{eff}. This is treated as a reminder that the values for K_{eff} must be considered as mere approximations. Nevertheless the K_{eff}-values show some reasonable agreement with literature: For example, $K_{eff} \approx 48\,kJ/m^3$ is in the same order of magnitude as K_{eff} determined from ZFC-FC measurements ($K_{eff} \approx 89\,kJ/m^3$) and fitting to particle heating data ($K_{eff} \approx 26\,kJ/m^3$) for the same particles reported previously [245]. Due to this agreement with literature values, the K_{eff}-value for FFAM is used as an approximation throughout this thesis, even though it must technically be excluded since $\Delta_{FFAM} < 45\%$ (as discussed above).

The fact that K_{eff} increases remarkably for FFLA and MLs compared to FFAM seemingly contradicts the fact that smaller particles are assumed to have higher K_{eff}-values due to increasing surface anisotropy effects, as discussed in Section 2.1.2 (eq. (2.19)). However, an explanation can be given by the superimposed increase in K_{eff} induced by magnetic dipole-dipole particle interactions [14, 246]: Upon freeze-drying during sample preparation, the MNP protective coatings (lauric acid for FFLA and phospholipides for MLs) are distroyed and MNP are generally assumed to agglomerate. This agglomeration decreases the interparticle distance strongly and gives rise to increased magnetic dipole-dipole particle interactions. Such an increase in particle interactions can be qualitatively confirmed by ZFC-measurements, as seen here. For example, the same effect has also been reported from ZFC measurement comparing intracellular MNP and freely suspended MNP [247, 248]. From knowing that the magnetic sizes are $d_{M,FFLA} \approx d_{M,ML1} \approx d_{M,ML2} \approx d_{M,ML3}$ (cf. Table 4.7), and by comparing $K_{eff,FFLA} \approx K_{eff,ML3} < K_{eff,ML1} \approx K_{eff,ML2}$ (cf. Table 4.8), one can assume that the particle interactions for ML2 and ML3 are higher than for FFLA.

4.7. Vibrating Sample Magnetometry

Vibrating sample magnetometry (VSM) is utilized to measure the magnetic properties of the thermal decomposition samples SEA1 through SEA4. As performed with SQUID measurements (s. Section 4.6), the samples' $M(H)$ curves will be measured, from which the saturation magnetization, M_S, initial susceptibility, χ_0, and the particle magnetic size, d_M, are determined. VSM achieves almost as good sensitivity for magnetic moments of $\leq 10^{-9}\,\text{A}\cdot\text{m}^2$ as a SQUID ($\leq 10^{-11}\,\text{A}\cdot\text{m}^2$) [179], while not being affected by constant magnetic fields.

General Principal

A change in the magnetic flux, $\Phi(t)$, induces a voltage, V, in a conductor according to

$$V = \frac{d\Phi(t)}{dt} = \left(\frac{d\Phi(z)}{dz}\right)\left(\frac{dz}{dt}\right), \tag{4.17}$$

where the time-varying flux can be converted to a position-dependent flux $\Phi(z)$. When a magnetic sample is enclosed by a pick-up coil and oscillated in the z-direction, the induced voltage V is induced in the pick-up coil: For a sinusoidal oscillation with angular frequency ω the magnetization, $M(t)$, of the sample reads

$$M(t) = \frac{V}{C \cdot A \cdot \omega \sin(\omega t)} \tag{4.18}$$

where A describes the oscillation amplitude and C the coupling constant, unique to the VSM setup. Usually, A is kept at small values ($A \sim 1\,\text{mm}$) and within the region of homogeneous excitation field provided by a tunable electromagnet (s. Fig. 4.17). Further details on VSM are

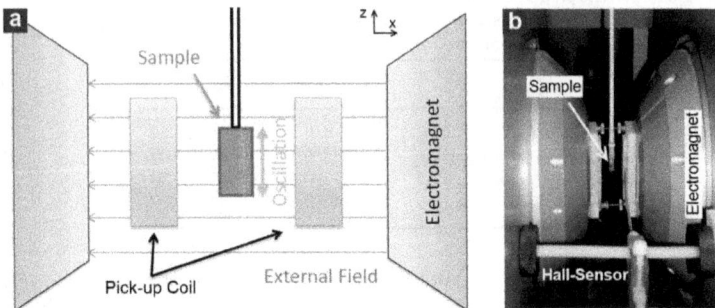

Fig. 4.17.: General principle of a VSM device. (a) schematic drawing of a VSM setup, where an external homogeneous field is generated using an electromagnet, while the sample is oscillated along the z-direction. The change in magnetic flux over time, $\Phi(t)$, caused by oscillating the magnetic sample, induces a voltage in the pick-up coil according to eq. (4.17). (b) shows the real VSM setup used for measurements (Lakeshore 7350), where a Hall-sensor was used to measure the applied field.

found in [249, 250].

Sample Preparation

$100\,\mu L$ of each SEA sample were pipetted in dedicated $100\,\mu L$ polycarbonate capsules (Quantum Design Inc., San Diego, CA, USA) with concentrations as listed in Table 4.6. Closed capsules were subsequently fixed inside of a polymer straw that suits as a sample holder.

Experimental Procedure

Magnetization curves in ambient conditions were measured using a Lakeshore 7350 VSM (Lakeshore Cryotronics Inc., Westerville, OH, USA). Two measurement runs with different field ranges were measured for all thermal decomposition samples: the high field ranging from $H = -160\,\mathrm{kA/m}$ to $H = 160\,\mathrm{kA/m}$ ($200\,\mathrm{mT}/\mu_0$) and the low field ranging from $H = -2\,\mathrm{kA/m}$ to $H = 2\,\mathrm{kA/m}$ ($2.5\,\mathrm{mT}/\mu_0$).

Experimental Results and Analysis

The magnetic moment [emu] data obtained from VSM measurements was normalized to the magnetite content per sample (using the MNP concentration, cf. Table 4.6) and expressed in [kA/m]. Furthermore, a linear function was subtracted from the resulting $M(H)$-curves to account for a diamagnetic portion of the sample, as described already for SQUID data in Section 4.6. In the same way, the data was fitted with the Langevin function, eq. (4.10) and Chantrell fitting was performed using, eqs. (4.11)-(4.15). The initial magnetic susceptibility, χ_0, was determined by linear fitting to the $M(H)$-data for low fields with $H = 2\,\mathrm{kA/m}$. Results are depicted in Fig. 4.18 and listed in Table 4.9 (details on the fitting parameters are found in Appendix A.2.5). All samples show no hysteresis at room temperature, indicating superpara-

Table 4.9.: Magnetic particle properties determined from VSM measurements via Langevin function and Chantrell fitting: magnetic (mass) susceptibility, χ_0, saturation magnetization, M_S and particle magnetic size, d_M. For details on the fitting parameters see Appendix A.2.5.

	χ_0	M_S [kA/m]	d_M [nm]
SEA1	64.70 ± 0.14	346.31 ± 0.93	21.14 ± 3.84
SEA2	94.61 ± 0.23	376.38 ± 0.32	23.20 ± 3.03
SEA3	125.78 ± 0.52	389.14 ± 0.43	25.55 ± 3.59
SEA4	151.51 ± 1.39	345.68 ± 0.88	27.44 ± 6.97

magnetic behavior. This is expected even for SEA4 particles that are as large as $d_C \approx 28\,\mathrm{nm}$, whose magnetic moments are typically block at ambient conditions. The reason for this is that liquid samples were measured, allowing the magnetic moments to relax with Brownian relaxation even if the internal rotation (Néel relaxation) is blocked for such large particles. As seen from Fig. 4.18, the applied field of $H = 160\,\mathrm{kA/m}$ is strong enough to saturate the magnetic moment of the particles.

For all samples the saturation magnetization M_S increases with core size up to approximately $82\,\%$ of the bulk magnetization ($476\,\mathrm{kA/m}$ [232]) for SEA3. The slightly lower M_S-values

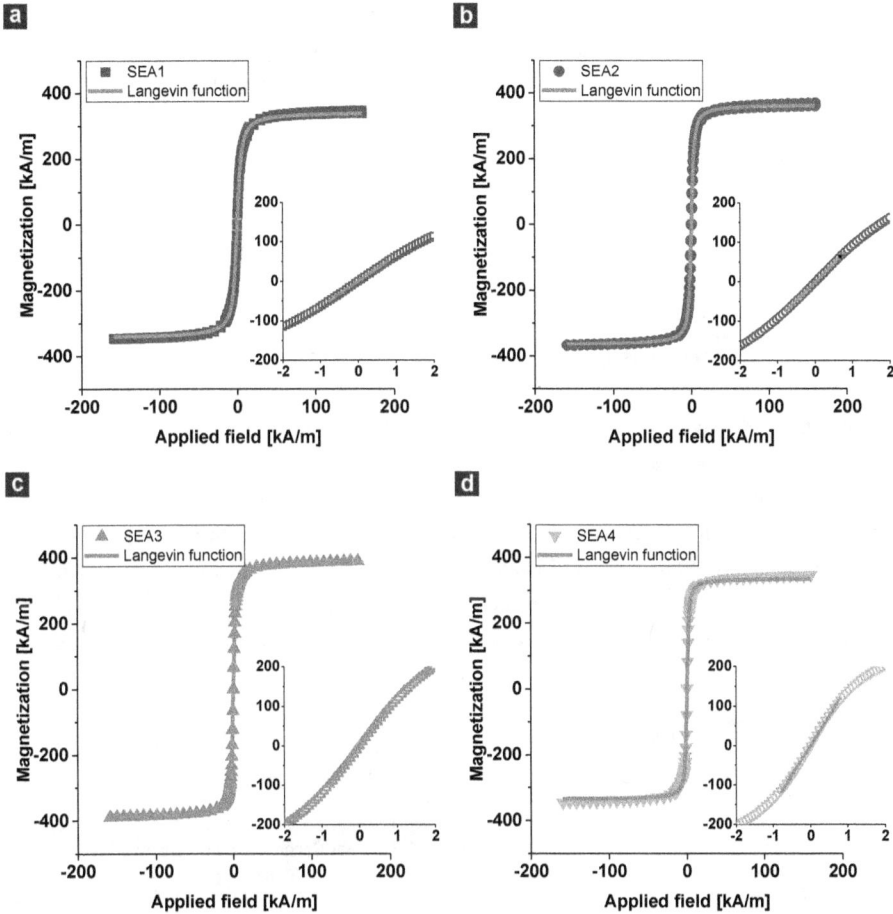

Fig. 4.18.: VSM magnetization measurements in ambient conditions for (a) SEA1, (b) SEA2, (c) SEA3, (d) SEA4, all fitted with the Langevin function. Magnetization curves for low fields $H = 2\,\text{kA/m}$ are shown in the insets. The slope of a linear fit through the origin is used to determine the initial magnetic susceptibility χ_0. Adapted from [251].

compared to those of co-precipitation particles (cf. Table 4.7) may be explained by the much higher applied field during SQUID measurements ($H \sim 4\,000\,\text{kA/m}$) compared to the field applied during VSM ($H \sim 160\,\text{kA/m}$): As outlined in detail in Section 4.6, such high fields in SQUID could possibly align canted spins at the surface of the particles, which otherwise do not contribute to the overall M_S. It may therefore be assumed here that a magnetic-dead layer of canted spins [235] lowers the M_S-value for SEA particles compared to co-precipitation particles. Sample SEA4 has a significantly lower saturation magnetization 73 % of bulk value, even though the particle size increases. Three reasons for this irregularity can be named: First, large particles (with e.g. $d_\text{C} \geq 40\,\text{nm}$ [23]) may split in multidomain particles, whose individual domain magnetization vectors will align antiparallel (in order to minimize the total magnetocrystalline and magnetostatic energy), leading to lower total magnetization values. Second, the PMAO-

PEG coating has been observed to sometimes "glue" particles together [214], decreasing their interparticle distance, which in turn decreases the saturation magnetization due to increased particle dipole interaction. Third, larger partices are more predisposed for antiferromagnetic wustite (FeO) portions[13], forming core(FeO)-shell(F_3O_4) nanostructures [238] which decrease the saturation magnetization [237], as discussed above (sec. 4.6).

In the same way, an increase in initial magnetic susceptibility, χ_0, is observed for increasing the particle magnetic size, d_M. This trend of increasing χ_0 with increasing magnetic particle size is observed in literature as well [252].

4.8. Particle Properties Summary

The physical and magnetic properties of MNP described in this chapter form the foundation of the subsequent analysis of particle relaxation and heating dynamics in experiments in the subsequent chapters. In the following, the main findings from characterization of each MNP property and its discussion are succinctly summarized. At the end of this section, the different particle sizes determined by XRD, TEM and magnetic measurements are compared to each other and briefly discussed. The MNP samples under investigation are listed in Table 4.10, while the MNP properties derived from characterization are summarized at the end of this section in Table 4.11. .

Table 4.10.: Summary of the MNP samples characterized.

Sample	Synthesis method	Coating	Sample	Synthesis method	Coating
FFLA	Co-Precipitation	Lauric Acid	SEA1	Thermal Decomposition	PMAO-PEG polymer
ML1	Co-Precipitation	Phospholipides	SEA2	Thermal Decomposition	PMAO-PEG polymer
ML2	Co-Precipitation	Phospholipides	SEA3	Thermal Decomposition	PMAO-PEG polymer
ML3*	Co-Precipitation	Phospholipides	SEA4	Thermal Decomposition	PMAO-PEG polymer
FFAM	Co-Precipitation	Sodium Citrate			

* ML3 was only characterized with SQUID magnetometry.

Based on diffraction patterns from XRD measurements (Section 4.2), a mix of magnetite and maghemite was estimated as the material of all co-precipitation MNP. The particle crystalline size, estimated from the Scherrer equation (eq. (4.2)), was of the order of $\sim 10\,nm$ for all MNP, which is in agreement with the particle sizes determined from TEM and Chantrell fitting (s. below). For thermal decomposition MNP, magnetite was assumed as the sample material, as described in literature [196, 207].

The mean particle core size, d_C, was determined from transmission electron microscopy (TEM) image analysis: the particle core sizes were measured, yielding the particle core size cumulative distribution. The distribution was fitted with the cumulative distribution function (CDF), cf. eq. (4.5), from which d_C was derived. The particle core sizes of co-precipitation particles

---

[13]During thermal decomposition synthesis, the MNP are left to oxidize to F_3O_4. Larger particles require longer times to fully oxidize, as their individual volume increases with $V_C \propto d_C$. For synthesizing larger particles it is challenging to find the right balance between oxigen flow and exact oxidation time.

were of the order of $\sim 10\,$nm, in accordance with the particle crystalline sizes, d_{XRD}, from XRD measurements. The SEA particles displayed an excellent size control for each individual sample batch. d_C of all four SEA samples together spanned a particle core size range of $d_C \sim (22 - 28)\,$nm. The size distribution width, σ, is relatively large for FFLA and MLs compared to FFAM and SEA particles (cf. Fig. 4.7), rendering the former MNP samples rather polydisperse, while the latter (FFAM and SEA particles) are considered monodisperse.

The mean particle hydrodynamic size, d_H, was determined from fitting the PDF of the log-normal distribution to scattering intensity data from DLS (Section 4.4). Generally, $d_H < 100\,$nm for all MNP samples under investigation, confirming the nanoscopic sizing of all samples. In particular, the particle hydrodynamic size depends on the MNP coating, as it was generally shown that $d_{\text{H,FFAM}} < d_{\text{H,FFLS}} < d_{\text{H,MLs}}$, which all have different coating materials (cf. Table 4.10. Additionally, thermal decomposition samples coating with PMAO-PEG showed some variation in particle hydrodynamic size, which are attributed to the coating efficiency during the necessary phase transfer to the aqueous phase: Further sonification could presumably equalize the sizes. Comparing the particle hydrodynamic size polydispersity, the SEA particles show good monodispersity with a PDI ≤ 0.13, while the co-precipitation particles were more polydisperse with PDI ≥ 0.19, showing the trend PDI$_{\text{FFAM}} \approx$ PDI$_{\text{FFLS}} <$ PDI$_{\text{MLs}}$. This is in line with the observations for trends observed for the particle core sizes from TEM analysis, described in the previous paragraph.

The iron concentration, c, of each MNP sample was determined either from photometric analysis (PA) of Fe^{3+}-complexation with Tiron in case of the co-precipitation particles or from inductively coupled plasma optical emission spectroscopy (ICP-OES) in the case of thermal decomposition particles. As expected from literature [23, 179], the co-precipitation synthesis yields generally higher iron concentrations compared to thermal decomposition synthesis.

Magnetic properties of the MNP were either characterized by superparamagnetic quantum interference device (SQUID) magnetometry in the case of co-precipitation particles or by vibrating sample magnetometry (VSM) in the case of thermal decomposition particles. Co-precipitation particles showed very good saturation magnetization with $M_S \approx (85 - 95)\,\%$ of bulk magnetite magnetization. A possible explanation for these high M_S-values is given by the alignment of surface spins within the MNP in the very high applied fields ($H \sim 4\,000\,$kA/m) during SQUID measurements [236]. These surface spins are otherwise canted and not align with an applied (lower) field, forming a surface layer that does not contribute to the MNP magnetization (sometimes denoted as the magnetic-dead layer). Thermal decomposition particles also demonstrated good saturation magnetization with $M_S \approx (73 - 82)\,\%$ of bulk magnetite magnetization. Their M_S is slightly lower than the M_S of co-precipitation particles, which was either presumed to be due to a magnetic-dead layer on the surface of the MNP in the comparatively low applied fields of the VSM (s. above) or caused by the formation of an antiferromagnetic wustite particle core (due to incomplete MNP oxidization during synthesis) [237]. In both cases, M_S is expected to reside below bulk magnetite value, as observed experimentally. The magnetic particles sizes, d_M, are generally comparable to the particle core sizes reported from TEM analysis above and will be compared in more detail in the last paragraph of this section. SQUID magnetometry also allowed to measure the zero-field-cooled field-cooled (ZFC-FC)-curves of the co-precipitation particles, from which the mean blocking temperature, T_B, and the effective anisotropy constant, K_{eff} could be estimated. Even though such an estimation is only an approximation, as it does not take the MNP sample core size and anisotropy value distributions into account [241], the comparison of K_{eff}-values allowed to qualitatively confirm an increasing magnetic dipole-dipole particle interaction for ML1 and ML2 MNP compared to FFLA and ML3.

The particle core size, d_C, particle crystalline size, d_{XRD}, and the particle magnetic size, d_M, all agree within one standard deviation. For co-precipitation particles, it is observed that $d_C \approx d_{XRD} < d_M$. However, SEA particles showed $d_C \approx d_M$. In general, $d_C < d_{XRD} < d_M$ is expected [253] and the particle magnetic size is considered the most reliable concerning statistics, as it is determined from the entire MNP sample magnetization, averaging over the contributions from the entire MNP sample population. In contrast to that, TEM analysis is restricted by the number and quality of the TEM images analyzed and XRD is limited to the resolution of the diffraction pattern, from whose intensity peaks the particle crystalline size is calculated.

Table 4.11.: Summary of MNP properties characterizing the samples used in this thesis. Details are found in the respective section marked in the footnotes. c denotes the sample iron mass concentration, d_{XRD} the particle crystalline size, d_C the particle core size, d_H the particle hydrodynamic size, PDI the particle hydrodynamic index of polydispersity, M_S the saturation magnetization, χ_0 the initial magnetic (mass) susceptibility, d_M the particle magnetic size and K_{eff} the effective anisotropy constant. The $*$ marks the sample material determined from literature. d_{XRD} and K_{eff} could only be determined for co-precipitation particles but not for thermal decomposition particles; respective slots are marked with –. ML3 was only characterized with SQUID magnetometry yielding the magnetic properties.

Synthesis:	Co-Precipitation				Thermal Decomposition				
MNP Property ↓	FFLA	ML1	ML2	ML3	FFAM	SEA1	SEA2	SEA3	SEA4
Material	$Fe_3O_4/$ $\gamma\text{-}Fe_2O_3$	$Fe_3O_4/$ $\gamma\text{-}Fe_2O_3$	$Fe_3O_4/$ $\gamma\text{-}Fe_2O_3$	–	$Fe_3O_4/$	$Fe_3O_4^*$	$Fe_3O_4^*$	$Fe_3O_4^*$	$Fe_3O_4^*$
c [mg(Fe)/mL]	11.023 ±0.083	1.184 ±0.027	2.771 ±0.020	–	4.444 ±0.037	1.498 ±0.012	1.315 ±0.010	1.513 ±0.009	1.499 ±0.012
d_{XRD} [nm] $\sigma_{d_{XRD}}$	10.71 ±0.41	10.05 ±0.22	10.55 ±0.26	–	7.97 ±0.24	–	–	–	–
d_C [nm] σ_{d_C}	9.68 ±3.10	11.03 ±3.67	10.63 ±3.35	–	8.93 ±1.31	21.92 ±0.88	23.13 ±1.16	25.38 ±2.03	27.77 ±1.95
d_H [nm] σ_{d_H}	57.67 ±19.51	67.31 ±21.13	93.00 ±35.19	–	14.35 ±03.69	75.17 ±21.33	39.39 ±11.26	41.50 ±09.19	80.04 ±23.47
PDI	0.194	0.248	0.214	–	0.189	0.130	0.127	0.104	0.121
M_S [kA/m] σ_{M_S}	385.18 ±1.76	423.35 ±1.60	422.13 ±1.43	386.07 ±1.39	445.86 ±1.08	346.31 ±0.93	376.38 ±0.32	389.14 ±0.43	345.68 ±0.88
χ_0 σ_{χ_0}	23.32 ±0.55	28.97 ±1.12	26.51 ±0.85	23.56 ±0.78	16.51 ±0.12	64.70 ±0.14	94.61 ±0.23	125.78 ±0.52	151.51 ±1.39
d_M [nm] σ_{d_M}	11.83 ±5.42	12.01 ±5.42	11.62 ±5.32	11.75 ±5.52	10.43 ±3.48	21.14 ±3.84	23.20 ±3.03	25.55 ±3.59	27.44 ±6.97
K_{eff} [kJ/m³]	≈ 78	≥ 108	≈ 108	≈ 80	≈ 48	–	–	–	–

5. Magnetic Particle Relaxation in Alternating Magnetic Fields

Besides the MNP properties described in Chapter 4, the relaxation of MNP in an alternating magnetic field (AMF) is of high importance for the biomedical application of magnetic fluid hyperthermia (MFH) and magnetic particle imaging (MPI), as both techniques (MFH and MPI) exploit the same physics of particle relaxation (as discussed in detail in Sections 2.2, 2.3 and 2.4). For this, the present chapter studies the MNP relaxation behavior of all MNP samples by means of magnetic particle spectroscopy (MPS), a zero-dimensional MPI (cf. Section 2.3) and MFH measurements: First, the MPI performance of the MNP used in this thesis is investigated via MPS in Section 5.1. Subsequently, a comparison of the MNP heating performance in MFH is presented in Section 5.2. Finally, the results of Monte-Carlo (MC-)simulation of particle heating are validated with experimental particle heating data in Section 5.3. These validated MC-simulations are then used to predict an optimized MNP heating as a function of the particle core size and by adjusting the applied AMF parameters under medical safety considerations.

5.1. Magnetic Particle Spectroscopy

Magnetic particle spectroscopy (MPS) measurements probe the magnetic relaxation of a MNP in response to a sinusoidal alternating magnetic field (AMF). The general concept of MPS is described in detail in Section 2.3. Throughout this work the MNP relaxation behavior is depicted in the MPS harmonic spectra, represented by the signal amplitude of the odd harmonics of the magnetic moment, A_k, in the frequency domain. $A_k := \hat{V}_k$ is the k-th Fourier coefficient derived from the Fourier transform of the received signal in time domain, as defined previously in eq. (2.54). MPS can be employed for the analysis of two MNP characteristics: First, the iron concentration, c, of a MNP sample is quantified by MPS, as the magnitude of the received signal is directly proportional to c (cf. eq. (2.57)). Usually, the thrid harmonic of the MPS spectra, A_3, is chosen as it has the highest spectral magnitude [14,254]. MPS was found to have superior sensitivity for iron quantification when compared to other quantification techniques such as ultra-violet spectrophotometry (UV/VIS) and atomic absorption spectroscopy (AAS) [255]. MPS has furthermore proven to of the iron content of MNP internalized inside cells [256]. Secondly, as MPS relies on intrinsic Néel and physical Brownian rotation of the MNP magnetic moment (cf. sec. 2.2), changes in MNP arrangement, microstructure, dipole-dipole interaction and different environemntal factors, such as binding to environemtal matrices, immobilization of MNP and viscostiy effects, can be studied via MPS [257].

Both characteristics of MPS, i.e. iron quantification and analysis of the MNP relaxation behavior, will be exploited throughout this theses: The iron content internalized inside pancreatic tumor cells over certain incubation times measured via MPS is presented in Section 6.3 and used throughout Chapter 8. Furthermore, the effect of agglomeration and immobilization of MNP upon intracellular internalization on the particle relaxation behavior is analyzed with MPS in Section 6.3, providing valuable information about the underlying changes in MNP arrangement and interaction with their immediate surrounding.

5.1.1. Magnetic Particle Spectrometers and Sample Preparation

MPS measurements presented in this thesis were conducted in cooperation with other research groups:

- **TU Braunschweig (TUB)**
  Samples FFLA, ML1, ML2 and FFAM were characterized in cooperation with the *Insitute für Elektrische Messtechnik und Grundlagen der Elektrotechnik* at the TU Braunschweig (Braunschweig, Germany) employing a home-built device [258, 259], operated at $f = 10\,\text{kHz}$ and $H_0 = 25\,\text{mT}/\mu_0$ at $T = 25\,°\text{C}$. $150\,\mu\text{L}$ of liquid MNP sample were prepared in $0.15\,\text{mL}$ microtiter vials containing with an iron concentration of $c = 300\,\mu\text{g(Fe)/mL}$. Final results are averages of five independent MPS runs.

- **Physikalische-Technische Bundesanstalt (PTB)**
  Sample ML1 and all cell samples and intracellular MNP (in Chapters 6 and 8) have been analyzed at the Physikalisch-Technische Bundesanstalt (Berlin, Germany) employing a BioSpin device (Bruker BioSpin MRI GmbH, Ettlingen, Germany), operated at $f = 25\,\text{kHz}$ and $H_0 = 25\,\text{mT}/\mu_0$ at $T = 25\,°\text{C}$. $50\,\mu\text{L}$ of liquid MNP sample were prepared in $0.2\,\text{mL}$ PCR tubes with an as-prepared iron concentration of $c = 1.184\,\text{mg(Fe)/mL}$ (s. Table 4.6). Cell samples were prepared embedded in agarose ($\mu_{\text{agar}} = 0.3\,\%$) as described in Section 6.3.1. Final results are averages of three independent MPS runs.

- **University of Washington (UW)**
  Samples SEA1 through SEA4 have been measured at the University of Washington (Seattle, WA, USA) employing a home-built device [39], operated at $f = 25\,\text{kHz}$ and $H_0 = 20\,\text{mT}/\mu_0$ at $T = 25\,°\text{C}$. $150\,\mu\text{L}$ of liquid MNP sample were prepared in $0.5\,\text{mL}$ PCR-tubes with an as-prepared iron concentration $c \approx 1.5\,\text{mg(Fe)/mL}$, as specified in Table 4.11. Final results are averages of three independent MPS runs.

5.1.2. Results of MPS Characterization

The MPS spectra of all samples were normalized to the third harmonic, $A_i[\text{norm}] = A_i/A_3$, in order to compare the MPS spectra indepdendent of the sample iron concentration. Results for FFLA, MLs and FFAM are plotted in Fig. 5.1a and those for SEA1 through SEA4 in Fig. 5.1b. Note that A_1 of the first harmonic f_1 is removed from all plots, since it is covered by the excitation signal, as discussed in Section 2.3. A direct comparison of all samples was impossible due to the different field amplitudes, H_0, and freuqencies, f, used: As shown in Appendix A.3.1, higher field amplitudes H_0 increases the spectral magnitude of higher harmonics even for A_3-normalized MPS spectra, as expected from literature [32]. While differences in the AMF frequency f render the harmonics, $f_k = k \cdot f$ cf. eq. (2.51), plotted on the x-axes, incompatible (here: k-th-fold of $f = 10\,kHz$ for TUB measurements and k-th-fold of $f = 25\,kHz$ for UW measurements). As shown in Appendix (A.3.1), the detection limit of the TUB setup is reached

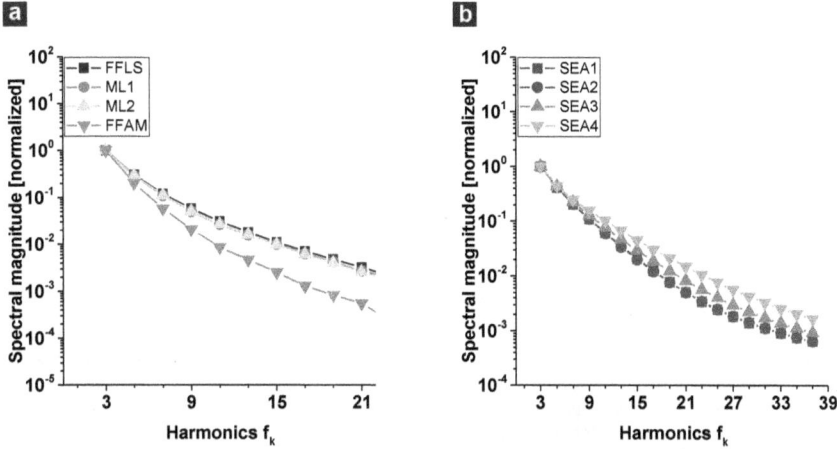

Fig. 5.1.: MPS spectra normalized to the third harmonic for (a) FFLA, MLs and FFAM measured at $f = 10\,kHz$ and $H_0 = 25\,mT/\mu_0$ and (b) SEA1 through SEA4 measured at $f = 25\,kHz$ and $H_0 = 20\,mT/\mu_0$.

for higher harmonics $f_k > 21$ for sample FFAM, wherefore the plots are limited to $f_k \leq 21$ for a meaningful comparison of the MNP samples in Fig. 5.1a. Generally, it can be seen that FFLA, ML1 and ML2 perform equally well in MPS measurements, while FFAM spectral magnitude decays faster. This is well expected, as FFLA and ML samples have the same MNP core, with a particle core size of $d_C \approx (10-11)\,nm$, while FFAM samples are smaller with $d_C \approx 9\,nm$ (c. f. Table 4.11).

The same trend is observed for SEA MNP systems, where larger particle core sizes are associated with higher intensity harmonics. Ergo, larger particle core sizes perform better in MPS, which is in agreement with literature [39, 260, 261]. According to current research, MNP with a particle core size of $d_C \approx (25-27)\,nm$ performs best in MPI [262, 263].

The size dependency of MPS performance can be conveniently visualized by comparing the relative magnitude of higher harmonics, defined by the ratio A_5/A_3, as summarized in Fig. 5.2. Note that interpretation of the MPS performance is allowed only for the same AMF settings and samples measured with the same MPS device, therefore the group of FFLA, ML1, ML2 and FFAM and the group of SEA1 through SEA4 can be compared among each other, respectively. From comparing the MPS measurements of sample ML1, performed once with the PTB setup ($f = 25\,kHz$ and $H_0 = 25\,mT/\mu_0$) and then with the TUB setup ($f = 10\,kHz$ and $H_0 = 25\,mT/\mu_0$) one observes the following: The ratio, A_5/A_3, drops for the measurement with the TUB setup as expected, since the A_3 and A_5 harmonics are proportional to the AMF frequency, which is higher for the PTB setup. However, taking into account that higher f will decrease A_5/A_3 and higher H_0 will increase A_5/A_3 (s. Appendix A.3.1) for the same sample, these effects balance out. Therefore, the qualitative comparison between all samples

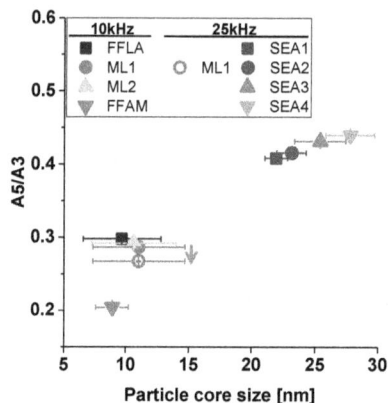

Fig. 5.2.: MPS spectral ratio A_5/A_3 versus particle core size for all MNP systems. Note that FFLA, MLs and FFAM have been measured at a different frequency than SEA1 through SEA4. ML1 has been measured twice, once with the TUB setup (solid circle; $f = 10$ kHz, $H_0 = 25$ mT$/\mu_0$) and once with the PTB setup (open circle; $f = 25$ kHz, $H_0 = 25$ mT$/\mu_0$). The green arrow indicates the drop in A_5/A_3 due to the increased frequency.

summarized in Fig. 5.2 is possible, confirming the size-dependence of MPI performance: Larger particles with $d_C \approx (25 - 28)$ nm perform best.

These results will be used as references further on in this thesis to identify changes in MPS signal due to MNP cellular internalization (Chapter 6) and MNP agglomeration and immobilization (Chapter 7).

5.2. Magnetic Fluid Hyperthermia

Alternating magnetic fields (AMF) with $f \sim 100$ kHz frequencies stimulate MNP to generate heat via magnetic relaxation processes (cf. Section 2.4), which can be applied for therapeutic tumor treatment in magnetic fluid hyperthermia (MFH) therapy. In this section, the MNP used in this thesis are characterized for their heating performance by means of direct calorimetric hyperthermia measurements.

5.2.1. Measuring the Particle Heating Efficiency

Characterization of the Measurement Setups

Two different hyperthermia setups have been employed throughout this thesis, a custom-build Hüttinger setup and a commercially available magneTherm setup. Their specifications are described in the following:

- **Hüttinger Setup**

  For all heating experiments performed with samples FFLA, MLs and FFAM a custom-built Hüttinger hyperthermia setup (Trumpf Hüttinger GmbH, Freiburg, Germany) was used. The setup consists of a TIG 5/300 DC generator, an AC-resonant oscillator with exchangeable capacitor banks and a hollow water-cooled copper coil, as depicted in Fig. 5.3a. The coil, designed to hold a 4 mL cylindrical sample vial wrapped in 1 mm styrofoam insulation, had the following dimensions: outer diameter $d_{out} = 30$ mm, inner diameter $d_{in} = 20$ mm, $N = 7$ turns, turn-to-turn distance $a = 3.5$ mm and turn thickness $t = 3$ mm (s. Fig. 5.3b). The coil's cooling water was kept in a sealed circulatory system constantly held at $17 °C$[1]. The setup allowed for field amplitudes in the range of, $H_0 = (15 - 90)$ mT$/\mu_0$, and frequencies ranging from $f = (83 - 270)$ kHz in non-uniform increments determined by choosing specific combinations of inductors and capacitors installed in the AC-resonant oscillator. The magnetic field inside the coil was accurately determined with the specific AC magnetic field sensor 'Probe' (NanoScience Labs, Staffordshire, UK), measuring AMF amplitude and frequency simultaneously.

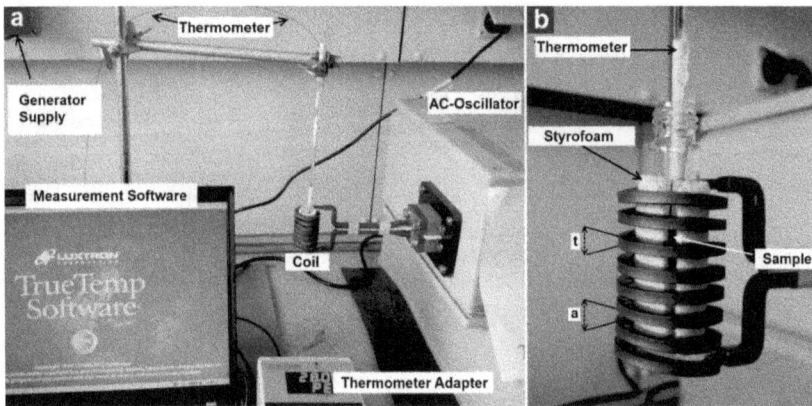

Fig. 5.3.: The measurement setup used in Aachen. (a) An AC-oscillator circuit features a cooper coil, which holds the sample wrapped in styrofoam. The current flowing through the coil, inducing the magnetic field, is provided by a generator (not shown). For measurements a fiber optic thermometer is immersed in the sample. (b) shows the coil with the turn-to-turn distance a and turn thickness t, along with the MNP sample, the styrofoam insulation and the fiber optic thermometer with the plastic tube guiding. (b) adapted from [264].

- **magneTherm Setup**

  Samples SEA1 through SEA4 were heated in a magneTherm 1.0 hyperthermia device (nanoTherics Ltd., Staffordshire, UK) consisting of a water-cooled copper coil (outer and inner diameters $d_{out} = 54$ mm and $d_{in} = 44$ mm, $N = 17$ turns, turn-to-turn distance $a = 1$ mm and turn thickness $t = 1$ mm), an AC-resonant circuit, a DC power generator and a waveform generator (s. Fig. 5.4). This setup is capable of generating frequencies ranging from $f = 176$ kHz up to $f = 992$ kHz by changing the AC-resonant circuit capacitance. Due to the limitation of maximum power applied to the setup, the maximum

---

[1]The actual temperature in the sample is a balance of the current applied to the coil, causing Joule heating, and the cooling temperature. These settings were chosen to keep 1 mL of cell culture medium at exactly $37 °C$ during in vitro MFH experiments using the AMF parameters $H_0 = 50$ mT$/\mu_0$ and $f = 270$ kHz (cf. Chapter 8). For general comparability, the cooling setting are kept constant for all heating experiments.

applicable field amplitude is limited for each individual frequency as follows for the four pre-selected frequencies used in this thesis: $20\,mT/\mu_0$ at $176\,kHz$, $15\,mT/\mu_0$ at $373\,kHz$, $9\,mT/\mu_0$ at $744\,kHz$ and $6\,mT/\mu_0$ at $992\,kHz$. The AMF of the coil is measured with a digital oscilloscope. The AMF amplitude, H_0, was calculated from the peak-to-peak value of the applied voltage, V, measured with the oscilloscope. For that, a calculation software supplied by nanoTherics was used, which calculates the current, I, flowing through the coil, from V. Then, $H_0 = I \cdot N/l$ with the number of turns, N, and the length of the coil, l, was used for calculating the AMF amplitude of a coil [265].

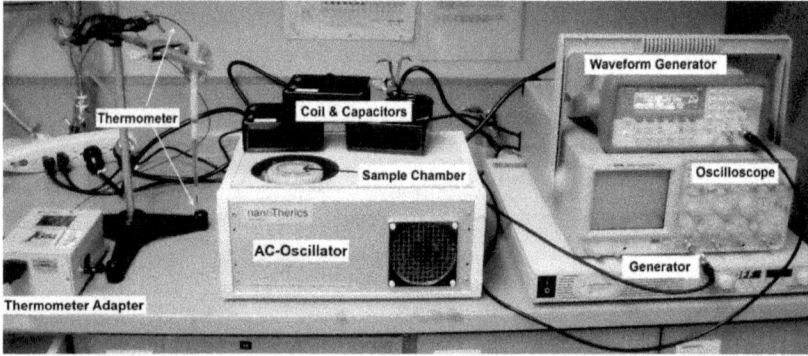

Fig. 5.4.: The magneTherm 1.0 setup consists of a waveform generator and a DC-power generator, which generate an AMF within the coil of the AC-oscillator. The AMF frequency is measured using an oscilloscope. The frequency is set by a specific coil and capacitor combination. The temperature of the MNP sample placed in the sample chamber is measured with a fiber optical thermometer.

Determination of the Particle Heating Efficiency

A fiber optic thermometer (LumaSense Inc., Santa Clara, CA, USA) measured the temperature, T, within the MNP sample upon application of the AMF. The temperature sensor was guided through a NMR-glass tube and placed approximately $5\,mm$ deep into the sample. In this way, T was measured in the center of the sample volume in accordance with [266]. The experimental specific loss power (SLP) can be calculated from the initial heating rate, $\frac{dT(t)}{dt}\big|_{t\to 0}$ according to [267]:

$$SLP = c_h \cdot \frac{m_{Fe_3O_4} + m_{CM}}{m_{Fe_3O_4}} \cdot \frac{dT(t)}{dt}\bigg|_{t\to 0} \approx c_h \cdot \frac{m_{CM}}{m_{Fe_3O_4}} \cdot \frac{dT(t)}{dt}\bigg|_{t\to 0} \qquad (5.1)$$

where $c_h = 4.187\,J/(g\cdot K)$ is the specific heat capacity of water (or the respective carrier matrix), $m_{Fe_3O_4}$ the mass of the dissolved MNP and m_{CM} the mass of the carrier matrix. The approximation is valid for $m_{Fe_3O_4} \ll m_{CM}$, which is the case for all following measurements.

In both setups the particle heating is considered non-adiabatic, as the sample is in thermodynamic exchange with its surrounding (s. discussion below). Furthermore, as the MNP sample vial is not sealed during MFH measurements to allow insertion of thermometer, convection is also possible. As a result, the temperature during particle heating saturates at T_{max} within tens of minutes.

For experiments carried out with the Hüttinger setup the temperature during heating was measured for 30 min every 0.5 s for each sample. The resulting $T(t)$-curve was fitted with the Box-Lucas function, which has proven to be most reliable for non-adiabatic measurements over 20 min that reach saturation temperature T_{max} [268, 269]:

$$T(t) = \Delta T \cdot (1 - \exp(-b \cdot t)) + T_0, \qquad (5.2)$$

with the relative temperature increase $\Delta T = T_{max} - T_0$, the rise rate b and the initial temperature T_0. In this way, also the saturation (maximum) temperature during heating, T_{max}, is accessible from the fit, which is an important parameter for assessing in vitro MFH effects in Chapter 8. Using eq. (5.2), the initial heating rate $\frac{dT}{dt}\big|_{t\to 0} = \Delta T \cdot b$ was calculated and used to determine the experimental SLP from eq. (5.1).

For measurements with the magneTherm device the temperature was recorded for 3 min of heating every 0.5 s. With such a short measurement time, the initial heating rate $\frac{dT}{dt}\big|_{t\to 0}$ was determined from the slope of a linear fit of the first 60 s of heating from the recorded $T(t)$-curve [267]. From this, the SLP value was determined as described above in eq. (5.1). Consequently, neither the Box-Lucas function was fitted nor the saturation temperature $T_{max} = T_0 + \Delta T$ could be extracted from the measurements and analysis is limited to SLP value.

It was observed that water cooling and current-induced Joule heating effects influenced the measured temperature in the sample despite sample insulation in both setups. Consequently, the setups are considered non-adiabatic. Depending on the applied field strength and frequency (i. e. current within the coil) and the temperature of the cooling water of the coil, the effect could either be a temperature increase or decrease. In order to account for this, reference measurements of the pure carrier matrix/ liquid not containing MNP was performed (e. g. water for FFLA and FFAM and TES buffer for ML1 and ML2). The resulting $T(t)$-curves, denoted as background, were subtracted from the heating curves to obtain the heating data generated solely by the MNP. Exemplary heating curves for FFAM particles performed with the Hüttinger setup and the temperature rise after background-subtraction are shown in Fig. 5.5. As can be seen in Fig. 5.5a, there are temperature oscillations in both $T(t)$-curves, caused by the cooling aggregate: The aggregate is set to keep the cooling water temperature constant at $T = 17\,^{\circ}C$ but its control system has only a precision of $\delta T = 1\,^{\circ}C$. Consequently, as soon as the temperature of the cooling water reaches $T = 18\,^{\circ}C$, the cooling water is cooled down to $T = 16\,^{\circ}C$. These temperature oscillations is superimposed on the recorded $T(t)$-curves. By appropriately starting the temperature measurements of the sample and the background with the same temperature cooling phase, the superimposed temperature oscillations cancel each other out after background subtraction (as demonstrated in Fig. 5.5b).

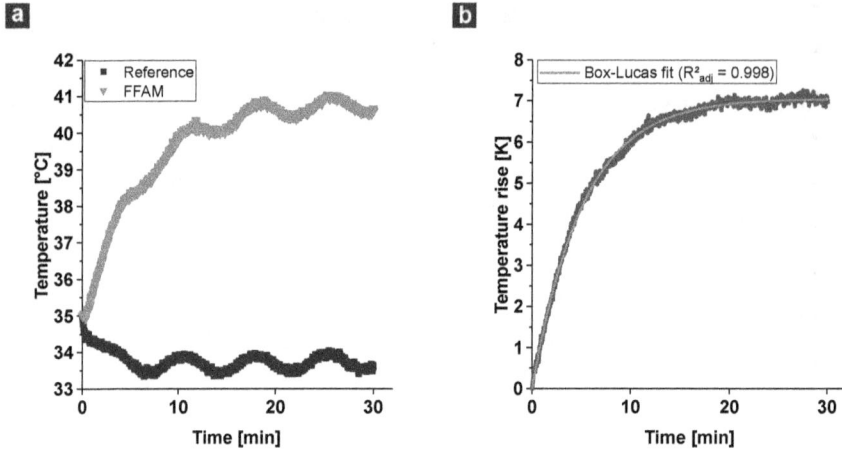

Fig. 5.5.: Exemplary temperature curves of FFAM and background measured at $H_0 = 50\,\text{mT}/\mu_0$ at $f = 270\,\text{kHz}$ at initial temperature $T_0 = 35\,°\text{C}$: (a) temperature measured for a reference sample (1 mL DI-H_2O) and FFAM particles dispersed in DI-H_2O (1 mL with $c = 300\,\mu\text{g(Fe)/mL}$). (b) Temperature rise $\Delta T(t)$ after background subtraction of the reference data from the FFAM sample data, fitted with the Box-Lucas function (cf. eq. (5.2)). Note that the temperature oscillations in (a) disappear upon background subtraction (s. text).

Such an issue did not arise for the magneTherm device, which was cooled with a constant flow of tap water.

5.2.2. Field-Dependence of Particle Heating Efficiency

Across literature, there is a firm agreement that the AMF parameters, i.e. amplitude, H_0, and frequency, f, are key contributors to particle heating [14, 42, 60, 269, 270], which does not need to be further confirmed. However, in this section the experimental particle heating results for $\text{SLP}(H_0, f)$ are compared to the predictions of $\text{SLP}(H_0, f)$ from MC-simulations from Section 3.3. From this, the characterization of the field-parameter-dependent SLP value offers additional insight in particle anisotropic behavior, as will be demonstrated in the following.

Sample Preparation and Data Acquisition

Hüttinger Setup using Co-Precipitation Particles
1 mL of FFAM with an iron concentration of $c = 300\,\mu\text{g(Fe)/mL}$ was transferred in a 4 mL Rotilabo glass vial (Carl Roth GmbH + Co. KG, Karlsruhe, Germany). Samples were prepared and measured in triplicate, along with one reference sample containing 1 mL of DI-H_2O for background subtraction. When not measured, samples were stored on a hotplate at a constant temperature of $T_0 = 35\,°\text{C}$. As the experimental results for $\text{SLP}(H_0, f)$ should be interpreted in terms of MC-simulation predictions from Section 3.3, with special concern to the change in the field-amplitude dependence of SLP at $H_0 = H_K/$, FFAM particles were chosen, since these showed the smallest effective anisotropy constant $K_{\text{eff}} = 48\,\text{kJ/m}^3$ of all co-precipitation

particles (cf. Table 4.8. This results in $H_K/4 = 54.3\,\text{mT}/\mu_0$ (using eq. (2.15) with values from MNP characterization from Table 4.11), which lies well within the range of field amplitudes accessible with the Hüttinger setup. For SLP(H_0) measurements, the frequency was fixed to $f_1 = 190\,\text{kHz}$ or $f_2 = 270\,\text{kHz}$, while the field amplitude was varied in the range of $H_0 = (16 - 86)\,\text{mT}/\mu_0$ or $H_0 = (17 - 72)\,\text{mT}/\mu_0$, respectively. For SLP($f$) measurements, the field amplitude was fixed to $H_0 = 50\,\text{mT}/\mu_0$, while the frequency was varied in the range of $f = (82 - 270)\,\text{kHz}$. The SLP value was calculated as described in Section 5.2.1 and final results are averages of triplicate measurements.

magneTherm Setup using Thermal Decomposition Particles

$250\,\mu\text{L}$ of liquid MNP sample was prepared from SEA1 through SEA4 in $0.5\,\text{mL}$ PCR-tubes (Eppendorf Vertrieb Deutschland GmbH, Wesseling-Berzdorf, Germany) with an as-prepared iron concentrations $c \approx 1.5\,\text{mg(Fe)}/\text{mL}$, as specified in Table 4.11. Samples were prepared and measured in triplicate with the magneTherm device, along with one reference sample containing $250\,\mu\text{L}$ of DI-H_2O for background subtraction. When not measured, samples were stored at a constant temperature of $T_0 = 25\,°\text{C}$. For SLP(H_0) measurements, the maximum field amplitude was set within the limits for the four frequencies available: $20\,\text{mT}/\mu_0$ at $176\,\text{kHz}$, $15\,\text{mT}/\mu_0$ at $373\,\text{kHz}$, $9\,\text{mT}/\mu_0$ at $744\,\text{kHz}$ and $6\,\text{mT}/\mu_0$ at $992\,\text{kHz}$. The smallest field amplitude was $H_0 = 3\,\text{mT}/\mu_0$. From here, H_0 was varied in increments of $3\,\text{mT}/\mu_0$ (except going from $18\,\text{mT}/\mu_0$ to $20\,\text{mT}/\mu_0$ at $f = 176\,\text{kHz}$). The SLP value was calculated as the average of triplicate measurements as described in Section 5.2.1.

Results and Discussion

Field Amplitude Dependency of SLP

The field amplitude dependent specific loss power values, SLP(H_0), are plotted for SEA particles in Fig. 5.6 and for FFAM in Fig. 5.7.

Generally, the SLP values are proportional to H_0^2 for all samples analyzed. However, for high field amplitudes, $H_0 \geq H_K/4$, associated with the MNP anisotropy field $H_K = 2K_{\text{eff}}/(\mu_0 \cdot M_S)$ (s. eq. (2.15); discussed below), the SLP values increase less steeply, dropping to a linear dependency with SLP$\propto H_0$. Besides, FFAM shows saturating SLP values for the highest field amplitudes of $H_0 \geq 60\,\text{mT}/\mu_0$. The following discussion of the SLP(H_0)-behavior is divided in the regions with SLP$\propto H_0^2$ and the region where SLP saturates. The intermediate region, where SLP$\propto H_0$, is discussed last in the framework of MC-simulation results on particle heating.

The relation SLP$\propto H_0^2$ is well described throughout literature: It has been proven for poly-disperse ferrofluid in low field amplitudes ($H_0 \approx 8\,\text{mT}/\mu_0$) [271] and further for higher fields ($H_0 \approx 20\,\text{mT}/\mu_0$) for small MNP ($d_C \leq 10\,\text{nm}$) [272]. However, field amplitude-dependencies diverging from the square-dependency have been reported as well, ranging from SLP$\propto H_0^{1.7}$ [273] to SLP$\propto H_0^3$ [177]. Reasons for this have been assumed to lie within the superparamagnetic to ferromagnetic (SPM-FM) transition: numerical hysteresis simulations for high field amplitudes

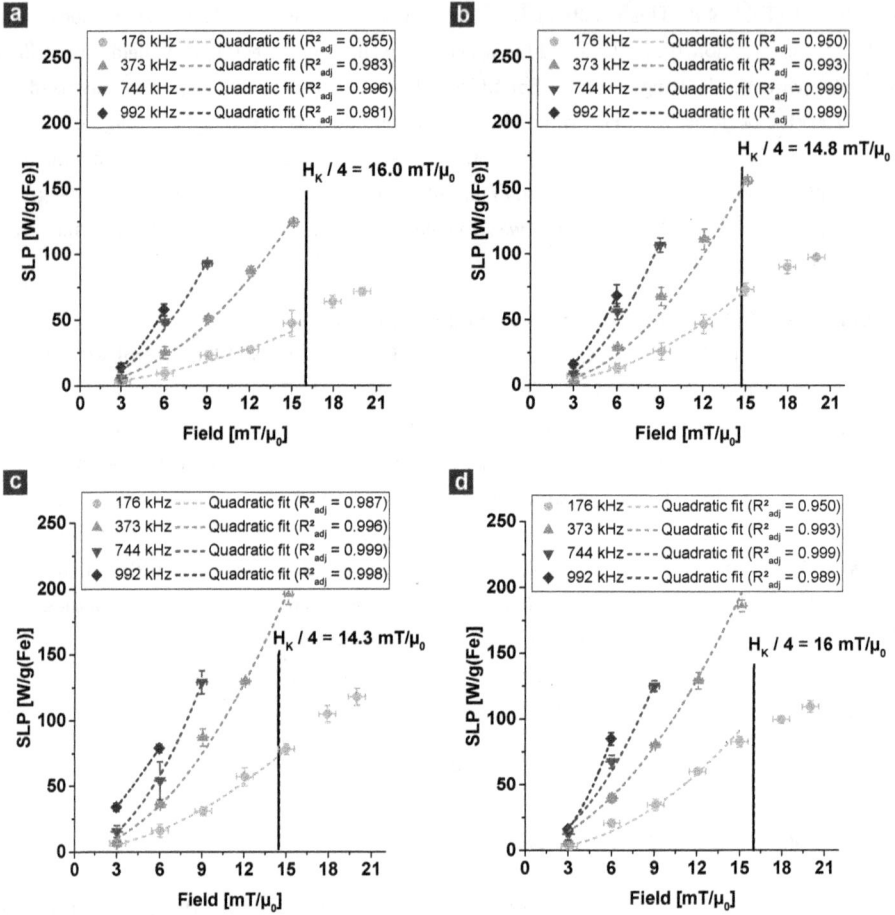

Fig. 5.6.: SLP in dependence of field amplitude for (a) SEA1, (b) SEA2, (c) SEA3 and (d) SEA4 at various frequencies. Up to a value of $H_K/4$ a quadratic function, $\mathrm{SLP} \propto H_0^2$, was fitted to the data for $H_0 \leq H_K/4$.

confirmed that the exponent $\mathrm{SLP} \propto H_0^x$ is $x < 2$ for SPM and ranging between $x = 2-6$ for FM [42].

The behavior of saturating SLP at high field amplitudes is also known well in the literature [47, 274, 275]. The current understanding of SLP saturation-behavior is closely related to the role of anisotropy and interparticle dipole-dipole interactions in particle heating [12]: E. g. Dennis et al. measured the SLP saturation trends for MNP at $f = 150\,\mathrm{kHz}$ and $H_0 \leq 60\,\mathrm{mT}/\mu_0$ [276]. The authors demonstrated by means of small angle neutron scattering structural analysis that the clustered micromagnetic arrangement of MNP gives rise to interparticle dipole-dipole interactions, which in turn largely affect the $\mathrm{SLP}(H_0)$-behavior. Likewise, numerical simulations emphasized the importance of effective anisotropy on the $\mathrm{SLP}(H_0)$-behavior [25, 42, 277]. Here, Carrey et al. suggest that the shape of the $\mathrm{SLP}(H_0)$-curve depends on a complex interplay of

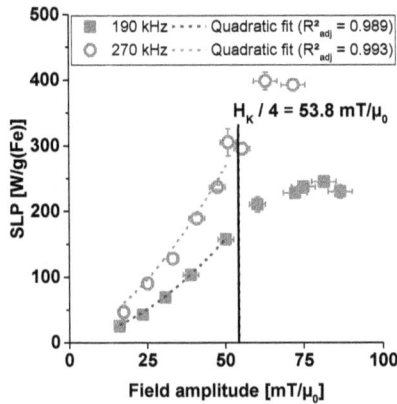

Fig. 5.7.: SLP in dependence of field amplitude for FFAM at two different frequencies. Up to a value of $H_K/4$ a quadratic function, SLP$\propto H_0^2$, was fitted to the data for $H_0 \leq H_K/4$.

the particle core size (esp. the SPM-FM transition-size), the effective anisotropy and applied field amplitude [42].

The intermediate region with SLP$\propto H_0$ can be best interpreted using the MC-simulation results presented in Section 3.3, Figs. 3.4b and 3.5, and summarized in Table 3.2: Here it was shown that the SLP field-dependence can be approximated as SLP$\propto H_0^2$, up to $H_0 \approx H_K/4$, then SLP$\propto H_0$ for $H_0 \geq H_K/4$, and SLP approximates saturation-like behavior for $H_0 > H_K/2$. Therefore the anisotropy field of the MNP samples was estimated from the MNP properties known from sample characterization (s. Chapter 4, esp. Table 4.11) and is listed in Table 5.1. For SEA particles the bulk anisotropy value of $K_{eff} = 11 \, \text{kJ/m}^3$ was assumed [20]. Using

Table 5.1.: MNP properties used for calculating the anisotropy field H_K (cf. eq. (2.15)). The bulk anisotropy value was assumed for SEA particles.

Sample	M_S [kA/m]	K_{eff} [kJ/m³]	H_K [mT/μ_0]
SEA1	346.3	11	64.0
SEA2	376.4	11	59.2
SEA3	389.1	11	57.2
SEA4	345.7	11	64.0
FFAM	445.9	48	215.3

$H_K/4$ to divide the SLP(H_0)-curves of the SEA particles in two regions, as depicted in Fig. 5.6, shows SLP$\propto H_0^2$ up to $H_0 \approx H_K/4$ and then SLP$\propto H_0$ in the experimental SLP. This is in good agreement with the predictions of MC-simulations mentioned above. However, the regime saturating SLP, predicted by MC-simulations for $H_0 > H_K/2$, remains inaccessible for SEA particles with the given setup.

Such a saturation in SLP is observed for FFAM at field amplitudes $H_0 > 60 \, \text{mt}/\mu_0$ (Fig. 5.7). For the given value of $H_K/4 = 53.8 \, \text{mT}/\mu_0$ this behavior does not fit of MC-simulation, which predict SLP$\propto H_0$ for $H_0 > H_{K/4}$. Nevertheless, the SLP(H_0)-trend measured for FFAM at $f = 190$ kHz fits the trends predicted by MC-simulations (cf. Fig. 3.5c) remarkably well, if one assumes lower effective anisotropy values for FFAM: Assuming e.g. $K_{eff} = 26$ kJ/m^3, as estimated for FFAM from a different batch but otherwise prepared identically [245], one calculates $H_K = 116.6$ mT/μ_0. Now using $H_K/4 = 29.2$ mT/μ_0 and $H_K/2 = 58.4$ mT/μ_0 to device the SLP(H_0)-behavior in Fig. 5.7 fits much better to the predictions of MC-simulations mentioned above. As discussed in Section 4.6, the K_{eff} derived from ZFC-FC measurements with eq. (4.16) must only be treated as an estimate. The discrepancy of the measured ($K_{eff} = 48$ kJ/m^3) and the literature ($K_{eff} = 26$ kJ/m^3) effective anisotropy constants suggests further measurements to clarify the exact value of K_{eff}, even more so when taking into account the strong influence of K_{eff} on predicting the SLP(H_0)-dependence, as demonstrated above. Nevertheless, the general qualitative agreement of experiment and MC-simulation for the SLP(H_0)-behavior of SEA and arguably with FFAM particles confirms that MC-simulation predictions are realistic, which will be further investigated in Section 5.3 for size-dependent heating characterization.

Frequency-Dependency

The frequency dependent specific loss power values, SLP(f), are plotted in Fig. 5.8.

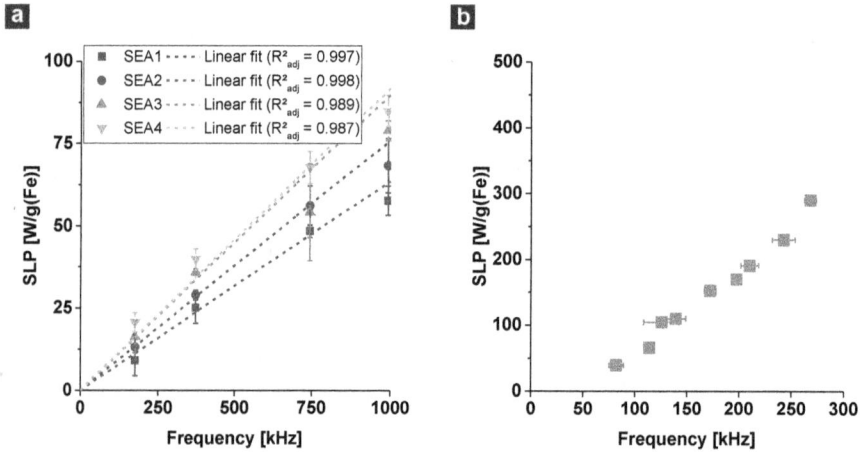

Fig. 5.8.: SLP in dependence of field frequency for (a) SEA particles at $H_0 = 6$ mT/μ_0 and (b) FFAM at $H_0 = 50$ mT/μ_0. A linear fit through the origin is shown in (a).

The SLP(f) values of SEA particles are well described by a linear dependency, SLP$\propto f$ for all cases studied. This linear dependence agrees well with literature [14, 278], as well as with predictions from all theories (LRT, SWMBT, MC-simulations) summarized in Table 3.2 and shown in Figs. 3.6 and 3.7.

For FFAM, there is a general linear dependency in SLP(f) observed (s. Fig. 5.8b), however, it does not go through the origin. This is not expected from common theories (LRT, SWMBT) and the reason why the SLP(f) was not fitted with a linear fit. This deviation from common theory is explained by the high magnetic field amplitude of $H_0 = 50\,\mathrm{mT}/\mu_0$, for which the MC-simulation also shows non-linear SLP(f)-behavior at small frequencies $f < 100\,\mathrm{kHz}$, as exemplarily shown for $d_C = 15\,\mathrm{nm}$ and $H_0 = 20\,\mathrm{mT}/\mu_0$ in Fig. 3.6.

5.2.3. Size-Dependence of Particle Heating Efficiency

In order to compare the SLP of the differently sized MNP samples measured throughout this Capter, the SLP-dependence on the AMF parameters must first be removed. For this, Kallumadi et al. introduced the intrinsic loss power (ILP) [279], which is definded by [269, 280]:

$$ILP = \frac{SLP}{f \cdot H_0^2}. \tag{5.3}$$

ILP is limited to the following conditions: Firstly, SLP must be linear in f, and the H_0-dependency of SLP must be quadratic, simultaneously being at a sufficiently low magnitude that does not saturate the MNP. As shown in Section 5.2.2, these AMF conditions are well met by for the MNP systems used for $H_0 < H_K/4$. Secondly, the experiments to be compared should be conducted under similar thermodynamic conditions. This condition is presumed to be true, as all experiments were carried out by the same experimenter using the same methods and comparable setups. In this way, the ILP allows for a comparison of the particle heating efficiency of all MNP used throughout the thesis, despite the different AMF parameters used.

Sample Preparation and Data Acquisition

Triplicates of FFLA and FFAM in DI-H_2O, ML1 and ML2 in TES-buffer were prepared and measured as described above for field-dependent particle heating analysis. Field parameters were fixed to $f = 270\,\mathrm{kHz}$ and $H_0 = 50\,\mathrm{mT}/\mu_0$.

Likewise, SEA1 through SEA4 were prepared and measured as described above and the field parameters were fixed to $f = 373\,\mathrm{kHz}$ and $H_0 = 15\,\mathrm{mT}/\mu_0$.

These AMF parameters were chosen to maximize the SLP, while staying within the restrictions of ILP to be valid, as presented above.

Results and Discussion

The results of ILP (calculated with eq. (5.3)) versus particle core size are plotted for all eight samples together in Fig. 5.9.

It can generally be seen that the ILP increases with particle core size. Towards larger core sizes, $d_C > 25\,\mathrm{nm}$, ILP presumably saturates, which will be further analyzed by MC-simulation predictions in the Section 5.3. Note that the errors of the ILP for FFLA, MLs and FFAM are much smaller compared to those for SEA particles. This is due to the different methods of calculating the SLP, as the linear-slope-fitting method used for SEA particles is more error-prone, which was also reported in the literature [269].

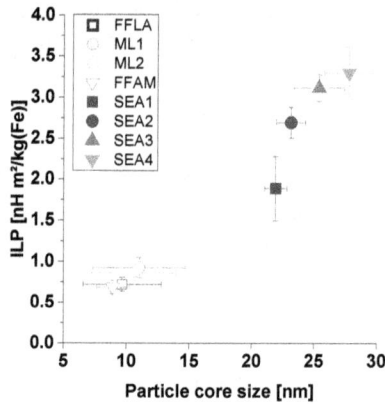

Fig. 5.9.: Intrinsic loss power (ILP) in dependence of the particle core size for all particles studied in this thesis. Open symbols represent measurements taken with the Hüttinger setup, closed symbols those taken with the magneTherm 1.0 setup.

The size-dependent particle heating is also known from literature, however, different results are reported: An increase in particle heating with increasing particle core size was observed for mangetite MNP with $d_C \approx (5-14)$ nm by Gonzales-Weimueller et al. [267] or likewise for maghemite MNP with $d_C \approx (8-17)$ nm by Fortin et al. [194]. An increase in particle heating with size was furthermore observed by Kallumadil et al. [279], reporting on a comprehensive study of 16 commercially available iron oxide MNP with $d_C \approx (7-13)$ nm. Consequently, these studies are in line with the ILP(d_C)-behavior presented in Fig. 5.9.

In contrast, for larger particle core sizes ($d_C > 20$ nm) Lima et al. reported an increasing particle heating up to $d_C \approx 18$ nm for magnetite MNP, followed by a decrease for MNP with $d_C = 23$ nm [281]. Moreover, a similar trend was found by Mehadoui et al. for iron nanostructures (spherical and cubic MNP) with $d_C \approx (5-28)$ nm, demonstrating a dramatic decrease in particle heating for $d_C = 28$ nm [47]. These differences observed for the ILP(d_C)-behavior by Lima et. al. and Mehdaoui et al. compared to the general increase in ILP with d_C observed in Fig. 5.9 can be explained by a closer look at the intrinsic effective anisotropy characteristics of each sample: In the former study of Lima et al., the MNP mass fraction was $\mu_m = 1\%$ (c. f. eq. (7.1)), compared to $\mu_m = (0.13-0.15)\%$ for SEA particles (cf. Table 4.11). Consequently, the average interparticle distance of these MNP was presumably strongly reduced to the regime, where magnetic dipole-dipole interactions dominate, leading to an increase in the effective anisotropy constant [246]. In the same way, iron nanostructures in the latter case of Mehdaoui et al. have a much higher effective anisotropy constant than magnetite ($K_{Fe} = 48$ kJ/m^3 vs. $K_{Fe3O4} = 11$ kJ/m^3 for bulk material [14, 20]). Here, K_{eff} was presumably even enhanced further by shape anisotropy contributions arising from the cubic shape of the nanostructures (cf. Section 2.1.2). Such a considerably higher effective anisotropy value can be the cause for dramatic changes in the size-dependency of particle heating [12]: As shown in Section 3.3,

Fig. 3.3 and summarized in Table 3.2, the size-dependent particle heating predicted by MC-simulations shows highest heating for a distinct particle size, denoted as d_M^*. As shown in Fig. 3.3, d_M^* shifts to smaller particles sizes for increasing K_{eff}-values. Therefore, high effective anisotropy constants assumed for the samples of Lima et al. and Mehdaoui et al. could possibly explain the difference to the ILP(d_C)-behavior measured for the SEA particles (Fig. 5.9; indeed it will be shown that in Section 5.3 that the SLP values for SEA particles show best agreement with MC-simulations for K_{eff}-values remarkably below bulk magnetite value.

5.3. Optimizing Particle Heating by Combining Experiment and Simulation

The SLP(d_C)-data measured for SEA particles is matched to SLP(d_C)-predictions from MC-simulations for various field amplitudes and frequencies in Section 5.3.1. In this way, the MC-simulations are verified against experimental data and the effective anisotropy constant, K_{eff}, and damping parameter, α', are estimated for the SEA particles. Using these estimates for K_{eff} and α' in MC-simulations, the ideal combination of AMF parameters, H_0 and f, and particle core size, d_C, for maximizing particle heating is predicted from MC-simulations in Section 5.3.2. This optimization of SLP(H_0, f, d_C) is performed under additional consideration of the medical safety limitations for the AMF parameters.

This section is based on an original publication by the writer [251] (s. Appendix C). Figures adapted from [251] are marked in the respective figure's caption.

Due to their high monodispersity and narrow size distribution (s. Section 4.8, Table 4.11), SEA particles were selected for an in-depth analysis of the size-dependent SLP(d_C)-behavior. The frequency-dependent SLP values for $H_0 = 6\,\text{mT}/\mu_0$ and the field amplitude-dependent SLP values for $f = 176\,\text{kHz}$ from Fig. 5.7 were selected for this study and are replotted in Fig. 5.10.

SLP value increases with particle core size independent of the applied AMF parameters; however, the relative difference between SLP values decreases for larger sizes ($d_C = 25.4\,\text{nm}$ (SEA3) and $d_C = 27.8\,\text{nm}$ (SEA4)), possibly suggesting a saturating behavior in SLP.

5.3.1. MC-Simulations in Comparison to Experimental Data

MC-simulations were performed within the framework described in Chapter 3. The particle core size, d_C, was varied between $10\,\text{nm}$ and $30\,\text{nm}$ in increments of $0.5\,\text{nm}$. At the same time the effective anisotropy constant, K_{eff}, was varied between $4\,\text{kJ/m}^3$ and $7\,\text{kJ/m}^3$ in increments of $1\,\text{kJ/m}^3$, and $K_{\text{eff}} = 11\,\text{kJ/m}^3$. For each constellation of d_C and K_{eff}, the damping parameter, α', was additionally varied between 0.5 and 1.0 in increments of 0.1. Temperature was fixed to $T = 300\,\text{K}$. The other input parameters were chosen as: Particle core size distribution width, $\sigma_{d_C} = 0.06$, particle hydrodynamic size, $d_H = 75\,\text{nm}$, with distribution width, $\sigma_{d_H} = 0.12$, and saturation magnetization, $M_S = 375\,\text{kA/m}$. These values are the average of the respective

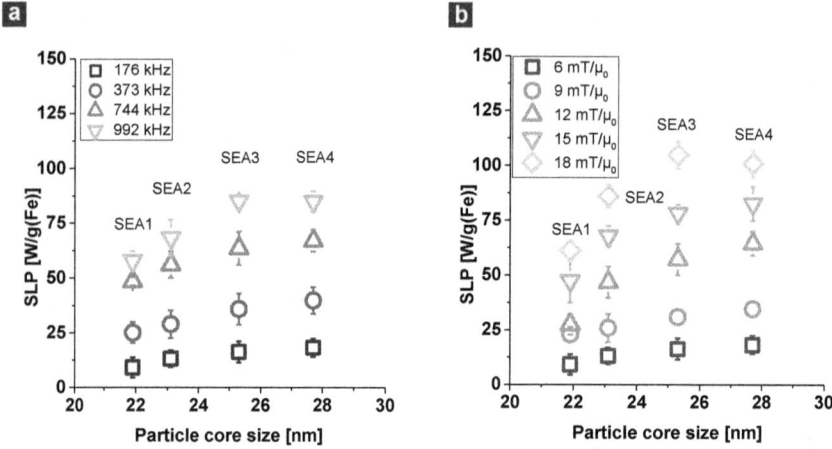

Fig. 5.10.: Particle core size dependent SLP values determined (a) for various frequencies at fixed $H_0 = 6\,\mathrm{mT}/\mu_0$ and (b) for various field amplitudes at fixed $f = 176\,\mathrm{kHz}$. Adapted from [251].

MNP properties of SEA particles derived from materials characterization (s. Section 4.8, Table 4.11).

Furthermore, magnetic dipole-dipole interparticle interactions were neglected for the MC-simulations. This assumption is based on an estimation of the average interparticle distance of approx $(240 - 300)\,\mathrm{nm}$ for all SEA particles. This estimation is based on the iron concentrations of $c = (1.31 - 1.51)\,\mathrm{mg(Fe)/mL}$ used during MFH measurements; details on are discussed in Appendix A.3.2. At such average interparticle distances no significant contributions of magnetic dipole-dipole interaction are expected; details on the calculation are also found in Appendix A.3.2. Besides, the fact that magnetic dipole-dipole interactions do not affect MC-simulations at iron concentrations of $c = 1\,\mathrm{mg(Fe)/mL}$ was also confirmed in Section 3.4, where SLP values simulated with and without magnetic dipole-dipole interparticle interactions were compared, resulting in identical predictions.

In order to find the best match of the simulated data (henceforth *Sim-Data*) to the experimental data (henceforth *Exp-Data*), the quadratic deviation of the SLP value of Sim-Data and the SLP value of Exp-Data, denoted as Γ, was calculated in the following way: Firstly, the Exp-Data were normalized to the highest SLP value (SEA4 at $f = 992\,\mathrm{kHz}$ for Fig. 5.10a and SEA3 at $H_0 = 18\,\mathrm{mT}/\mu_0$ for Fig. 5.10b). Secondly, $\mathrm{SLP}_{\mathrm{Sim\text{-}Data}}$ was averaged over the three $\mathrm{SLP}(d_C)$-values closest to the experimental $d_C = (21.9; 23.1; 25.4; 27.8)\,\mathrm{nm}$ for samples SEA1, SEA2, SEA3 and SEA4, respectively. I. e., for the example of SEA1 with $d_C = 21.9\,\mathrm{nm}$, the mean was calculated from $\overline{\mathrm{SLP}_{\mathrm{Sim\text{-}Data}}}(\mathrm{SEA1}) = \frac{1}{3}(\mathrm{SLP}_{\mathrm{Sim\text{-}Data}}(21.5\,\mathrm{nm}) + \mathrm{SLP}_{\mathrm{Sim\text{-}Data}}(22.0\,\mathrm{nm}) + \mathrm{SLP}_{\mathrm{Sim\text{-}Data}}(22.5\,\mathrm{nm}))$. The same procedure was used for SEA2, SEA3 and SEA4. This averaging was necessary to adjust the experimental particle core sizes to the simulated particle core sizes, which did not match perfectly. As a beneficial side effect, the averaging also accounted for statistical fluctuations in the simulated $\mathrm{SLP}_{\mathrm{Exp\text{-}Data}}$-values. Thirdly, the quadratic deviation of the SLP values of Sim-Data and Exp-Data, henceforth denoted as the individual fitting quality parameter, was normalized to the experimental error, σ_{SLP}^2, was calculated and summed over the results of each sample SEA1, SEA2, SEA3 and SEA4 according to:

$$\Gamma = \sum_{\text{SEA samples}} \frac{(SLP_{\text{Exp-Data}} - \overline{SLP_{\text{Sim-Data}}})^2}{\sigma_{SLP}}, \tag{5.4}$$

By normalizing to the experimental error, the statistical uncertainty is taken into account; weighing $SLP_{\text{Exp-Data}}$-values with high uncertainties less in the overall Γ. Finally, Γ was calculated for each set of $(K_{\text{eff}}, \alpha', f, H_0)$ and summed for the AMF-parameters (f, H_0) of the respective set for $(K_{\text{eff}}, \alpha')$, resulting in the fitting quality parameter, $\overline{\Gamma}(K_{\text{eff}}, \alpha')$, defined as:

$$\overline{\Gamma}(K_{\text{eff}}, \alpha') = \sum_{f, H_0} \Gamma_{f, H_0}(K_{\text{eff}}, \alpha'). \tag{5.5}$$

The parameter $\overline{\Gamma}$ provides information about the quality of the match between the SLP-values from Sim-Data and Exp-Data for a set of fixed effective anisotropy constant and damping parameter. Choosing the smallest $\overline{\Gamma}$ yields the best match of K_{eff}, α' to the experimental SLP data. As seen from Table 5.2, the minimal $\overline{\Gamma}$, i.e. the best agreement in matching Sim-Data with Exp-Data, is found for $K_{\text{eff}} = 4\,\text{kJ/m}^3$ and $\alpha' = 0.7$. The SLP values derived from MC-simulation employing these parameters are compared to the experimental SLP values in Fig. 5.11.

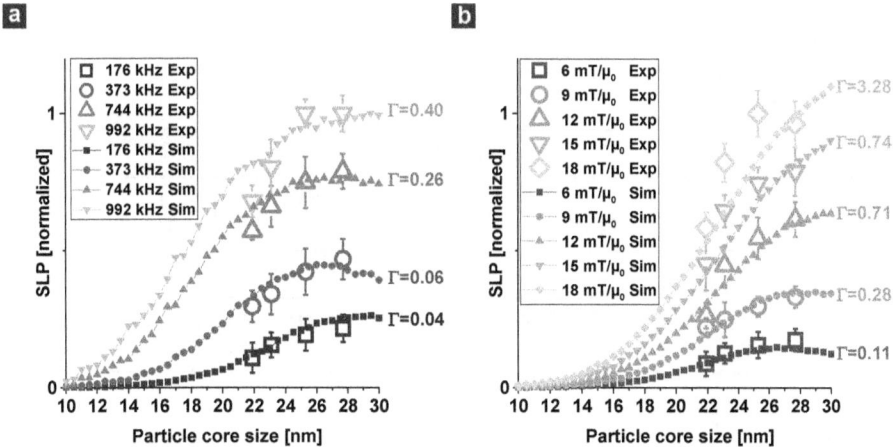

Fig. 5.11.: Normalized SLP values as a function of particle core size from experiment (open symbols) and MC-simulation (closed symbols) for various (a) frequencies with $H_0 = 6\,\text{mT}/\mu_0$ and (b) field amplitudes with $f = 176\,\text{kHz}$. Simulations were performed with an effective anisotropy constant $K_{\text{eff}} = 4\,\text{kJ/m}^3$ and dampling paramter $\alpha' = 0.7$. The individual fitting quality parameter Γ is shown for each simulated curve (the sum of all Γ shown yields the overall fitting quality parameter $\overline{\Gamma}$, as defined in eq. (5.5). Adapted from [251].

---

[2]The experimental error on SLP was calculated from the standard deviation of the triplicate MFH measurements.

Table 5.2.: Quality fit parameter, $\overline{\Gamma}$, eq.(5.5), describing the deviation of simulated and experimental data points for each set of $(K_{\text{eff}}, \alpha')$. The set of $(K_{\text{eff}}, \alpha')$ minimizing $\overline{\Gamma}$ is marked in bold.

$\alpha' = 0.5$

$K_{\text{eff}} \left[\frac{kJ}{m^3}\right]$	$\overline{\Gamma}$	$K_{\text{eff}} \left[\frac{kJ}{m^3}\right]$	$\overline{\Gamma}$	$K_{\text{eff}} \left[\frac{kJ}{m^3}\right]$	$\overline{\Gamma}$	$K_{\text{eff}} \left[\frac{kJ}{m^3}\right]$	$\overline{\Gamma}$	$K_{\text{eff}} \left[\frac{kJ}{m^3}\right]$	$\overline{\Gamma}$
4	14.79	5	15.63	6	18.19	7	21.59	11	23.76

$\alpha' = 0.6$

$K_{\text{eff}} \left[\frac{kJ}{m^3}\right]$	$\overline{\Gamma}$	$K_{\text{eff}} \left[\frac{kJ}{m^3}\right]$	$\overline{\Gamma}$	$K_{\text{eff}} \left[\frac{kJ}{m^3}\right]$	$\overline{\Gamma}$	$K_{\text{eff}} \left[\frac{kJ}{m^3}\right]$	$\overline{\Gamma}$	$K_{\text{eff}} \left[\frac{kJ}{m^3}\right]$	$\overline{\Gamma}$
4	6.75	5	10.37	6	14.29	7	17.83	11	39.58

$\alpha' = \mathbf{0.7}$

$K_{\text{eff}} \left[\frac{kJ}{m^3}\right]$	$\overline{\Gamma}$	$K_{\text{eff}} \left[\frac{kJ}{m^3}\right]$	$\overline{\Gamma}$	$K_{\text{eff}} \left[\frac{kJ}{m^3}\right]$	$\overline{\Gamma}$	$K_{\text{eff}} \left[\frac{kJ}{m^3}\right]$	$\overline{\Gamma}$	$K_{\text{eff}} \left[\frac{kJ}{m^3}\right]$	$\overline{\Gamma}$
4	**5.88**	5	9.72	6	12.39	7	19.28	11	46.44

$\alpha' = 0.8$

$K_{\text{eff}} \left[\frac{kJ}{m^3}\right]$	$\overline{\Gamma}$	$K_{\text{eff}} \left[\frac{kJ}{m^3}\right]$	$\overline{\Gamma}$	$K_{\text{eff}} \left[\frac{kJ}{m^3}\right]$	$\overline{\Gamma}$	$K_{\text{eff}} \left[\frac{kJ}{m^3}\right]$	$\overline{\Gamma}$	$K_{\text{eff}} \left[\frac{kJ}{m^3}\right]$	$\overline{\Gamma}$
4	6.18	5	9.64	6	17.65	7	28.34	11	94.15

$\alpha' = 0.9$

$K_{\text{eff}} \left[\frac{kJ}{m^3}\right]$	$\overline{\Gamma}$	$K_{\text{eff}} \left[\frac{kJ}{m^3}\right]$	$\overline{\Gamma}$	$K_{\text{eff}} \left[\frac{kJ}{m^3}\right]$	$\overline{\Gamma}$	$K_{\text{eff}} \left[\frac{kJ}{m^3}\right]$	$\overline{\Gamma}$	$K_{\text{eff}} \left[\frac{kJ}{m^3}\right]$	$\overline{\Gamma}$
4	13.93	5	14.87	6	16.31	7	17.66	11	100.20

$\alpha' = 1.0$

$K_{\text{eff}} \left[\frac{kJ}{m^3}\right]$	$\overline{\Gamma}$	$K_{\text{eff}} \left[\frac{kJ}{m^3}\right]$	$\overline{\Gamma}$	$K_{\text{eff}} \left[\frac{kJ}{m^3}\right]$	$\overline{\Gamma}$	$K_{\text{eff}} \left[\frac{kJ}{m^3}\right]$	$\overline{\Gamma}$	$K_{\text{eff}} \left[\frac{kJ}{m^3}\right]$	$\overline{\Gamma}$
4	8.43	5	13.48	6	27.21	7	44.91	11	131.74

For the frequency-varied case (Fig. 5.11a), the individual fitting quality parameters of $\Gamma = (0.04..0.40)$ confirm a good agreement of MC-simulated and experimental data. The agreement is similarly good for the amplitude-dependent case, resulting in $\Gamma = (0.11 - 0.74)$ (Fig. 5.11b). An exception is the comparison of the highest field amplitude, $H_0 = 18 \, \text{mT}/\mu_0$, showing $\Gamma = 3.28$. This irregularity is mainly attributed to the experimental SLP value of SEA4, which does not follow the general trend observed for other samples and field amplitudes. One reason for this could be the caused by a change in the MNP saturation magnetization, M_S, upon heating, as it has been observed that MNP may continue an incomplete oxidation under the influence of heat [223]: As SEA4 particles have the largest particle core size, $d_C = 27.8 \, \text{nm}$, and the SLP in question was measured for a high field amplitude, $H_0 = 18 \, \text{mT}/\mu_0$, the effective temperature rise was highest for this sample and could thus have lead to a continued oxidation of magnetite to maghemite. This would lower M_S (as bulk magnetite shows $M_S = 476 \, \text{kA/m}$ compared to $M_S = 394 \, \text{kA/m}$ for bulk maghemite [232]) and with it also the SLP value [42, 282]. However, a clear correlation cannot be confirmed here.

The results from MC-simulations are shown to agree *quantitatively* with the experimental data and thereby brigde the gap between experiment and theoretical prediction, especially for large particles with $d_C > 20 \, \text{nm}$, unlike previous models and simulations discussed in Section 3.1 [25, 42, 48, 163, 165]. Since the MC-simulations are not limited to a particluar range of particle core sizes, it could advance the understanding of partilce heating, where previous studies reached a limit using common theories: For example, the linear response theory (LRT) has been demonstrated to accurately experimental data only for $d_C \leq 13.5 \, \text{nm}$ and $H_0 = 6.8 \, \text{mT}/\mu_0$ for cobalt ferrite MNP [283]. Stoner-Wohlfarth-model-based-theory (SWMBT) on the other hand, was limited in correctly predict the size-dependent SLP values measured for iron nanostructures to $d_C \leq 20 \, \text{nm}$ [47]. More advanced dynamic relaxation hysteresis simulations have emphasized the importance of adjusting the effective anisotropy of MNP in order to optimize heating, but experimental validation is missing [25, 163].

The effective anisotropy constant, $K_{eff} = 4 \, \text{kJ/m}^3$, showing the best match of MC-simulation to experimental data agrees well with $K_{eff} \approx 5 \, \text{kJ/m}^3$ from earlier experimental work, derived from mangetorelaxometry, magnetic particle spectroscopy, and AC-susceptibilty measurements perforemd with similarly synthesized thermal decomposition particles with $d_C \approx 20 \, \text{nm}$ [284]. Furthermore, similar K_{eff}-values have been reported recently from qualitative fitting of kinetic Monte Carlo (MC-) hysteresis simulations to experimental data: Niculaes et al. report $K_{eff} = 5 \, \text{kJ/m}^3$ for interacting magnetite nanocubes (edge length $d_C \approx 20 \, \text{nm}$) [285]; Cabrera et al. observed best agreement between MC-simulation data and AC-susceptibility measurements for $K_{eff} = 4.1 \, \text{kJ/m}^3$ for magnetite MNP ($d_C = 21 \, \text{nm}$) [286].

Altogether, the MC-simulations agree very well with the experimental data for reasonably small effective anisotropy values, $K_{eff} = 4 \, \text{kJ/m}^3$ and a damping constant of $\alpha'' = 0.7$. MC-simulations with these parameters are used to predict optimized particle heating under medically safe conditions in the next Section 5.3.2.

5.3.2. Optimized Particle Heating under Medical Safety Constraints

In this section, MC-simulations are used to predict an optimal combination of particle core size and AMF-parameters to maximize particle heating. The assessment is performed under the additional constraints of medically tolerable conditions, as any AMF applied to the body must be limited for safety reasons in order not to induce non-selective body heating and undesired nerve stimulation due to eddy currents, as discussed in Section 2.5. In the following, particle heating is calculated under consideration of the two medical limits for the AMF parameters derived from clinical trials (s. Section 2.5), one for treatment of the body, $Z_{med}^{body} = H_0 \cdot f = 628\,\text{kHz} \cdot \text{mT}/\mu_0$, and one for treatment of the head, $Z_{med}^{head} = H_0 \cdot f = 1\,758\,\text{kHz} \cdot \text{mT}/\mu_0$. MC-simulations were carried out for $K_{eff} = 4\,\text{kJ}/\text{m}^3$, $\alpha' = 0.7$, $d_C = (10-30)\,\text{nm}$ and for various $f = (10-1\,000)\,\text{kHz}$, while adjusting H_0 in order to not exceed the limits Z_{med} mentioned above. Varying f and adjusting H_0 accordingly was done for practical reasons, as typical hyperthermia setups allow for a continuous adjustment of H_0, whereas f is usually specified by the specific coil and capacitor combinations of the resonant circuit. The results are demonstrated as a heat map in Fig. 5.12.

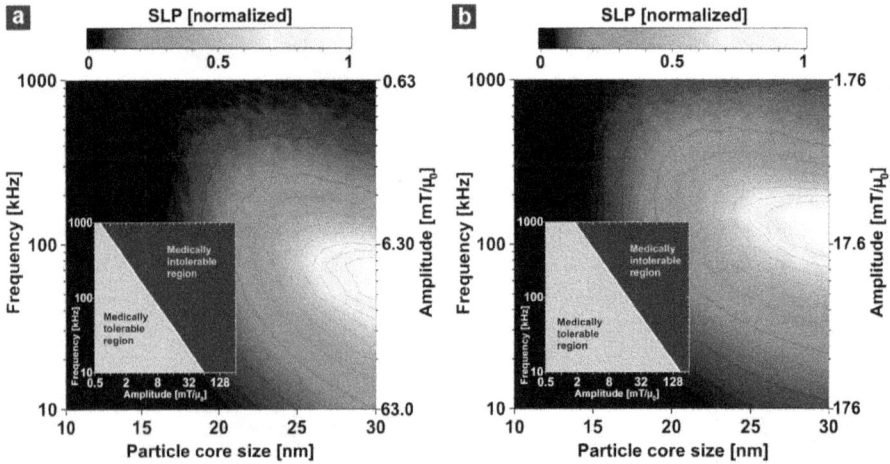

Fig. 5.12.: Heat map of normalized SLP values depending on AMF parameters and particle core size under medically safe condictions: (a) for the body limit and (b) for the head limit. Inset graphs show the corresponding linear relationship of field amplitude, H_0, and frequency, f, of the AMF, separating the medically tolerable region from the medically intolerable region. Adapted from [251].

Independent of applying the body or head limit, the highest heating occurs for the largest core sizes, $d_C = (27-30)\,\text{nm}$, a result already expected from the preceding results of Section 5.3.1, esp. Fig. 5.11. Such a trend was also reported by Mamiya and Jeyadevan for dynamic relaxation hysteresis simulations of rotable MNP within the 2-level-approximation (s. Section 3.1 for details) under similiar medical constraints [164]. The set of AMF parameters generating highest heating is $f_{body} \approx 75\,\text{kHz}$ and $H_{0,body} = 8.4\,\text{mT}/\mu_0$ for the body limit and shifts to higher frequencies with $f_{head} \approx 150\,\text{kHz}$ and $H_{0,body} = 11.7\,\text{mT}/\mu_0$ for the head limit. Under the assumption of similar K_{eff}-values for the iron oxide MNP ($d_C \approx 15$ nm), this implies that the AMF-frequency $f = 100$ kHz applied in the above mentioned clinical trials [64, 65] was optimal for maximizing SLP. Carrying this assumption further within the framework of the MC-simulations, these trials could have been even more effective by increasing the particle core size from $d_C \approx 15$ nm to e. g. $d_C \approx 25$ nm, increasing the predicted SLP value by a factor of approximately 12. Or in other words, the same heating could have been achieved in these trials when using MNP with $d_C \approx 25$ nm, while reducing MNP dosing by the same factor (since the SLP value is directly proportional to the MNP concentration, cf. eq. (5.1)). Please note that such high MNP concentrations that are used currently for clincial MFH ($c \approx 112$ mg(Fe)/ml) will most likely cause MNP agglomeration, resulting in a collectively-coupled MNP relaxation behavior [287]. The heating behavior of such collectively relaxing MNP is substantially harder to control and predict than that of monodisperse, single MNP, which is the topic of Section 7.4. Consequently, a lower MNP dosing also helps to reduce uncertainty of accurately predicting the particle heating for the clinical application of MFH.

In fact, provided that well characterized MNP systems are used (esp. with known effective anisotropy constant and damping paramter), such MC-simulation-based predictions could help to optimally adjust AMF-parameters for MFH-therapy planning and reduce MNP dosing for existing MNP. One step further, new MNP heating agents could be produced that are optimized for maximizing SLP with the AMF parameters of existing clinical AMF-setups, such as the MFH 300F clinical MFH device (MagForce AG, Berlin, Germany) [74]. Here, MC-simulations could be employed to predict the optimal combination of MNP particle core size, effective anisotropy and saturation magnetization for a given set of AMF parameters. Either way, such improvements of the effective particle heating can help reduce the effective MNP dosing, making MFH therapy more safe for the patients, and also economically more affordable. Always provided that appropriate MNP amounts can be delivered in vivo to the targeted site.

Please note that the MC-simulations presented here, albeit being a powerful tool for the prediction of particle heating, are still falling short when predicting the SLP for realistic in vivo settings: As demonstrated in the next Chapter 6, MNP agglomerate (cluster) and immobilize inside tissue and cells upon interacting with the body in vivo. Those effects of agglomeration and immobilization greatly influence particle relaxation behavior and, consequently, also particle heating, but cannot be modeled by the MC-simulations correctly at present. Moreover, bulk temperature rise due to particle heating as measured in calorimetric hyperthermia experiments is not the only contributing factor to cytotoxicity during MFH application: Nanoheating effects and cell membrane damage (caused by mechanical rotation of MNP) must also be considered when assessing MFH effectiveness for tumor therapy (s. Section 2.5.3 and Chapter 8).

6. Magnetic Particle-Cell-Interaction

During in vivo applicability in biomedicine, MNP will inevitably interact with biological media, cells and tissue and changes in MNP mobility, morphology and arrangement are expected. In order to predict these changes, this chapter studies the MNP-cell-interactions in vitro as a necessary step towards in vivo application of MNP in MPI and MFH[1]. As this thesis is part of a treatment approach for the pancreatic ductal adenocarcinoma (s. Section 2.5.4), this study is performed using pancreatic tumor cells MiaPaCa-2 and BxPC-3 as well as murine fibroblasts L929 as a healthy control. The main goal of this chapter is to identify changes in MNP morphology, arrangement and relaxation behavior upon interaction with cellular material and moreover to describe the MNP uptake kinetics inside cells, using ML particles. For this, the chapter first introduces the cell culture protocols necessary to grow cells appropriately in Section 6.1. Second, experiments on ML internalization inside cells are conducted and analyzed for morphological changes via TEM analysis in Section 6.2. Third, MPS is used to quantify the ML uptake inside cells, which is used to model the uptake kinetics of ML inside cells in Section 6.3. In this section also analyzes the magnetic relaxation behavior of intracellular ML. Furthermore, the changes of the ML magnetic properties caused by internalization are investigated via SQUID magnetometry in Section 6.4. Finally, the main findings are generalized in a brief concluding remark (s. Section 6.5).

For all experiments ML were used. They were chosen as the most suitable particle system for in vitro application, as their liposome coating mimicks the phospholipid bilayer of cell membranes [288] and it was demonstrated in literature before that liposome coating facilitates uptake inside cells [289, 290].

This Chapter is partly based on an original publication by the writer [291] (s. Appendix C).

6.1. Cell Culture Protocol

Human pancreatic tumor cell lines *MiaPaCa-2* and *BxPC-3* were used throughout this thesis as well as murine fibrobast cell line *L929* as a healthy control. All three cell lines were obtained from the German Collection of Microorganisms and Cell-Cultures (DSMZ, *Deutsche Sammlung von Mikroorganismen und Zellkulturen GmbH*, Braunschweig, Germany). The cell culture procedure described in the following is the standard protocol used for all cell experiments in this thesis, unless stated otherwise in the respective chapters and sections.

The three cell lines have the following characteristics:

- **MiaPaCa-2**
  This pancreastic cancer cell line was established by A. Yunis et al. in 1977 and obtained from a PDAC of a 65-year old Caucasian male [292]. It has a doubling time of approx.

---

[1]In vitro conditions provide comparatively well-controlled and reproducible situations and are thus well suited for investigating MNP-cell-interactions.

$t_{doub} \approx 40\,$h. MiaPaCa-2 shows fairly poor differentiation behavior of grade GIII[2] [294]. MiaPaCa-2 was cultured in Dulbecco's Modified Eagle Medium (DMEM) (gibco Life Technologies, Fisher Scientific GmbH, Schwerte, Germany) treated with penicillin and streptomycin (both with volume fraction $\mu_v = 1\,\%$) and fetal calf serum (FCS, $\mu_v = 10\,\%$).

- **BxPC-3**

  BxPC-3 display a high to moderate differentiation behavior of grade GI-GII [294]. This cell line was first established by Tan et al. in 1986 [295] and has a doubling time of $t_{doub} = (48-60)\,$h. It was obtained via biopsy from tissue of a primary tumor from a 61-year old female.

  BxPC-3 was cultured in Roswell Park Memorial Institute (RPMI) cell culture medium RPMI 1640 (gibco) treated with Na-Pyruvate, penicillin and streptomycin (all three with volume fraction $\mu_v = 1\,\%$) and fetal calf serum (FCS, $\mu_v = 10\,\%$).

- **L929**

  These murine fibroblasts has been developed from normal subcutaneous areolar adipose tissue of a 100-day-old C3H/An mouse (first obtained by Earle and Voegtlin in 1940 [296]) via cloning. L929 referes to the 929-th clone of the parent L strain first to be established successfully in continuous culture in 1948 by Sanford et al. [297]. Doubling time varries from $t_{doub} = 14\,$h [298] to $t_{doub} = 42\,$h [299] depeonding on the cell passage.

  BxPC-3 was cultured in Roswell Park Memorial Institute (RPMI) cell culture medium RPMI 1640 (gibco) treated with penicillin and streptomycin (both with volume fraction $\mu_v = 1\,\%$) and fetal calf serum (FCS, $\mu_v = 10\,\%$).

Cells were thawed after freezing ($-196\,°$C, liquid nitrogen) by prewarming in a water bath set to $37\,°$C for $30\,$s and subsequently being suspended in $9\,$mL cell culture medium prewarmed to $37\,°$C. The suspension was centrifuged for $5\,$min ($1\,500\,$rpm, $37\,°$C) and the supernatant was cast away. This step was necessary to remove the antifreeze agent dimethyl sulfoxide, which is cytotoxic under ambient conditions, but necessary to prevent the formation of ice crystals during freezing that would otherwise harm the frozen cells. After resuspending in $10\,$mL of cell culture medium, the cells were transferred in a T75-cell culture flask ($75\,$cm^2; Nunc EasYFlask, ThermoFisher Scientific) and stored in an Heracell 150i incubator (ThermoFisher Scientific) at $37\,°$C and $5\,\%$ CO_2. After approx. $24\,$h the cell medium was exchanged to remove dead cells, which are suspended in the cell medium as they have lost adhesion to the living cells growing at the bottom of the flask.

---

[2]Tumor cells and tissue are graded by their amount of abnormality, also called *differentiation behavior*, compared to healthy cells and tissue under the microscope: grade *GX* tumors cannot be assessed (undetermined differentiation behavior), grade *GI* tumors show a high differentiation behavior (i. e. they are similar to normal cells and tissue), grade *GII* tumors show a moderate differentiation behavior, and grade *GIII* and *GIV* tumors show a poor differentiation behavior [293]. Due to their poor differentiation behavior GIII and GIV tumor cells grow and spread quickly.

To grow the experiment-specific amount of cells (s. Table 6.1), the cells were passaged at a ratio 1:2 every 3-4 days, but never below 70% confluency (checked via light microcopy). For cell harvesting, medium was removed and 2.5 mL Typsin/EDTA solution (0.05 w%/0.02 v% in PBS (1x, w/o Ca$^+$ or Mg^{2+}; Biochem GmbH, Karlsruhe, Germany) was added. To allow full detachment of living cells from the bottom of the flask, the cells were then stored in the incubator for approx. 15 min. 10 mL of medium was added and cells were counted using an automated cell counting machine *LUNA* (Logos Biosystem Inc., Annandale, VA, USA). Afterwards, cells were seeded in experiment-specific amounts to either 6-/24-well plates (CELLSTAR, Greiner Bio-One International GmbH, Kremsmünster, Österreich) or in T25 flasks (ThermoFisher Scientific) in their medium and stored in the incubator for 24 h (cf. Table 6.1). Subsequently, medium was removed and cells were incubated with MNP mixed in cell culture medium (always ML in this thesis) for 24 h (standard value) or a specific incubation time as stated in the respective section. Every MNP sample was passed through a 0.22 µm syringe filter for sterilization.

After incubation with MNP, the cell harvesting is performed with the general protocol as described above, but in between, cells are washed with 1x DPBS (gibco) to remove excess MNP. The amounts of Typsin/EDTA solution and medium added subsequently for harvesting depends on the specific experiment, as listed in Table 6.1.

Table 6.1.: Specifications used for cell culture during incubation and for cell harvesting after incubation. The experiment-specific number of cells was seeded in the respecitve container with the amount of cell medium specified. After incubation with MNP, the cells were washed in 1x DPBS, harvested with Trypsin/EDTA and resuspended in the specified amount of medium.

		Cell growth and incubation			Cell harvest after incubation		
Experiment	**Section**	**Cells seeded**	**Container**	**Medium [mL]**	**DPBS [mL]**	**Trypsin/EDTA [mL]**	**Medium [mL]**
TEM	6.2	$5 \cdot 10^6$	T25 flask	3.5	2.5	–	–
MPS	6.3	$2 \cdot 10^4$	24-well plate	1.0	0.5	0.2	0.5
SQUID	6.4	$1 \cdot 10^6$	6-well plate	1.0	2.5	0.8	3.0
MPS/MFH	8.1	$1 \cdot 10^6$	6-well plate	1.0	2.5	0.8	3.0

6.2. Morphology of Magnetic Particles inside Cells

This section studies the arrangement, localization and distribution — defined as *morphology* in the following — of ML upon internalization inside cells via transmission electron microscopy (TEM; s. Section 4.3 for a general description of TEM). The cell lines MiaPaCa-2, BxPC-3 and L929 were incubated with ML and afterwards investigated with TEM; the respective experimental procedure is described in Section 6.2.1. Following, the various aspects of the morphology of MNP inside cells are studied: First, Section 6.2.2 studies the internalization of ML over the initial 2 h of incubation with TEM images taken at fixed time points to assess the different states of ML uptake in a chronological manner. Second, in Section 6.2.3, the localization and distribution of ML inside cells is assessed after 24 h of incubation and the differences between the ML arrangement inside the three cell lines are discussed. Finally, the results on ML morphology are briefly summarized in Section 6.2.4.

TEM measurements were carried out in cooperation with the Electron Microscopy Facility at the Institute of Pathology from the RWTH Aachen University Hospital (Aachen, Germany).

6.2.1. Experimental Procedure

Standard Procedure for TEM Cell Imaging

Each cell line was cultivated as described in Section 6.1 until $5 \cdot 10^6$ cells could be seeded in a T25 flask ($25 \, cm^2$; Thermo Fisher Scientific). Cells were stored in the incubator for approx. 24 h, then the medium was removed and the cells were subsequently incubated with 3.5 mL fresh cell-specific medium (cf. sec. 6.1) including ML1 particles with various concentrations of $c = (0.150, 0.225, 300) \, mL(Fe)/mL$. One flask was prepared for each cell line and each individual ML concentration. The standard incubation time with ML1 was 24 h. After incubation, the ML-treated medium was removed and cells were washed once in 2.5 mL DPBS (1x; gibco) immediately afterwards to remove excess ML. Subsequently, 1.5 mL Sørenses phosphate buffer (Merck KGaA, Darmstadt, Germany) mixed with glutaraldehyde (volume fraction $\mu_v = 3\%$; GA) was added to fix the cells. The GA-treated cells were stored at $4\,°C$ for 4 h before GA excess was cast away and the cells fixed in the residual GA were scraped off the bottom of the flask. From here, they were transferred into 1.5 mL PCR tubes (Eppendorf) and centrifuged for 5 min (1 200 rpm; ambient conditions). The supernatant was removed and the remaining cell pellet was embedded in agarose (mass fraction $\mu_m = 2.5\%$) preheated to $60\,°C$. The mixture was stored at $4\,°C$ for 10 min to allow agarose solidification. Afterwards, the agarose pellet was cut in blocks, which were individually immersed in GA ($\mu_v = 3\%$) and stored at $4\,°C$ overnight (approx. 12 h).

Then, the agarose blocks were immersed in Sørensen buffer for 15 min, followed by immersion in a mixture of osmium tetroxide (mass fraction $\mu_m = 1\%$; OsO_4; Sigma Aldrich) and sucrose buffer (mass fraction $\mu_m = 8.5\%$) for 1 h. Afterwards, the agarose blocks were washed with sucrose buffer (mass fraction $\mu_m = 8.5\%$) and then with pure DI-H_2O. Subsequently, the samples were dehydrated by repetitive washing with increasing concentrations of ethanol (volume fraction $\mu_v = 30\%$, 50%, 70%, 90% and 100%) each for 10 min ($\mu_v = 100$ was applied three times). The dehydrated agarose blocks were subsequently cast in propylene oxide (C_3H_6O; Sigma Aldrich) for 30 min, in a 1:1 mixture of epoxy EPON resin (Sigma Aldrich) and propylene oxide for 1 h and finally in pure EPON resin for 1 h. The final sample was cured at $90\,°C$ for 2 h.

The solid sample blocks were cut into slices of $(70 - 90)$ nm thickness with a Leica EM UC-6 microtome (Leica Microsystems, Wetzlar, Germany) and affixed onto 3 mm diameter Cu/Rh-150-mesh maxtaform grids (Electron Microscopy Sciences). To enhance contrast during TEM imaging, the samples were stained with a mixture of uranyl acetate (mass fraction $\mu_m = 0.5\%$) and lead citrate (mass fraction $\mu_m = 1\%$; both from Electron Microscopy Sciences). TEM imaging was performed with a Zeiss LEO 906 TEM at 60 kV as described in Section 4.3.

TEM Procedure for Chase Experiments

In order to image the process of ML internalization, beginning with the binding of ML to the cell membrane, special cell experiments, called *chase* experiments (as defined in literature by Wilhelm et al. [300]), were conducted for MiaPaCa-2 and L929 cells. Here, cells were seeded with $5 \cdot 10^6$ cells in T25 flasks as described above for the TEM standard procedure. Before incubation with ML-treated medium, however, the cells were stored at $4\,^\circ$C for 1 h. This short period of cold passivated the cells, only allowing ML to attach to the cell membrane (s. Section 2.5.2). Then, the cells were treated with medium containing ML1 ($c = 0.15\,\text{mg(Fe)/mL}$) as described above in the TEM standard procedure and incubated for another hour at $4\,^\circ$C to allow ML membrane attachment. Afterwards, the cells were washed and resuspended in fresh medium before incubation at $37\,^\circ$C. Cells are reactivated at this temperature and the membrane-bound ML start to internalize. The incubation time was varied for $t_\text{chase} = (0, 10, 20, 45, 60$ and $120)$ min. After the specific incubation times, the cell samples were further prepared for TEM cell imaging following the steps presented in TEM standard procedure above.

6.2.2. Chase Experiments Reveal ML Uptake Kinetics

The process of ML internalization begins with the binding of ML to the cell membrane, which is shown in exemplary TEM images from chase experiments for MiaPaCa-2 and L929 in Figs. 6.1 and 6.2, respectively. Here, ML (visible as agglomerates of black dots in the TEM images) are first allowed to attach only to the membrane of cells passivated at $T = 4\,^\circ$C, and are allowed subsequently to be internalized by reactivated cells at $T = 37\,^\circ$C for up to 2 h. The results for $t_\text{chase} = 0$ min for either cell line verifies that ML were only bound to the membrane (s Figs. 6.1a and 6.2a), however ML are grouped in small agglomerates of several tens of nanometers here. With restored endocytic activity at $T = 37\,^\circ$C incubation, one observes that these membrane-bound ML agglomerates are internalized over a few tens of minutes (s. Figs. 6.1(b,c) and 6.2(b,c)). Note the display of an endocytosis internalization process in Fig. 6.1b, which is discussed in detail in the following section. ML are fully internalized inside cells after 45 min (s. Figs. 6.1d and 6.2d). Consequently, the ML uptake process takes less than one hour for full internalizatoin of ML particles inside both cell lines. No more membrane-bound ML are observed for $t_\text{chase} \leq 45$ min, as all ML are internalized. Even though the TEM images in Figs. 6.1 and 6.2 show only a part of the whole cell, the absence of membrane-bound ML was verified from a multitude of chase TEM images at each the given chase time t_chase. The time scale for the internalization of ML of approx. 1 h is in line with the findings of Wilhelm et al., who reported full internalization of DMSA[3]-coated MNP inside HeLa tumor cells within 1 h of incubation [300]. The vesicle size increases to approx. 100 nm to 200 nm in diameter with increasing chase times ($t_\text{chase} = (1, 2)$ h, s. Figs. 6.1(e,f) and 6.2(e,f)), indicating the evolution of smaller early endosomes into larger endosomes by merging of ML-carrying vesicles, as described in Section 2.5.2.

---

[3] meso-2,3-dimercaptosuccinic acid (HOOC-CH-(SH)-CH(SH)-COOH)

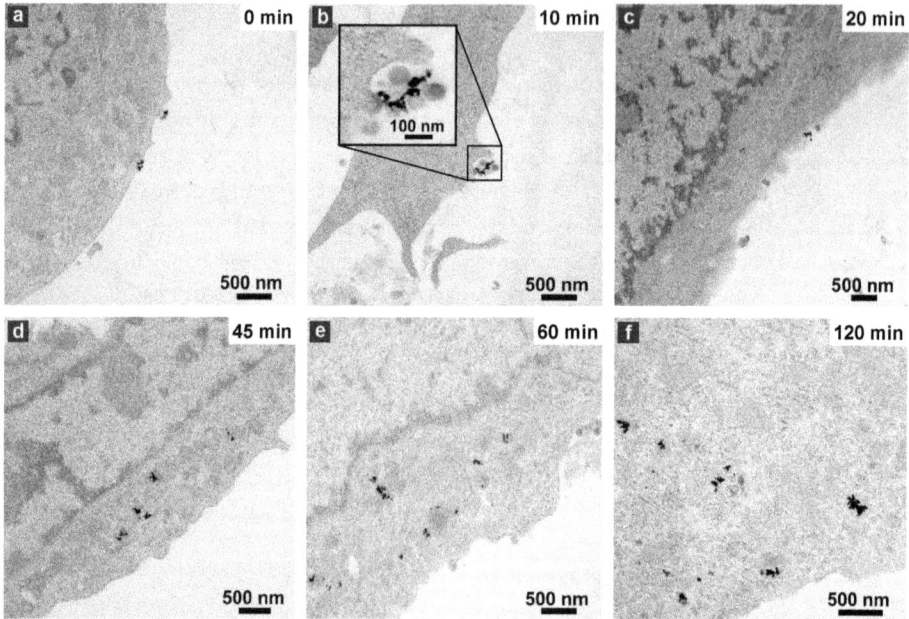

Fig. 6.1.: TEM images of MiaPaCa-2 cells incubated with ML1 particles ($c = 0.15\,\text{mg(Fe)/mL}$) during chase experiments for chase times and magnifications: (a) 0 min, 21 560 x, (b) 10 min, 21 560 x, (c) 20 min, 12 930 x and (d) 45 min, 21 560 x, (e) 60 min, 21 560 x and (f) 120 min, 27 800 x. Note the capture of an endocytosis internalization process in the inset in (b).

6.2.3. Morphology of Intracellular ML

Localization of ML after 24 h of Incubation with Cells

After an incubation time of 24 h the internalization of ML inside cells is expected to have saturated [11] and the arrangement and distribution of ML can be assumed to have reached a stable state. Please note that the results presented in the following are exemplary demonstrated for an ML incubation concentration of $c = 0.15\,\text{mg(Fe)/mL}$. The results for higher incubation concentrations tested ($c = 0.225$ and 0.3)mg(Fe)/mL are found in Appendix A.4 and yield no differences to the findings obtained for $c = 0.15\,\text{mg(Fe)/mL}$. Therefore, the trends described in the following hold for all incubation concentrations used.

An exemplary TEM image for each of the cells MiaPaCa-2, BxPC-3 and L929 after 24 h of incubation with ML ($c = 0.15\,\text{mg(Fe)/mL}$) is shown in Fig. 6.3. Generally, ML are internalized in all three cell lines. The majority of ML is grouped and clustered inside cells internalized in lysosomes, while a few ML are attached to the cell membrane. In agreement with the observations from chase experiments, the membrane-bound ML are also grouped in agglomerates, but smaller ones than the agglomerates observed intracellularly. Interestingly, no ML are observed at the membrane of BxPC-3 cells (cf. Fig. 6.3b), which possibly reflects less adhesion of the

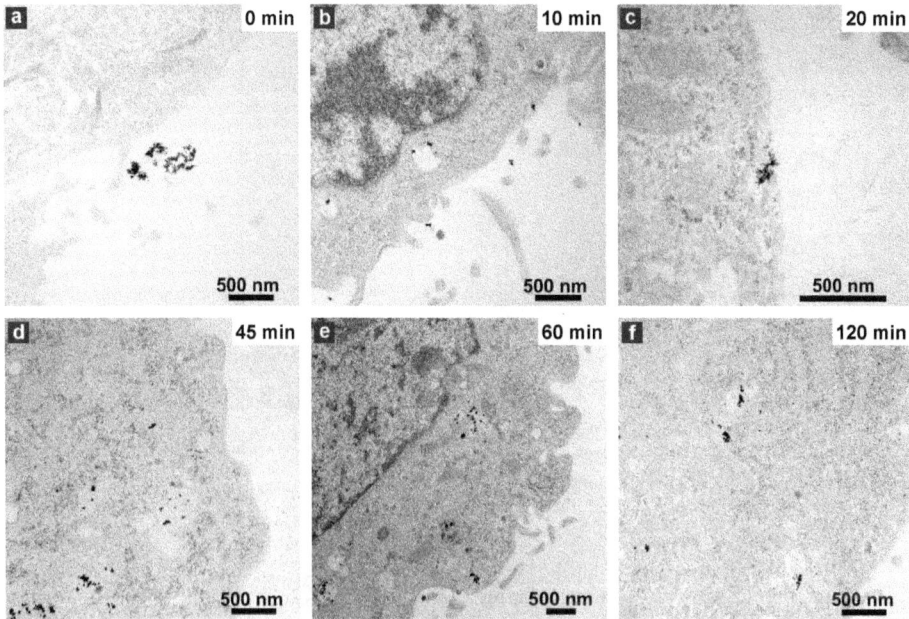

Fig. 6.2.: TEM images of L929 cells incubated with ML1 particles ($c = 0.15\,\mathrm{mg(Fe)/mL}$) during chase experiments for chase times and magnifications: (a) 0 min, 27 800 x, (b) 10 min, 27 800 x, (c) 20 min, 60 000 x and (d) 45 min, 27 800 x, (e) 60 min, 12 930 x and (f) 120 min, 27 800 x. (a), (b), (d) and (f) adapted from [291].

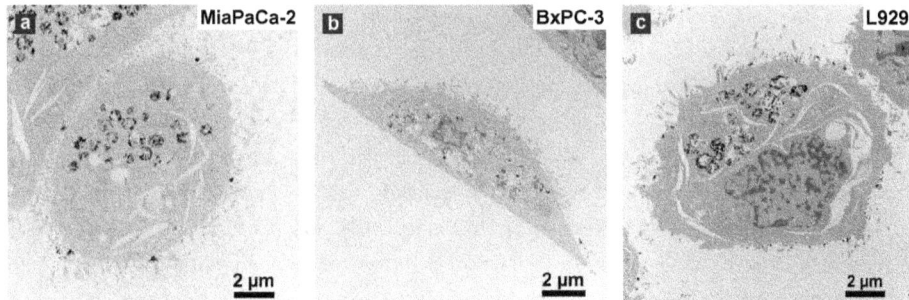

Fig. 6.3.: Exemplary TEM images of (a) MiaPaCa-2, (b) BxPC-3 and (c) L929 cells after incubation with ML1 particles ($c = 0.15\,\mathrm{mg(Fe)/mL}$) for 24 h; magnification 6 000 x. ML are mostly located inside cells clustered in lysosomes; some ML are found also clustered at the membrane of MiaPaCa-2 (a) and L929 (c) cells.

ML to the BxPC-3 cell membrane, resulting in ML being washed away by washing in PBS during sample preparation.

Membrane transport processes such as endocytosis and exocytosis are displayed in for MiaPaCa-2 in Fig. 6.1b and for BxPC-3 and L929 in Fig. 6.4. Here, ML agglomerates at the cellular membrane were engulfed by membrane invaginations. Since TEM images always show snapshots of the otherwise continuous processes, it cannot be distinguished whether endo- or exocytosis was captured here after 24 h of ML incubation (s. Fig. 6.4(a,b)). However, such an invagination

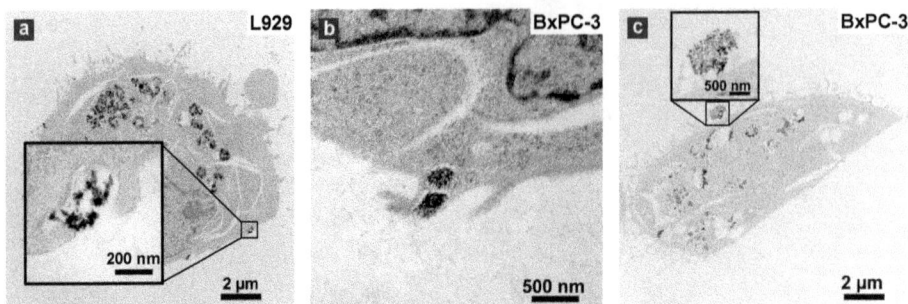

Fig. 6.4.: Membrane transport processes after 24 h of incubation: Cellular membrane invaginations engulf an agglomerate of ML for (a) L929 (inset shows a zoom); magnification: 6 000x; and (b) BxPC-3; magnification: 27 800x. (c) shows a large agglomerate of ML bound to the cell membrane of a BxPC-3 cell (inset shows a zoom), supposedly as a result of exocytosis (s. text); magnification: 6 000x. (b) and inset in (c) adapted from [291].

process was also observed for MiaPaCa-2 after a chase time of $t_{chase} = 10$ min (s. Fig. 6.1b) with an agglomeration size of ML of approx. 150 nm. Given the short chase time, this is doubtlessly an endocytotic process. The ML agglomerates are approximately $(50-250)$ nm in diameter when being engulfed for endocytosis by the cellular membrane independent of the cell type. These sizes of endocytosed ML agglomerates are in line with the assumption of a minimum diameter necessary for effective endocytic internalization and further in agreement with observations from other groups: Chithrani et al. demonstrated the higher uptake rates for transferrin-coated Gold nanoparticles (Au-NP) with particle sizes $d \approx 50$ nm in HeLa ovarian cancer cells, compared to Au-NP with $d \approx 14$ nm [301]. Similarly, carboxyl-modified nanoparticles were internalized best for a particle size $d \approx 40$ nm in a variety of cells (including HeLA tumor cells and RAW 264.7 macrophages), with substantially decreasing amount of particles internalized for larger sizes of $d \approx (100, 200, 500)$ nm [302] as reported by Dos Santos et al.. Using ceria (CeO_2) nanoparticles (CeO_2-NP), Limbach et al. demonstrated almost no uptake of CeO_2-NP with $d = (20-50)$ nm inside human lung fibroblasts ATCC and MRC-9, however, a remarkably increased uptake was observed for the same CeO_2-NP clustered in agglomerates of $d \approx (100-500)$ nm [303]. The observation of a minimum MNP (agglomerate) size is supported by theorecical modeling of the uptake kinetics, derived from the system free energy, reported by Chaudhuri et al., who demonstrated an improved uptake for interacting (potentially agglomerated) MNP, but also predicting a minimum particle size of $d \approx 40$ nm below which no particle uptake could take place [304].

In Fig. 6.4c a large ML agglomerate bound to the cellular membrane of a BxPC-3 cell is depicted, which could either be an ML agglomerate forming in the RPMI medium and about to be internalized or on the other hand be a ML agglomerate after exocytosis. The latter hypothesis is supported by the fact that the size of the ML agglomerate, $d_C \approx 800$ nm, is much larger than any other membrane-engulfed ML agglomerate about to be internalized observed in

Figs. 6.1b and 6.4. Furthermore, the ML agglomerate is shaped like a typical lysosome[4]. Even though an unambiguously conclusive statement as to whether endo- or exocytosis is observed cannot be given here, exocytosis seems more likely. Exocytosis will also be considered as one possible membrane transport process for the modeling of the internalization uptake kinetics of ML inside cells in Section 6.3.4.

Intracellular Arrangement of ML

The arrangement of ML inside lysosomes after 24 h of incubation is compared between the three cell lines in Fig. 6.5. The ML packing density inside lysosomes differs for each cell line,

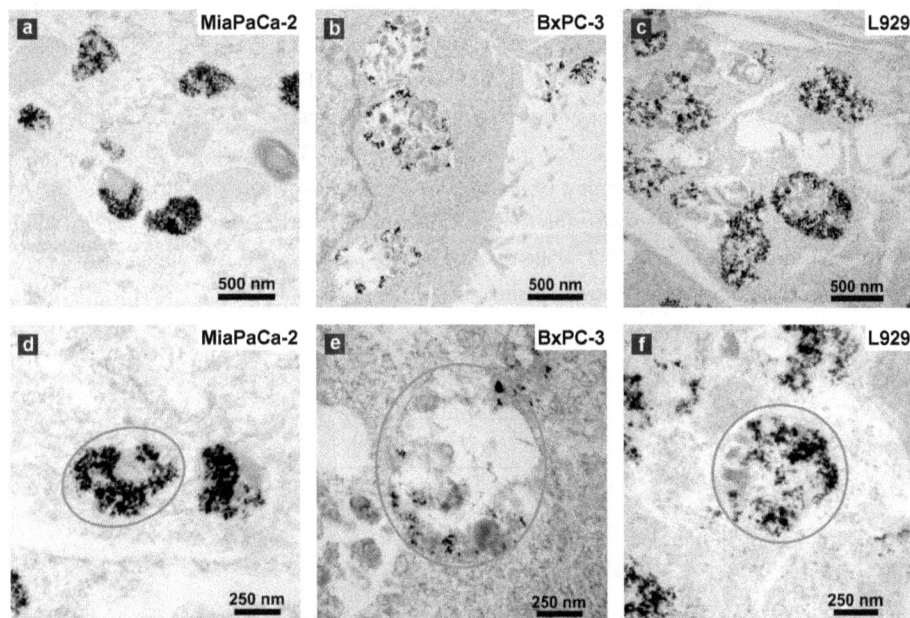

Fig. 6.5.: ML-loaded lysosomes inside (a,d) MiaPaCa-2, (b,e) BxPC-3 and (c,f) L929 cells after 24 h of incubation. Single lysosomes are circled in red in (d,e,f). Magnifications: (a,b,c) 35 870 x; (d,e,f) 60 000 x. (a), (b) and (c) taken from [291].

where MiaPaCa-2 shows the tightest packing density (s. Fig. 6.5(a,d)), L929 have a still tight packing density (s. Fig. 6.5(c,f)) and BxPC-3 have a relatively low packing density (s. Fig. 6.5(b,e)). Although this observation is only qualitatively and one must keep in mind that this information is derived from 2-dimensional slices, while in reality the intracellular lysososmes are 3-dimensional structures, it is an indicator that the uptake and intracellular processing of ML is different for each individual cell line. This observation is in line with literature, reporting on individual uptake morphology and kintetics for different cell types [11, 302]. The size of lysosomes is substantially larger after 24 h of incubation compared to the size of endosomes observed for shorter incubation times of $(1 - 2)$ h during chase experiments (cf. Figs. 6.1(e,f)

---

[4]As will be shown below, the average lysosome size for BxPC-3 is $d_{lys}^{BxPC-3} = 646$ nm after 24 h of incubation, which is another indicator that this particular ML agglomerate has been exocytosed.

and 6.2(e,f)). This confirms that ML are intracellularly processed by the cell, merging smaller (early) endosomes into larger (late) lysosomes over time in order to group foreign cargo (ML) together, which is in accordance with literature [98, 105]. It is furthermore assumed that the mobility of ML in the densely packed lysosomes is severely restricted. Therefore, the ML are considered as immobilized upon internalization in accordance with literature [245, 305, 306].

Size Distribution Analysis of ML-loaded Lysosomes

The average size of ML-loaded lysosomes after 24 incubation, d_{lys}, was determined from a multitude of TEM images for each cell line. Lysosomes of all three incubation concentrations, $c = (0.150, 0.225, 0.300)\,mg(Fe)/mL$, were analyzed together as no difference was observed in the average lysosome sizes with concentration (cf. Appendix A.2.2 for exemplary TEM images for incubation concentrations $c = (0.225, 0.300)\,mg(Fe)/mL$). For spherical lysosomes d_{lys} was determined by measuring the concentric diameter, while asymmetrically shaped lysosomes were measured along their maximum concentric length a and along the maximum length b perpendicular to a and averaged according to $d_{lys} = (a + b)/2$, as demonstrated in Fig. 6.6d. The distribution of d_{lys} was fitted with the CDF (cf. eq. (4.5) for each cell line as shown in Fig. 6.6. Results are listed in Table 6.2, while the CDF fitting parameters can be found in Appendix A.4, Table 6.2. MiaPaCa-2 show the smallest average lysosome size, $d_{lys}^{MiaPaCa\text{-}2} = 410\,nm$,

Table 6.2.: Average lysosome sizes, d_{lys}, calculated from the cumulative size distribution for each cell line from N analyzed lysosomes.

Cell line	N	d_{lys} [nm]
MiaPaCa-2	151	410 ± 166
BxPC-3	69	646 ± 273
L929	168	542 ± 223

followed by L929 with $d_{lys}^{L929} = 542\,nm$ and finally BxPC-3 with $d_{lys}^{BxPC\text{-}3} = 646\,nm$. An average estimate for the final size of late lysosomes is given with $d_{lys} \approx 500\,nm$ [106]. As a certain degree of variation is expected for d_{lys} for each specific cell line, the averages lysosome sizes calculated above can be considered final, i.e. d_{lys} is not expected to change significantly for longer incubation times. The observation that MiaPaCa-2 displayed the highest ML packing density fits well to the finding that MiaPaCa-2 also has the smallest average lysosome size. This could indicate that MiaPaCa-2 cells try to store their ML cargo most energetically efficient, as merging smaller lysosomes into one larger (late) lysosome happens at the dispense of energy, which is proportional to the final size of the late lysosome [307]. It could be hypothesized further that MiaPaCa-2 cells use the energy saved to internalize more ML. This will be further investigated in Section 6.3, where the uptake of ML inside cells is quantified. BxPC-3 show the highest average lysosome size, while at the same time displaying the lowest packing density. Taken together with the observation that BxPC-3 show no membrane-bound ML after $24\,h$ of incubation, this might indicate that BxPC-3 cells generally internalize less ML compared to MiaPaCa-2 and L929. Furthermore, BxPC-3 had the lowest number of lysosomes available for

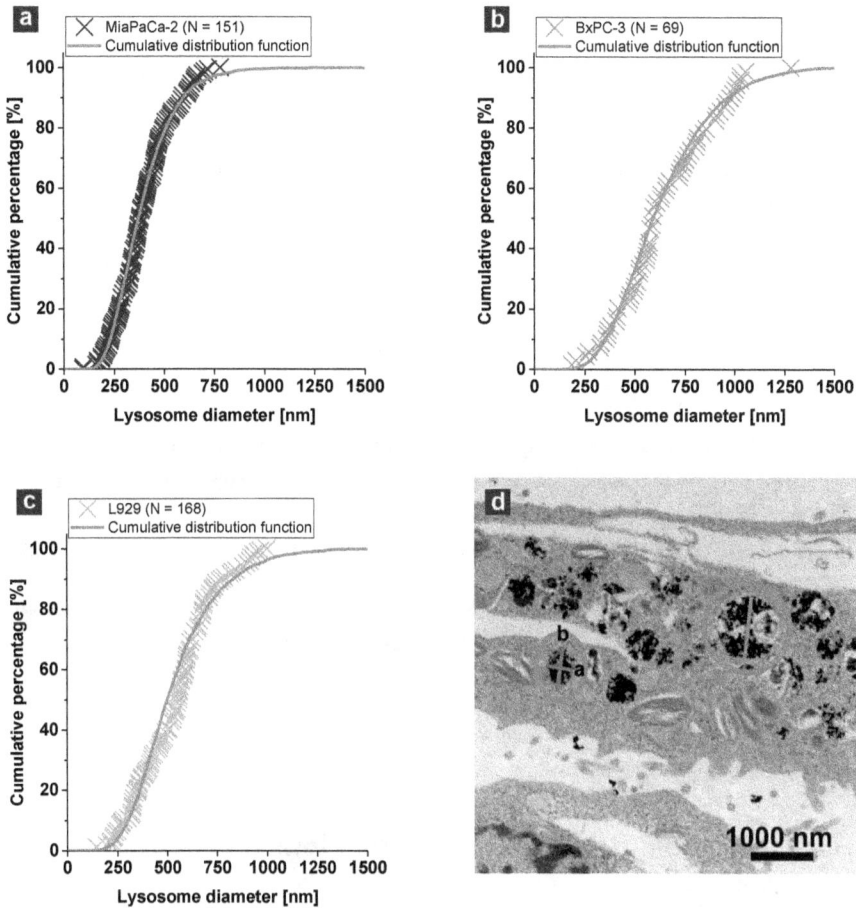

Fig. 6.6.: Cumulative lysosome diameter (size) distribution counting N individual lysosomes, determined from TEM images after incubation for 24 h with ML1 particles. The distributions were fitted with the CDF (cf. eq. (4.5)) for (a) MiaPaCa-2 ($R^2_{adj} = 0.963$, $N = 151$), (b) BxPC-3 ($R^2_{adj} = 0.824$, $N = 69$) and (c) L929 cells ($R^2_{adj} = 0.881$, $N = 168$). (d) demonstrates the size measurement for spherical lysosomes (marked in red; $d_{lys} = c = 945\,\text{nm}$) and for asymmetrical lysosomes (marked in blue; $d_{lys} = (a + b)/2 = (644\,\text{nm} + 418\,\text{nm})/2 = 531\,\text{nm}$; s. text for explanation). (d) shows L929 with incubation concentration $c = 0.300\,\text{mg(Fe)/mL}$. Fitting parameters for (a,b,c) are found in Appendix A.4.

assessment ($N = 69$), possibly indicating the fusion of several endosomes / early lysosomes into large late lysosomes. This could also point towards an increased lysosomal digestive activity inside BxCP-3 cells [307, 308]. L929 cells display a mixture of the observations described above: an intermediate late endosome / lysosome size of $d_{lys}^{L929} = 542\,\text{nm}$ and high ML packing density.

6.2.4. Summary of Morphological Changes of ML upon Internalization in Cells

To summarize, the following morphological characteristics of ML internalization are derived from TEM image analysis:

- The internalization process requires a minimum size of ML agglomerated at the cellular membrane of $d_C \geq 50\,\text{nm}$.

- The internalization process across the cellular membrane is completed within several tens of minutes, but surely after $1\,\text{h}$.

- The endocytosis of ML happens by membrane invaginations engulfing a cluster of ML. In turn, exocytosis was hypothesized to be observed as well after $24\,\text{h}$ of incubation.

- ML-loaded endosomes merge with other endosomes, evolving finally into (late) lysosomes with final sizes in the order of $d_{\text{lyo}} \sim 500\,\text{nm}$ after $24\,\text{h}$ of incubation. The particular average lysosome sizes are cell-specific, with the smallest size for MiaPaCa-2, $d_{\text{lys}}^{\text{MiaPaCa-2}} = 410\,\text{nm}$, followed by L929 with $d_{\text{lys}}^{\text{L929}} = 542\,\text{nm}$ and finally BxPC-3 with $d_{\text{lys}}^{\text{BxPC-3}} = 646\,\text{nm}$.

- The ML packing density inside lysosomes is inversely proportional to the average lysosome size, showing highest packing density for MiaPaCa-2, still high packing density for L929 and low packing density for BxPC-3.

- The agglomeration of ML is maintained inside the cells at all times. It is assumed that ML are also immobilized inside lysosomes.

6.3. Quantification of Magnetic Particle Uptake Kinetics

In the following, the ML uptake kinetics inside cells are quantified as a function of the incubation time with MPS and compared to a particle internalization model. Therefore, MiaPaCa-2, BxPC-3 and L929 cell lines were incubated with ML for incubation times of up to $24\,\text{h}$ at two different temperatures ($T = 4\,°\text{C}$ and $T = 37\,°\text{C}$) that allows to distinguish between the amount of ML attached to the surface membrane of passivated cells ($T = 4\,°\text{C}$) and internalized ML inside active cells ($T = 37\,°\text{C}$). MPS is employed here due to its high sensitivity, capable of quantifying even small amounts of ML inside cells based on the ML non-linear magnetization response to the applied sinusoidal field (as discussed in detail in Section 2.3). The cell sample preparation is described in Section 6.3.1. The preliminary calibration of MPS for iron quantification measurements is discussed in Section 6.3.2. Section 6.3.2 also presents preliminary cell growth experiments that are used to calculate the amount of ML per cell. In Section 6.3.3 the ML uptake kinetics measured by MPS are discussed and considered for an approach to model the ML uptake kinetics. Such a model based on simple mathematical-physical assumptions on the endo- and exocytosis process is developed in Section 6.3.4 and compared to the experimental uptake data in Section 6.3.5. Here, the quantitative ML uptake kinetics are also interpreted and discussed with the results from the morphological study of

ML internalization from the previous Section 6.2. Finally, Section 6.3.6 discusses the magnetic relaxation behavior of ML after being fully internalized.

6.3.1. Experimental Procedure

Cell Sample Preparation for MPS Measurements

Each cell line was cultivated as described in Section 6.1: 20 000 cells were seeded per well in 24-well plates (CELLSTAR) and stored in the incubator for 24 h. Subsequently, medium was replaced by 1 mL cell-specific medium mixed with ML1 particles with an iron concentration of $c = 0.15\,\text{mg(Fe)/mL}$ and the cells were stored in the incubator $(37\,^\circ\text{C})$ for various incubation times: $t_{inc} = (10, 20, 30, 60)\,\text{min}$ and $t_{inc} = (2, 4, 6, 12, 16, 24)\,\text{h}$. For each incubation time, samples were prepared in triplicate (i. e. three identical wells were prepared). During incubation time, the cells were active and able to internalize ML. A second batch of samples was prepared in the same way but stored at $4\,^\circ\text{C}$ 1 h prior to incubation with ML and subsequently during incubation for incubation times of $t_{inc} = (10, 20, 30, 45, 60)\,\text{min}$ and $t_{inc} = (2, 4, 6)\,\text{h}$. These cells were passivated, only allowing ML adsorption at the cell membrane. Reference cell samples not treated with ML were prepared in triplicate as well with every incubation time at both temperatures.

After incubation with ML, cells were harvested as described in Section 6.1 (quantities: 0.5 mL DPBS, 0.2 mL Trypsin, 0.5 mL cell-specific medium). Subsequently, cells were transferred to 1.5 mL PCR-tubes (Eppendorf) and centrifuged for 5 min (1, 300 rpm, 25 $^\circ$C). Supernatant was removed and the cell pellet resuspended in 0.1 mL formalin (10 % buffered solution; Sigma Aldrich) for cell fixation. The mixture was transferred to 0.2 mL PCR tubes, centrifuged for 3 min (6 000 rpm, ambient conditions) and the supernatant was removed. The final cell pellet was homogeneously mixed with 0.05 mL agarose (weight fraction $\mu_{agar} = 0.3\,\%$). All samples were measured in the PTB MPS setup (cf. Section 5.1.1). Final results were averaged over the results from triplicate measurement.

Cell Sample Preparation for Cell Growth Experiments

To obtain a meaningful result, the amount of iron per sample quantified with MPS must be normalized to the number of cells per sample. For this reason, reference samples for cell counting experiments were prepared as described in the subsection above (20 000 cells seeded, triplicates, $c = 0.15\,\text{mg(Fe)/mL}$ ML1 treatment). Cell samples were stored either at 37 $^\circ$C in the incubator $(t_{inc} = (10, 30, 60)\,\text{min}$ and $t_{inc} = (2, 3, 4, 6, 12, 24)\,\text{h})$ or at 4 $^\circ$C in the fridge $(t_{inc} = (10, 20, 30, 45, 60)\,\text{min}$ and $t_{inc} = (2, 4, 6)\,\text{h})$. Afterwards, the cells were harvested as described above and live cell counting was performed with the LUNA cell counter. Final results were averaged over the results from triplicate measurement.

6.3.2. Preliminary Considerations

Sample Iron Quantification with MPS

The quantification of the amount of iron taken up by cells were carried out with the PTB setup (cf. Section 5.1.1). In order to determine a calibration curve needed for the calculation of the absolute amount of iron in cell samples, ML1 samples of known iron content were prepared: ML were homogeneously dispersed in agarose (weight fraction $\mu_{agar} = 0.3\%$) with a total sample volume of $0.05\,\mathrm{mL}$ and various absolute iron contents: $(0, 0.25, 5, 25, 50, 250, 500, 2\,500, 5\,000, 7\,500, 15\,000, 30\,000$ and $50\,000)\,\mathrm{ng(Fe)}$. The $0\,\mathrm{ng(Fe)}$ sample was used for background subtraction. The MPS spectral magnitude of the 3rd harmonic[5], A_3, versus the iron content per sample is plotted in Fig. 6.7, demonstrating a linear across the range of iron content measured. The theoretical detection limit for ML in

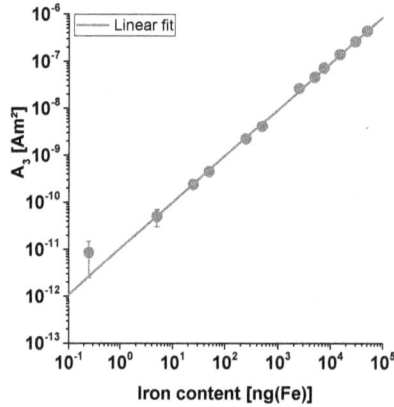

Fig. 6.7.: MPS spectral magnitude of the 3rd harmonic (A_3) versus iron content per sample, m_{iron}. The data was fitted with a linear function $A_3(m_{\mathrm{iron}}) = \alpha \cdot m_{\mathrm{iron}} + A_0$, as shown in red. The fitting parameters are $\alpha = (8.730 \pm 0.005) \cdot 10^{-12}\,\mathrm{ng(Fe)/(Am^2)}$, $A_0 = (7.86 \pm 1.27) \cdot 10^{-13}\,\mathrm{Am^2}$, $R^2_{\mathrm{adj}} = 0.999$. Adapted from [291].

this specific MPS setup can be assumed following the guidelines of the International Union of Pure and Applied Chemistry (IUPAC) [309]: $A_{\mathrm{limit}} = A_{\mathrm{back}} + 3 \cdot \sigma_{A_{\mathrm{back}}}$, with the mean and standard deviation of the background MPS amplitude signal, A_{back} and $\sigma_{A_{\mathrm{back}}}$, respectively. With $A_{\mathrm{back}} \approx 8 \cdot 10^{-12}\,\mathrm{Am^2}$ and $\sigma_{A_{\mathrm{back}}} \approx 3 \cdot 10^{-12}\,\mathrm{Am^2}$ the detection limit yields $A_{\mathrm{limit}} \approx 1.7 \cdot 10^{-11}\,\mathrm{Am^2}$. According to the calibration curve in Fig. 6.7, this corresponds to a sensitivity of the MPS setup for ML particles down to approximately $1\,\mathrm{ng(Fe)}$.

In order to produce a reference sample for iron quantification that mimics intracellular agglomerates, a highly agglomerated sample of ML with concentration $c = 0.133\,\mathrm{mg(Fe)/mL}$ was produced by adding $4\,\mathrm{M}$ of sodium chloride (NaCl) to increase the ionic strength of the carrier

---

[5] Using here the spectral magnitude of the harmonics of the MPS spectrum is directly proportional to the MNP (ML) iron content, as discussed in Section 2.3; cf. eq. (2.57).

liquid and break the liposome shell. Consequently, the MNP cores agglomerated as seen from intensity-weighted DLS measurements[6] (Fig. 6.8a): the hydrodynamic size, d_H, increases from approx. 94 nm to approx. 900 nm. The normalized MPS spectra[7] show a substantial reduction

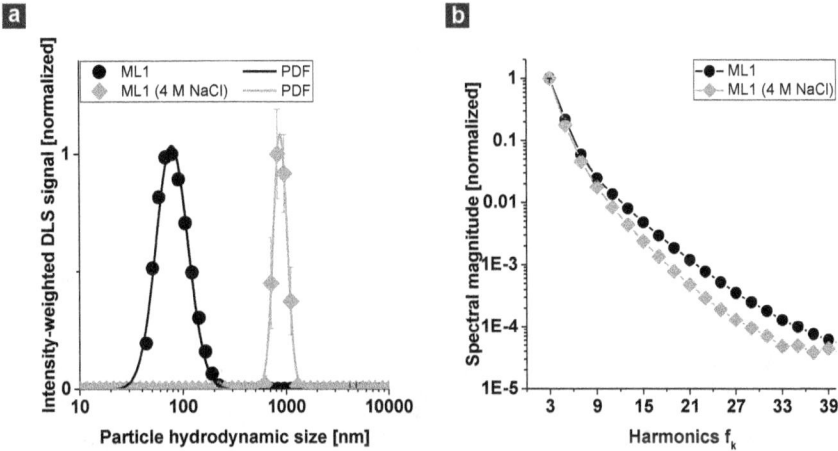

Fig. 6.8.: (a) Intensity-weighted DLS signal for ML1 and ML1 treated with 4 M of NaCl fitted with the log-normal probability density function (PDF) (cf. eq.(3.22)), yielding hydrodynamic sizes of $d_H = (94 \pm 35)$ nm and $d_H(4\,M\,NaCl) = (909 \pm 140)$ nm, with $R^2_{adj} = 0.990$; and $R^2_{adj} = 0.997$, respectively. (b) MPS spectral magnitude normalized to the third harmonic for ML1 and ML1 treated with 4 M of NaCl.

in the spectral magnitude of higher harmonics for agglomerated ML (Fig. 6.8b), presumably due to the blocking of Brownian rotation due to particle immobilization and demagnetization effects arising from increased interparticle dipole-dipole interaction upon particle agglomeration (this assumption for immobilized and agglomerated MNP will be discussed (and verified) in detail in Chapter 7). ML treated with 4 M NaCl serve as a second reference next to untreated ML1 (cf. Fig. 6.7) for analyzing MPS spectra of intracellular ML in the following. The conversion factor for this sample corresponds to $6\,650$ ng(Fe)$\hat{=}(2.671 \pm 0.086) \cdot 10^{-8}$ Am^2; derived from the sample iron content and the A_3 of the MPS spectrum.

Cell Growth Experiments

The results for the cell count as a function of incubation time with $t_{inc} = (1/6 - 6)$ h at $4\,°C$ and with $t_{inc} = (1/6 - 24)$ h at $37\,°C$ are shown in Fig. 6.9. The time-dependent cell count was fitted with a linear function, assuming that the presence of ML unspecifically inhibits the cell proliferation by influencing the metabolic activity of the cells [310, 311]. The (live) cell count of cells stored at $4\,°C$ shows high fluctuation but no clearly identifiable trend (s. Fig. 6.9a) and was thus approximated as constant over time (linear fit with zero slope in Fig. 6.9a). Compared to the initially seeded 20 000 cells for each cell line, L929 shows an unchanged average cell count

---

[6]DLS measurements were performed as described in Section 4.4. Intensity-weighted DLS was chosen for analysis as this detects large particle agglomerates best.

[7]MPS measurements were performed as described in Section 5.1 with ML1 samples dispersed in agarose (mass fraction $\mu_{agar} = 0.3\,\%$). Results are averaged over three independent cell samples measured.

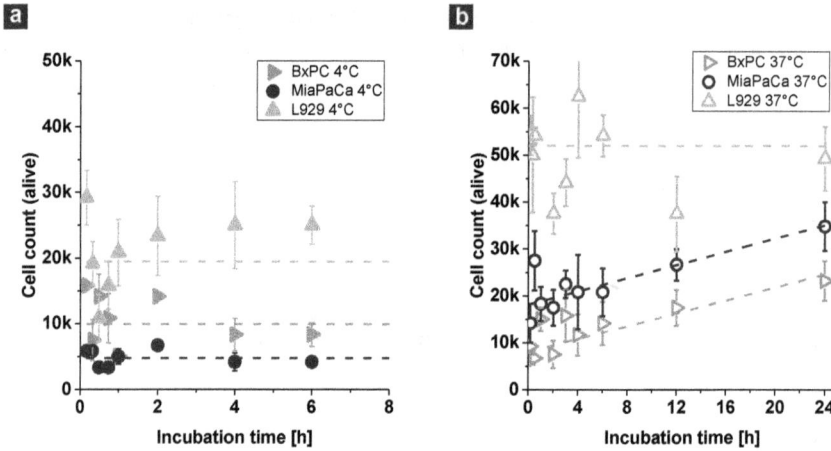

Fig. 6.9.: (Live) cell count versus incubation time, t_{inc}, for cell lines BxPC-3, MiaPaCa-2 and L929 incubated at (a) $4\,^{\circ}\text{C}$ and (b) $37\,^{\circ}\text{C}$. The data was fitted with a linear function, $G(t_{inc}) = \alpha \cdot t_{inc} + G_0$. Since no substantial increase over incubation time was measured in (a) and for L929 in (b), $\alpha = 0$ was assumed here. Fit results are (a) $G_{0,\text{MiaPaCa-2}} = (4\,753 \pm 422)$ with $R^2_{adj} = 0$ for MiaPaCa-2, $G_{0,\text{BxPC-3}} = (9\,920 \pm 2\,209)$ with $R^2_{adj} = 0$ for BxPC-3 and $G_{0,\text{L929}} = (19\,468 \pm 2\,060)$ with $R^2_{adj} = 0$ for L929 and (b) $\alpha_{\text{MiaPaCa-2}} = (11.7 \pm 2.7)\,1/\text{min}$ and $G_{0,\text{MiaPaCa-2}} = (18\,181 \pm 1\,360)$ with $R^2_{adj} = 0.692$ for MiaPaCa-2, $\alpha_{\text{BxPC-3}} = (11.5 \pm 3.7)\,1/\text{min}$ and $G_{0,\text{BxPC-3}} = (8\,089 \pm 809)$ with $R^2_{adj} = 0.524$ for BxPC-3 and $G_{0,\text{L929}} = (52\,011 \pm 2\,376)$ with $R^2_{adj} = 0$ for L929.

of approx. $19\,500$ cells. In contrast, the cell count decreases to approx. $10\,000$ cells for BxPC-3 and even down to approx. $4\,700$ for MiaPaCa-2.

At $37\,^{\circ}\text{C}$, BxPC-3 and MiaPaCa-2 cells display a linear increase over time (s. Fig. 6.9b), indicating some proliferation activity. However, BxPC-3 total cell count is remarkably lower than MiaPaCa-2 total cell count. L929 cells on the other hand do not show a clearly identifiable trend in cell count over incubation time and the cell count was thus approximated as constant again as for the cell count at $4\,^{\circ}\text{C}$ before. However, the overall average cell number was approx. $52\,000$ and thus approx. 2.5-times higher than the initially seeded $20\,000$ cells.

6.3.3. ML Uptake Kinetics

The MPS harmonic spectrum of each individual sample was assigned to one of the two reference spectra for untreated ML and agglomerated ML (cf. Fig. 6.8b) by comparison and used to calculate the individual ML iron content from the per sample A_3: For cell samples associated with untreated ML the calibration curve (s. Fig. 6.7) was used and for cell samples associated with agglomerated ML the conversion factor given for the agglomerated ML (s. Section 6.3.2) was used, respectively. In this way the agglomeration state of the intracellular ML is taken into account for the calculation of the ML iron content. After background subtraction, the ML iron content per sample was then averaged over triplicate measurements for each cell line at each respective incubation time and incubation temperature. Finally the result was divided by the cell number from cell growth experiments (cf. Fig. 6.9), yielding the ML iron content per cell. The resulting for ML uptake kinetics of all three cell lines are shown in Fig. 6.10.

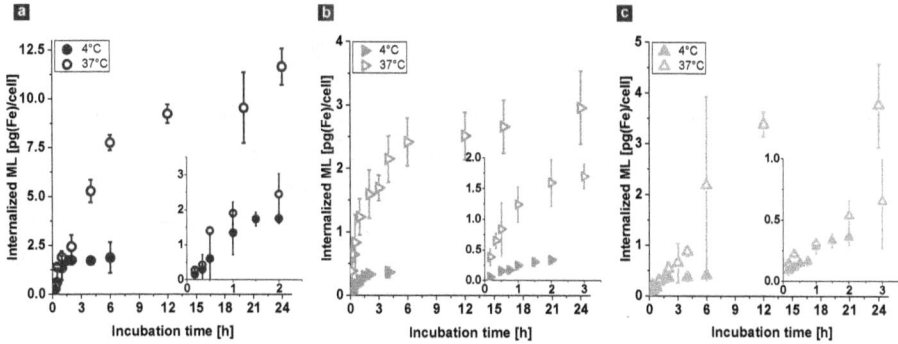

Fig. 6.10.: ML uptake kinetics at $4\,°C$ (passivated cells) and $37\,°C$ (cells actively internalizing ML) for (a) MiaPaCa-2, (b) BxPC-3 and (c) L929 cells. Insets show a zoom for the initial ML uptake at short incubation times. Adapted from [291].

Generally, the uptake kinetics differ substantially for passivated cells at $4\,°C$ and cells actively internalizing ML at $37\,°C$ for all three cell lines: the ML uptake for active cells was approx. 5-times higher for MiaPaCa-2 after $24\,h$ of incubation, and approx. 8-times higher for BxPC-3 and approx. 10-times higher for L929 cells. The ML uptake for passivated cells of all three cell lines saturates for $t_{inc} \approx (1.5 - 2)\,h$. During this time, passivated MiaPaCa-2 cells take up much more ML at their membrane compared to BxPC-3 and L929. In contrast to that the amount of uptaken ML inside active cells increases further, reaching saturation fastest for BxPC-3 after $t_{inc} \approx 6\,h$, followed MiaPaCa-2 saturating after $t_{inc} \approx (6 - 12)\,h$ and L929 saturating after $t_{inc} \approx 12\,h$. Here, MiaPaCa-2 reaches the highest saturation ML uptake with approx. $9\,pg(Fe)/cell$, followed by L929 with approx. $3.5\,pg(Fe)/cell$ and BxPC-3 with approx. $2.5\,pg(Fe)/cell$.

Please note that the ML uptake of approx. $11\,pg(Fe)/cell$ measured for MiaPaCa-2 after $t_{inc} = 24\,h$ suggests a potential further increase in ML uptake instead of a saturation. However, an additional ML uptake experiment performed under the same conditions[8] measured an ML uptake of approx. $8.5\,pg(Fe)/cell$ (cf. Fig. 8.1), which rather suggests a saturation of ML uptake for MiaPaCa-2 after $t_{inc} = 24\,h$ in Fig. 6.10a. For this reason, the value of approx. $11\,pg(Fe)/cell$ is excluded from the following modeling of the ML uptake kinetics in Section 6.3.5.

Please note further that the ML uptake measured for L929 after $t_{inc} = 6\,h$ shows a very high uncertainty (possibly due to contamination of the cells during growth). It is therefore also excluded from modeling of the ML uptake kinetics in Section 6.3.5.

---

[8] I. e. using the identical procedure, except seeding $1 \cdot 10^6$ cells in 6-well plates and using a different batch of ML (ML2, cf. Section 4.8).

It is important to note that the ML uptake is always higher for active cells compared to passivated cells for all cell lines. This indicates a 2-step process of ML internalization: first requiring ML to bind to the cellular membrane before, second, ML internalize inside cells. After intracellular internalization of ML, new extracellular ML can attach, in this way increasing the total amount of ML measured over (incubation) time, but only for cells actively internalizing ML. This finding is supported by TEM analysis in Section 6.2.4, observing that ML are first clustered at the cellular membrane before being internalized across the membrane via endocytosis from. Such a 2-step process, was also reported in literature for a multitude of healthy as well as tumor cells [312, 313]. It was suggested by Wilhelm et al. to model this 2-step process with a simple mathematical-physical model differentiating membrane bound particles and internalized particles [300]. This model is derived and adapted to the specific ML uptake kinetics found in this thesis in the following Section 6.3.4 and fit to the experimental data in Section 6.3.5.

6.3.4. Modeling of the ML Uptake Kinetics

As suggested in Section 6.3.3, the uptake (internalization) of nanoparticles can be described as a 2-step-process, where nanoparticles adsorb at the cellular membrane in groups first before being internalized. The model presented in the following is based on the 2-step-model of Wilhelm et al. (2002) [300, 313]. Here, the model will be expanded by a third step after internalization including exocytosis, for which the so-called compartment model of Rappaport et al. (2011) [314] was integrated. This step is reasonably necessary, since evidence for exocytosis of nanoparticles is mounting in the past decade [99, 315, 316] and was furthermore presumably observed from TEM imaging in the previous section as well (cf. Fig 6.4c).

For the model, several assumptions are made:

1. The adsorption capacity for nanoparticles on the cell membrane is limited per cell, expressed as the maximum adsorption mass m_0. This value is cell type-specific.

2. Only a finite fraction of the membrane surface, ϕ_0, is actively involved in the process of nanoparticle transport.

3. Endo- and exocytosis use the same active membrane surface fractions. In other words, the fraction of the membrane surface is exclusively active for either endocytosis or exocytosis simultaneously.

4. Exocytosed nanoparticles remain at the cellular membrane, from where they either detach or are endocytosed again.

5. Nanoparticles bind to the cell plasma membrane via electrostatic and van-der-Waals interactions [103, 317], allowing to describe the adsorption as a simple Langmuir process[9].

---

[9]A Langmuir process is defined by assuming that the adsorbate (here the nanoparticles) behaves like an ideal gas (randomly moving point-approximated particles that only interact with perfectly elastic collisions), while the adsorbent (here the cell) is composed of a distinct and limited sites capable of binding the adsorbate. Furthermore, only monolayer adsorption is considered.

For a Langmuir process (cf. assumption 5, above) a certain mass of nanoparticles adsorbed at the cellular membrane, $m_{mem}(t)$, can be expressed with the following differential equation:

$$\frac{dm_{mem}(t)}{dt} = c \cdot k_a \cdot (m_0 - m_{mem}(t)), \tag{6.1}$$

with the saturation (maximum) adsorption mass (cf. assumption 1, above), m_0, the global incubation nanoparticle concentration (considered as a reservoir), c, and the rate constant for adsorption at the cellular membrane, k_a[10] Assuming, that the reservoir, c, is inexhaustible compared to the maximum adsorption at the membrane and from $m_{mem}(t = 0) = 0$, eq. (6.1) can be solved by

$$m_{mem}(t) = m_0 \cdot (1 - \exp(-c \cdot k_a \cdot t)). \tag{6.2}$$

With the characteristic absorbance time $\tau = 1/(c \cdot k_a)$. Eq. (6.2) fully describes the time-dependent adhesion step of ML at the cellular membrane.

For modeling the step of internalization, the mass of nanoparticles absorbed at the cellular membrane, $m_{mem}(t)$, is distinguished from the mass of nanoparticles internalized, $m_{int}(t) = m_{endo}(t) - m_{exo}(t)$, which consist of the competing endocytosed and exocytosed masses, $m_{endo}(t)$ and $m_{exo}(t)$, respectively. For simplification, $m_{endo}(t)$ comprises all possible endocytotic pathways. The differential equation eq. (6.1) is modified to include the mass fraction of nanoparticles internalized (by endo- and exocytosis) over time:

$$\frac{dm_{mem}(t)}{dt} = c \cdot k_a \cdot (m_0 - m_{mem}(t)) - \frac{dm_{int}(t)}{dt}$$
$$= c \cdot k_a \cdot (m_0 - m_{mem}(t)) - \frac{dm_{endo}(t)}{dt} + \frac{dm_{exo}(t)}{dt} \tag{6.3}$$

It is assumed here that the maximum membrane-bound mass of nanoparticles, m_0, does not change during the process, meaning that the adsorption sites are constantly renewed after endo- or exocytosis. During internalization, the cellular membrane is actively bending and wrapping the particles [318, 319]. This is using a fraction of the active surface of the membrane, which is limited to the maximum ϕ_0 (cf. assumption 2, above). The active surface fraction involved in endocytosis, $\phi_{endo}(t)$, can be described by the differential equation:

$$\frac{d\phi_{endo}(t)}{dt} = k_{endo} \cdot (\phi_0 - \phi_{endo}(t)), \tag{6.4}$$

with the rate constant of endocytosis, k_{endo}. If exocytosis occurs across the same active membrane surface fraction as endocytosis (cf. assumption 3, above) and the total mass of nanoparticles exocytosed at time t is denoted as $m_{exo}(t)$ with the rate constant of exocytosis, k_{exo}, the

---

[10]The original model of Wilhelm et al. [300] also includes a rate constant of desorption of nanoparticles. The present model assumes an *effective* rate constant of adsorption, that also includes desorption fractions.

differential equation for the active surface fraction involved in exocytosis reads:

$$\frac{d\phi_{exo}(t)}{dt} = k_{exo} \cdot (\phi_0 - (\phi_{endo}(t) + \phi_{exo}(t))). \tag{6.5}$$

Knowing this, eq. (6.4) must include ϕ_{exo} as well (s. below, eq. (6.6). However, exocytosis occurs only after some delay time, t_{ex}, as nanoparticles first need to internalize. Therefore this delay is implemented by complementing eqs. (6.4) and (6.5) with the hyperbolic tangent function $\Theta(t) = 0.5 + 0.5 \cdot \tanh((t - t_{ex})/b')$[11]

$$\frac{d\phi_{endo}(t)}{dt} = k_{endo} \cdot (\phi_0 - (\phi_{endo}(t) + \phi_{exo}(t) \cdot \Theta(t))), \tag{6.6}$$

$$\frac{d\phi_{exo}(t)}{dt} = k_{exo} \cdot (\phi_0 - (\phi_{endo}(t) + \phi_{exo}(t))) \cdot \Theta(t), \tag{6.7}$$

where b' describes the time span across which exocytosis sets on. The total mass of nanoparticles endocytosed, m_{endo}, as well as that exocytosed, m_{exo}, during the time interval dt then depends on the product of the change in active surface fraction available for the respective process, $d\phi_{endo}(t)/dt$ or $d\phi_{exo}(t)/dt$ multiplied by the mass of nanoparticles available for the respective process:

$$\frac{dm_{endo}(t)}{dt} = \frac{d\phi_{endo}(t)}{dt} \cdot m_{mem}(t), \tag{6.8}$$

$$\frac{dm_{exo}(t)}{dt} = \frac{d\phi_{exo}(t)}{dt} \cdot m_{endo}(t). \tag{6.9}$$

Note that when inserting eqs. (6.8) and (6.9) in eq. (6.3), it is consequently assumed that nanoparticles exocytosed remain attached to the membrane after exocytosis and add their mass to the total mass at the membrane, as well as occupying a fraction of the active surface membrane (cf. assumption 4, above).

The total mass of nanoparticles bound to and internalized into the cell is determined by

$$m_{tot}(t) = m_{mem}(t) + m_{int}(t). \tag{6.10}$$

m_{tot} is accessible by MPS measurements (s. Section 6.3.3). Therefore, eq. (6.2) can be solved by fitting to experimental data of passivated cells, only able to adsorb nanoparticles to their membrane. Then, eqs. (6.3) and (6.6-6.10) can be fitted to experimental data of actively internalizing cells.

---

[11]Defined in this way, $\Theta(t)$ attains values between $[0, 1]$ and centered at $t = t_{ex}$.

6.3.5. Comparison of Experimental and Modeled ML Uptake Kinetics

The experimental ML uptake kinetics from Section 6.3.3, Fig. 6.10, were fitted with the 3-step model presented in Section 6.3.4. Results are shown for the three cell lines in Fig. 6.11 . The model fitting parameters are summarized in Table 6.3.

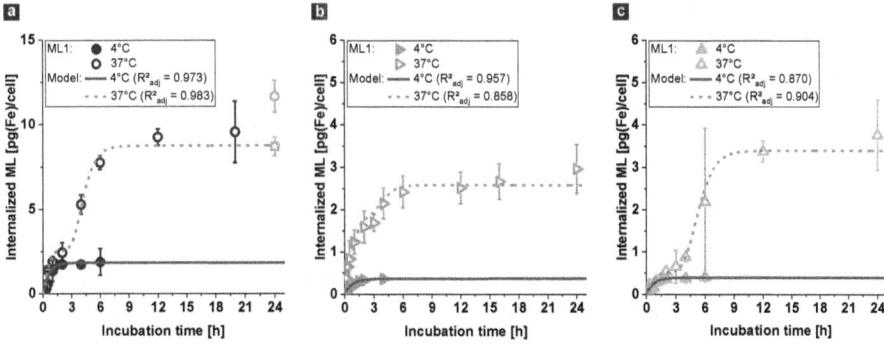

Fig. 6.11.: Experimental and modeled ML uptake kinetics at $4\,°C$ (passivated cells) and $37\,°C$ (cells actively internalizing ML) for (a) MiaPaCa-2, (b) BxPC-3 and (c) L929 cells. The values for MiaPaCa-2 after $t_{inc} = 24\,h$ and for L929 after $t_{inc} = 6\,h$ are grayed out since they are excluded from the modeling, as explained in Section 6.3.3; the open square for 24 h in (a) shows the ML uptake measured with $1 \cdot 10^6$ MiaPaCa-2 cells seeded for comparison. Adapted from [291].

The modeling was performed in the following way: First, the ML adsorption at the cell membrane was numerically modeled by fitting eq. (6.1) to the $4\,°C$ data of each cell line: Starting from an arbitrary initial input for the fitting parameters k_a and m_0, the model curve was iteratively fitted to the experimental data with a maximum of 400 iterations and tolerance of $\Delta\delta = 0.001$, maximizing the fitting quality parameter R^2_{adj}.

The fitting parameters k_a and m_0 were then input parameters for the second step of numerically modeling the ML uptake in active cells with the $37\,°C$ data. For this, eqs. (6.3) and (6.6-6.10) were used, again with arbitrary initial input for the fitting parameters k_{endo}, k_{exo} and ϕ_0. The R^2_{adj} value was maximized iteratively as described above. The time scale parameters for exocytosis, t_{ex} and b', are constant input parameters during one iterative fitting process but were varied over a multitude of fitting processes to maximize the R^2_{adj}.

Generally, the model is in good agreement with experimental data for all three cell lines ($R^2_{adj} > 0.85$) and describes the uptake of ML inside cells well. The model is furthermore clearly a 3-step-model, where ML need to be adsorbed at the membrane first before internalization. The third step of exocytosis is visible as a plateau of delayed ML uptake within the first hours of incubation at $37\,°C$, marking the onset of exocytosis (at time t_{ex}). The model adds substantially to the interpretation of the ML uptake kinetics started in Section 6.3.3 by quantifying the following relations with the fitting parameters:

Table 6.3.: Fitting parameters of the uptake kinetics modeling: k_a denotes the rate constant of effective adsorption of ML to the cellular membrane, m_0 the saturation mass of ML adsorbed at the cellular membrane, t_{ex} the time point for the onset of exocytosis, b' describes the time scale parameter for exocytosis, k_{endo} and k_{exo} the rate constants for endo- and exocytosis and ϕ_0 the maximum active cellular membrane surface fraction involved in the uptake of nanoparticles.

	k_a $[\frac{mL}{mg(Fe)\cdot min}]$	m_0 [pg(Fe)]	R^2_{adj}	t_{ex} [min]	b' [min]	k_{endo} $[\frac{1}{min}]$	k_{exo} $[\frac{1}{min}]$	ϕ_0	R^2_{adj}
MiaPaCa-2	0.184 ±0.004	1.83 ±0.14	0.973	240	100	0.0031 ±0.0007	0.0015 ±0.0007	9.1 ±3.5	0.983
BxPC-3	0.141 ±0.010	0.37 ±0.02	0.957	200	100	0.0183 ±0.0096	0.0010 ±0.0002	12.2 ±5.9	0.858
L929	0.154 ±0.054	0.39 ±0.07	0.870	310	120	0.0021 ±0.0024	0.0010 ±0.0006	11.9 ±8.0	0.904

1. Adsorption rates, k_a and active membrane surface fractions, ϕ_0, are of the same order of magnitude for all three cell lines. This indicates that ML adsorb equally fast to the cell membrane, independent of the specific cell type, which is expected since the ML's phospholipid shell promotes unspecific ML attachment to cell membranes [320, 321], independent of the cell type.

2. The saturation mass of ML adsorbed to the cellular membrane, m_0, is approximately 4.5-times higher for MiaPaCa-2 cells compared to BxPC-3 and L929 cells. As indicated in statement (1.) above, ML attach non-specifically to cell membranes. Consequently, the higher m_0 must be a characteristic of the MiaPaCa-2 cells. Taken together with the observation that MiaPaCa-2 also shows the highest absolute uptake of ML for active cells (cf. Section 6.3.3, Fig. 6.10a), MiaPaCa-2 display the highest affinity for ML uptake. This fits the hypothesis stated in Section 6.2.3 that MiaPaCa-2 might store ML most energetically efficient inside lysosomes in order to preserve energy for further ML uptake. Nevertheless, no conclusive statement can be given here, as further information on the specific MiaPaCa-2 endocytosis mechanism are missing.

3. The rate constant of endocytosis, k_{endo}, is approx. 6- or 8-times higher for BxPC-3 compared to that of MiaPaCa-2 or L929, respectively. This confirms a comparatively fast initial uptake of ML inside BxPC-3 that also reach ML uptake saturation quickest within 6 h of incubation time, which was also observed in Section 6.3.3. Nevertheless, the absolute amount of ML taken up in saturation is lowest for BxPC-3. This could possibly be linked to the observation that BxPC-3 display a low packing density of ML, indicating a low overall ML uptake.

4. The ratio of k_{endo} : k_{exo} is 2:1 for MiaPaCa-2 and L929 but 18:1 for BxPC-3. This indicates that exocytosis of ML is a comparatively slower process than the one of endocytosis, especially for BxPC-3, which internalize ML quickest. The actual endocytotic process generally happens within minutes after attachment to the cell [322, 323], as also confirmed for ML from chase experiments in Section 6.2.3.

5. BxPC-3 shows the shortest delay time for the onset of exocytosis ($t_{ex} = 200$ min), followed by MiaPaca-2 ($t_{ex} = 240$ min) and L929 ($t_{ex} = 310$ min). These delay time scales for ML are of the same order of magnitude as that of approx. 21 min reported for exocytosis of 8 nm quantum dots incubated with human cervical cancer cells (HeLa cells) [316], approx. 120 min for 14 nm-sized Au-nanoparticles [301] and even up to 360 min for 17 nm-sized Au-nanoparticles in endothelial cells [324]. Overall, this shows that smaller nanoparticles are more favorable for rapid exocytosis [99]. Some particles were even reported to be re-taken up by cells 4 h after exocytosis [325].

6. For $t_{inc} \geq 6$ h, however, all time-dependent effects of endo- and exocytosis are predicted to be at equilibrium and ML uptake saturates for the pancreatic tumor cell lines MiaPaCa-2 and BxPC-3, while L929 reaches equilibrium for $t_{inc} \geq 8$ h.

6.3.6. Magnetic Relaxation Study of Intracellular ML

The magnetic relaxation of intracellular ML after 24 h of incubation at 37 °C was also studied via MPS. Exemplary MPS harmonic spectra normalized to the third harmonic are shown in Fig. 6.12 for the three cell lines with $t_{inc} = 24$ h. Untreated and agglomerated ML references from Fig. 6.8 are shown for comparison.

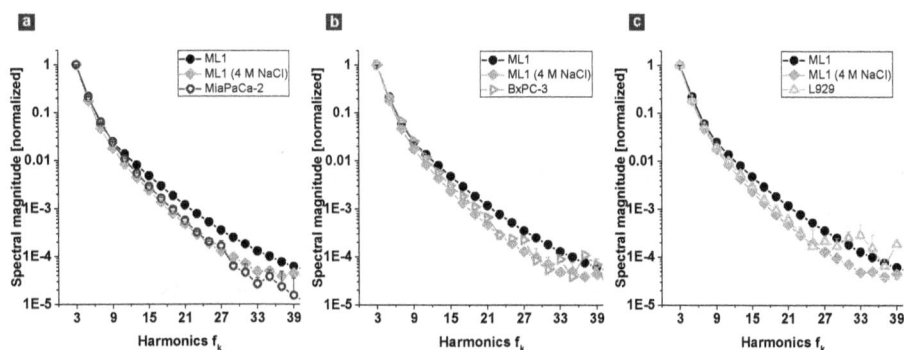

Fig. 6.12.: Normalized MPS harmonic spectra after 24 h of incubation for (a) MiaPaCa-2, (b) BxPC-3 and (c) L929. Untreated and (agglomerated) ML treated with 4 M NaCl (cf. Fig. 6.8) are shown for comparison as well. (a) adapted from [291].

The MPS spectra for intracellular ML ($t_{inc} = 24$ h) show an equivalent decrease in spectral magnitude as the agglomerated ML for all three cell lines. This decreased ability of relaxation for intracellular ML indicating a tendency of agglomeration for ML inside cells, as already confirmed in detail by morphological studies (cf. Section 6.2.4). A similar decrease in the MPS spectral magnitude upon MNP internalization was observed in literature [257, 326]. It was hypothesized from AC-susceptibility measurements that a decrease in the dynamic magnetic relaxation of intracellular MNP is caused by the inhibition of Brownian relaxation upon intracellular immobilization of MNP [305]. Moreover, the dynamic magnetic relaxation could be further decreased by demagnetizing effects of magnetically interacting MNP upon intracellular agglomeration [286]. Consequently, the observed decrease in the spectral magnitude of intracellular ML indicates ML agglomeration and immobilization, however, both effects occur indistinguishably superimposed. This superposition will be unraveled in Chapter 7, where the isolated effects of either agglomeration or immobilization on the magnetic relaxation of MNP are studied individually in more detail using dedicated model systems.

6.4. Magnetic Properties of Internalized Particles

TEM investigation revealed agglomeration and immobilization of intracellular ML inside lysosomes after 24 h of incubation (cf. Section 6.2) and the MPS spectral magnitude decreased for such intracellular ML as demonstrated in the last Section 6.3.6. To further elucidate the effects of intracellular uptake on the magnetic properties of ML, SQUID measurements were performed with intracellular ML as described in this section. Therefore, ML3 particles were incubated with MiaPaCa-2, BxPC-3 and L929 cells for 24 h as described in Section 6.4.1. The $M(H)$-curves and ZFC-FC curves are analyzed in Section 6.4.2 and discussed with regard to increased magnetic dipole-dipole particle interaction among ML upon internalization inside cells in dependence of the MNP packing density.

6.4.1. Experimental Procedure

Cell lines MiaPaCa-2, BxPC-3 and L929 were cultivated as described in Section 6.1 with the specifications given in Table 6.1. ML3 particles with a concentration $c = 0.3 \, \text{mg(Fe)/mL}$ were used for incubation for 24 h. After harvesting, the cells were transferred to 15 mL falcon tubes (Eppendorf) and centrifuged for 5 min (1 200 rpm, 20 °C). Supernatant was removed, the cell pellet resuspended in 0.03 mL cell suspension and the mixture was transferred to 0.1 mL polycarbonate capsules (Quantum Design) for SQUID measurements. Subsequently, samples were mixed with 0.03 mL of mannitol solution (mass fraction $\mu = 13.0 \, \%$) and freeze-dried as described in Section 4.6.

SQUID magnetometry was conducted as described in Section 4.6, however, the temperature range of ZFC-FC measurements was set from $T = 5 \, \text{K}$ to $T = 400 \, \text{K}$.

6.4.2. Results and Discussion

Fig. 6.13a shows the $M(H)$-curves from SQUID measurements for intracellular ML together with ML3 (cf. Fig. 4.14) for reference. As the final iron concentration for intracellular ML was inaccessible for these experiments since the ML uptake inside cells could not be quantified, Fig. 6.13a shows the magnetic measurement data in [Am^2] after subtraction of the diamagnetic portion (performed as described in Section 4.6). However, assuming that the saturation magnetization, M_s [Am^2], is proportional to the iron content per sample, C_{Fe}, a rough estimation of C_{Fe} can be given, since the iron content of ML3 is known from sample preparation

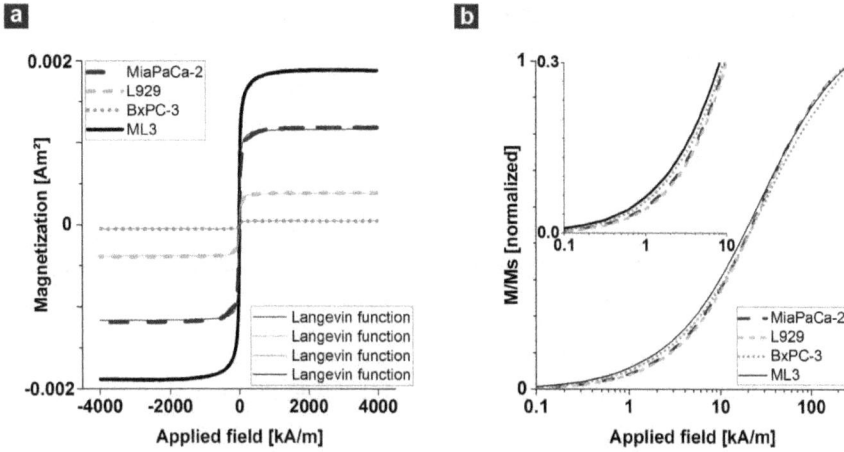

Fig. 6.13.: (a) Magnetization curve of intracellular ML3 for the three cell lines MiaPaCa-2, L929 and BxPC-3. ML3 is shown as a reference. Data was fitted with the Langevin function (cf. eq. (2.27)). (b) virgin curves of the same samples normalized to saturation magnetization (assumed from (a); cf. Table 6.4) with zoom at low applied fields (inset).

with $C_{Fe} = 18\,\mu g(Fe)$. For this, the $M(H)$-curves in Fig. 6.13a were fitted with the Langevin function (cf. eq. (2.27)), yielding M_S [Am^2], from which C_{Fe} is estimated for the cell samples by comparison to ML3. Results are listed in Table 6.4.

Table 6.4.: Saturation magnetization, M_s, derived from Fig. 6.13a with fitting the Langevin function with the fitting quality parameter, R^2_{adj}. An estimate for the iron content per sample, C_{Fe} follows from comparing to ML3, whose C_{Fe} is known from sample preparation.

Sample	M_S [mA m^2]	R^2_{adj}	C_{Fe} [μg(Fe)]
ML3	1.882 ± 0.052	0.998	18.0
MiaPaCa-2	1.174 ± 0.032	0.999	11.2
BxPC-3	0.387 ± 0.013	0.999	3.7
L929	0.050 ± 0.002	0.999	0.5

ML3 uptaken in MiaPaCa-2 show the highest M_S and consequently highest iron content, $C_{Fe} \approx$ 11.2 μg(Fe), followed by L929 with $C_{Fe} \approx 3.7\,\mu g$(Fe), and $C_{Fe} \approx 0.5\,\mu g$(Fe) for BxPC-3 (s. Fig. 6.13a and Table 6.4). Although C_{Fe} may only be interpreted as a rough estimate, this observation is line with the findings of highest saturation uptake of ML in MiaPaCa-2 (approx. 3.5-times higher than for L929 and BxPC-3; cf. Fig. 6.10), which, combined with the moderate cell proliferation rate of MiaPaCa-2 during incubation with ML (cf. Fig. 6.9b), results in higher absolute iron content uptaken. Arguing along the same lines, L929 proliferate more rapidly than MiaPaCa-2 but internalize less ML, resulting in intermediate absolute ML uptake. Lastly, BxPC-3 proliferate more slowly than both MiaPaCa-2 and L929 and also internalize lower amounts of ML per cell, resulting in an overall very low absolute amount of iron in the SQUID sample, which is just barely detectable by the SQUID magnetometer.

A closer look to the virgin curves normalized to the saturation magnetiaztion reveals lower magnetization for intracellular ML3 at low fields (Fig. 6.13b). This implies an increasing magnetic dipole-dipole particle interaction in intracellular ML due to particle agglomeration causing demagnetizing effects, as discussed before in literature [286, 327]. Indeed, ML agglomerates were confirmed to be tightly packed inside cells in lysosomes by TEM investigation (s. Section 6.2), for which the small interparticle distance can be assumed, inducing magnetic dipole-dipole particle interaction. For BxPC-3, the ML packing density was observed to be less dense, resulting in a slightly higher initial magnetization compared to MiaPaCa-2 and L929 (with high packing density; cf. Section 6.2.4) in Fig. 6.13b.

Further insight into the impact of ML internalization inside cells on magnetic particle interaction can be derived from ZFC-FC measurements (Fig. 6.14). Here, the peak temperature in the

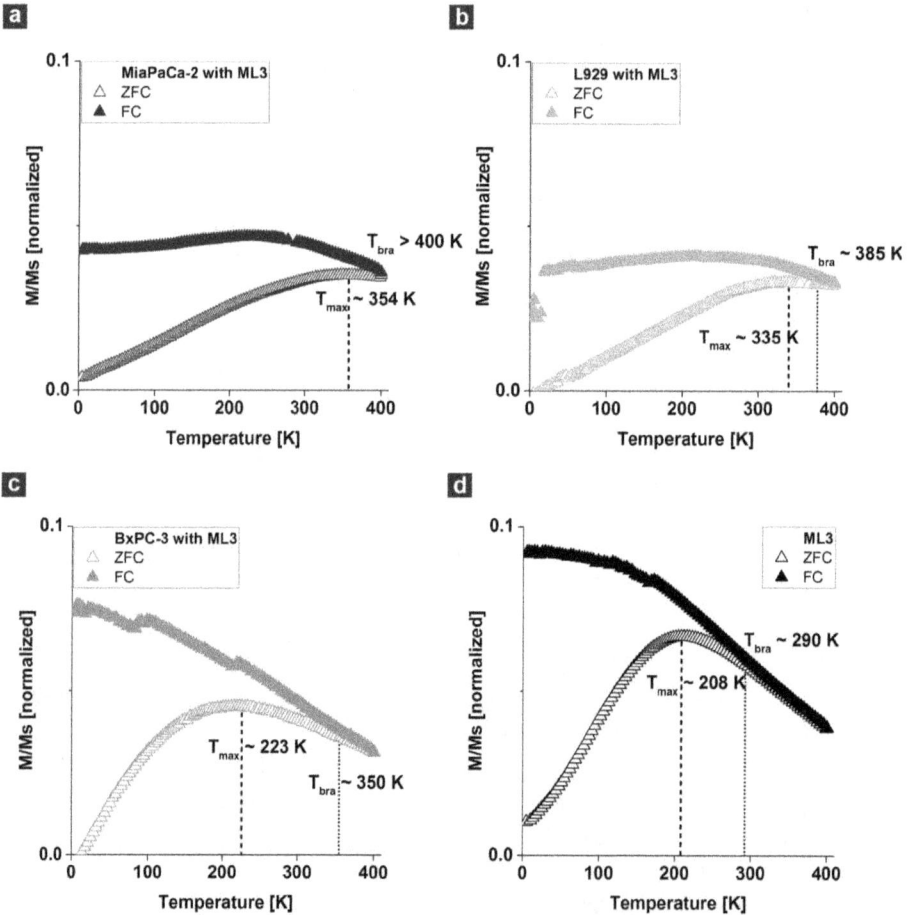

Fig. 6.14.: ZFC and FC magnetization curves normalized to the saturation magnetization, M_S (s. Table 6.4), for ML3 internalized in (a) MiaPaCa-2, (b) L929 and (c) BxPC-3. (d) shows ML3 for reference (cf. Fig. 4.16). The ZFC peak temperature, T_{max}, and the branching temperature, T_{bra} are marked.

ZFC curve, T_{max}, and the branching temperature, T_{bra}, both shift to higher temperatures for intracellular ML compared to homogeneously dispersed ML. From T_{max} the blocking temperature, T_B, and the effective anisotropy constant, K_{eff}, were estimated using eq. (4.16). Results are summarized in Table 6.5. As already discussed in Section 4.6, these results can only be used as rough approximation rather than reliable values.

Table 6.5.: ZFC-FC peak temperature, T_{max}, and branching temperature, T_{bra}, derived from Fig. 6.14. Furthermore, the mean blocking temperature, T_B, and effective anisotropy constant, K_{eff}, were estimated from eq. 4.16, using the magnetic particle size data derived from SQUID data for ML3 (cf. Table 4.11.

Sample	T_{max} [K]	T_{bra} [K]	T_B [K]	K_{eff} [kJ/m^3]
MiaPaCa-2	354 ± 20	> 400	334	136 ± 191
L929	335 ± 15	385 ± 10	316	128 ± 181
BxPC-3	223 ± 20	350 ± 10	211	85 ± 120
ML3	208 ± 10	290 ± 10	196	80 ± 112

As discussed above, MiaPaCa-2 exhibit the tightest ML packing density upon internalization inside lysosomes, followed by L929 and BxPC-3. This corresponds directly to the magnitude that T_B shifts towards higher temperatures for intracellular ML compared to ML3 (cf. Table 6.5), which is another consequence of the increased magnetic dipole-dipole particle interaction due to intracellular ML agglomeration. The same observation was reported by Lévy et al. for iron-oxide MNP ($d_C \approx 9\,nm$) uptaken in the liver, spleen and adipose tissue [247, 248].

In summary, $M(H)$ and ZFC-FC measurement on intracellular ML revealed an increased magnetic interparticle interaction upon internalization. The strength of these interactions corresponds directly to the packing density of the ML inside cells, showing strongest increase in interactions for ML uptaken in MiaPaCa-2, followed by L929 and lastly BxPC-3.

6.5. Concluding Remarks

In this chapter, the interaction of ML with the pancreatic tumor cell lines MiaPaCa-2 and BxPC-3 as well as with mouse fibroblasts L929 was analyzed. All three cell types internalized ML over incubation times up to 24 h, upon which the morphology[12], relaxation behavior and magnetic properties of ML changed substantially. The main characteristics are summarized in the following.

The morphology of intracellular ML was studied with TEM analysis, revealing that ML are first grouped at the outer cell membrane in clusters with sizes of approx. $(50 - 250)\,nm$, before internalizing (cf. Section 6.2). From here, ML are endocytosed within several tens of minutes for all cell lines (cf. Section 6.2.2) and internalized inside intracellular compartments, where ML are agglomerated and immobilized (i. e. ML interact with each other and are restricted in their mobility).

---

[12]Defined as the arrangement, localization and distribution of ML inside cells.

Over an incubation time of up to $24\,\text{h}$, ML are internalized inside early endosomes within the first hours, which merge with each other into larger (late) endosomes over time and are finally secreted via exocytosis or further develop in lysosomes (cf. Section 6.2.3). The final size and ML packing density of lysosomes after $24\,\text{h}$ of incubation is cell-specific (cf. Table 6.2 and Fig. 6.5): MiaPaCa-2 cells had the smallest average lysosome size with $d_{\text{lyo}} \approx 410\,\text{nm}$ but the qualitatively highest packing density. L929 lysosomes showed intermediate sizes of $d_{\text{lyo}} \approx 540\,\text{nm}$ with high ML packing density. The average lysosome size for BxPC-3 cell is largest with $d_{\text{lyo}} \approx 650\,\text{nm}$ but the packing density is comparatively much lower than for MiaPaCa-2 and L929.

The ML uptake kinetics of each cell line were quantified using MPS measurements to determine the iron content per cell as a function of incubation time. Here, the saturation ML uptake is substantially higher for MiaPaCa-2 cells with approx. $9\,\text{pg(Fe)/cell}$ compared to L929 with approx. $3.5\,\text{pg(Fe)/mL}$ and BxPC-3 with approx. $2.5\,\text{pg(Fe)/mL}$ (s. Section 6.3.3).

Moreover, the ML uptake kinetics were successfully modeled as a 3-step process, assuming ML adsorption at the cell membrane as the first step, followed by endocytosis of ML as the second step and exocytosis of ML as the third step (cf. Sections 6.3.4). The model allowed to determine cell-specific parameters such as the rate constants of endo- and exocytosis and enables the prediction of intracellular ML uptake for any given incubation time. E. g., saturation ML uptake is predicted after an incubation time of approx. $6\,\text{h}$ for pancreatic tumor cells and after approx. $8\,\text{h}$ for L929 (cf. Section 6.3.5). This may serve as an indicator for estimating a minimum incubation time for the application of intracellular MFH. As will be discussed in Chapter 8, the internalization of MNP is a key factor for amplifying the efficacy of MFH treatment.

The formation of intracellular ML agglomerates results in increased magnetic dipole-dipole particle interactions that affect the magnetic properties of intracellular ML: From SQUID magnetometry and ZFC-FC measurements, it was observed that the initial magnetization drops, while the blocking temperature and effective anisotropy constant increase upon ML internalization (cf. Section 6.4.2). The extent of changes in magnetic properties could even be assigned to a certain intracellular ML packing density (i. e. greatest shift in blocking temperature for highest packing density in MiaPaCa-2, indicating strongest magnetic particle interactions for ML internalized in MiaPaCa-2).

Moreover, also the magnetic relaxation of ML was affected by intracellular internalization: The MPS spectral magnitude was substantially decreased for ML internalized in all three cell lines (cf. Section 6.3.6). This decrease in MPS spectral magnitude is attributed to the superposition of ML agglomeration and immobilization inside cells, causing demagnetizing effects as well as inhibiting Brownian relaxation of ML. This indicates the need of quantifying the isolated effects of immobilization or agglomeration of MNP relaxation behavior and magnetic properties. Based on the ML morphology and magnetic behavior inside cells (i. e. intracellular agglomerate size, increased magnetic dipole-dipole particle interaction and ML immobilization) MNP systems mimicking the intracellular environment are developed and characterized for their magnetic particle heating performance in the next Chapter 7.

7. Magnetic Particle Heating in Model Systems

Magnetic particles internalized inside cells form tightly packed agglomerated arrangements and are simultaneously immobilized in lysosomes (as seen in Chapter 6). The present chapter addresses the question of how the states of MNP immobilization and MNP agglomeration can be modeled independently of each other and how each individual state affects the particle heating efficiency. For this, model hydrogel systems with incorporated MNP are synthesized and characterized in Section 7.1 and employed to study the isolated effects of gradual particle immobilization on particle heating in Section 7.2. Moreover, model systems of MNP clustered inside specifically tailored liposome shells as well as MNP agglomerated by damaging their coating upon adding sodium chloride are characterized in Section 7.3 and used to investigate the effects of MNP agglomeration on particle heating in Section 7.4.

Sections 7.1 and 7.2 are partly based on an original publication by the writer [264] (s. Appendix C). Where entire sentences are cited directly from the publication, this is marked with [‡] and figures adapted or taken directly from [264] are marked in the respective figure's caption. Sections 7.3 and 7.4 are partly based on two original publications by the writer [328, 329] (s. Appendix C). Where entire sentences are cited directly from [329], this is marked with [◇] and figures adapted or taken directly from [328, 329] are marked in the respective figure's caption.

7.1. Characterization of Ferrohydrogels

As described in Section 2.6, hydrogels display tissue-equivalent and biocompatible properties, which make them highly suitable for the use as tissue-mimicking model systems to immobilize MNP. Hydrogels have mesh sizes tunable to the particle hydrodynamic size of MNP and are therefore employed in the next Section 7.2 as in situ model systems to study the effect of gradual immobilization of FFAM particles on magnetic particle heating. Such hydrogels containing FFAM are denoted as ferrohydrogels (FHG) throughout this thesis. In the present section, the preparation of two types of FHG, in particular of low-melting (LM-)agarose and poly(acrylamide) (PAAm) FHG, is described in Section 7.1.1. The rheological characterization of these two hydrogels is presented in Section 7.1.2, assessing their mechanical properties and deriving an estimate for the mean mesh size of the polymer network. Following, additionally characterization via dynamic light scattering (DLS; Section 7.1.3), transmission electron microscopy (TEM; Section 7.1.4) as well as vibrating sample magnetometry (VSM; Section 7.1.5) is performed to assess potential agglomeration of the incorporated FFAM.

Rheological measurements, TEM and VSM experiments were performed in cooperation with the Institute of Physical Chemistry from the Universität zu Köln (Cologne, Germany).

7.1.1. Preparation of Ferrohydrogel Samples

LM-agarose and PPAm FHG were prepared from standard chemical within one day before the actual experiment was performed (rheological characterization, particle heating or VSM measurements (s. below)). The chemicals used to prepare the hydrogels are listed in Table 7.1.

Table 7.1.: Chemicals used for the preparation of hydrogels.

Chemical Name	Abbreviation	Purity	Supplier
Low melting agarose	LM-agarose	$\geq 99\,\%$	Carl Roth GmbH + Co. KG (Karlsruhe, Germany)
Acrylamide ($CH_2{=}CHCONH_2$)	AAm	$\geq 99\,\%$	Sigma-Aldrich Chemie GmbH (Taufkirchen, Germany)
N,N'-methylenebisacrylamide (($H_2C{=}CHCONH)_2CH_2$)	BIS	$\geq 99\,\%$	Sigma-Aldrich Chemie GmbH
N,N,N',N'-tetramethylethylene-diamine (($CH_3)_2NCH_2CH_2N(CH_3)_2$)	TEMED	$\geq 99\,\%$	Sigma-Aldrich Chemie GmbH
Ammonium persulfate (($NH_4)_2S_2O_8$)	APS	$\geq 98\,\%$	Sigma-Aldrich Chemie GmbH

Agarose Samples

Low melting (LM-)agarose FHG were prepared by dissolving a certain LM-agarose mass fraction in DI-H_2O. The mass fraction was varied according to the general definition:

$$\mu_m = \frac{m_m}{m_{tot}}, \tag{7.1}$$

with the mass of the substance fractioned, m_m, and the total mass of the mixture, m_{tot}. For LM-agarose it reads: $\mu_{agar} = \frac{m_{agar}}{m_{agar}+m_{H2O}+m_{FFAM}}$, with the mass of LM-agarose powder, m_{agar}, the mass of preheated DI-H_2O, m_{H2O}, in which agarose is dissolved and m_{FFAM} the mass of FFAM in the FHG. For preparation of FHG, DI-H_2O was heated on a hotplate set to $T = 70\,°C$. First, the desired amount of LM-agarose was dissolved in the preheated DI-H_2O on the hotplate, vortexed slowly and stored at $T = 70\,°C$ for 30 min to ensure full dissolving. Then, the liquid LM-agarose heated at $T = 70\,°C$ was mixed with FFAM. The FHG was immediately vortexed slowly again to disperse the FFAM homogeneously and cooled at $-18\,°C$ for 30 min directly afterwards. During this time, the gel rapidly and reproducibly thickens without freeze the samples. Samples must not be frozen to avoid inhomogeneous network structuring due to the formation of ice crystals from the water contained by the hydrogels.

LM-agarose FHG with the following mass fractions were prepared: $\mu_{agar} = (0.3, 0.5)\,\%$, $\mu_{agar} = (1.0 - 3.0)\,\%$ in increments of 0.5 % and $\mu_{agar} = (4.0 - 10.0)\,\%$ in increments of 1.0 %. The FFAM concentration was $c = 0.3\,\text{mg(Fe)/mL}$ all the LM-agarose FHG prepared.

Reference LM-agarose hydrogels not containing FFAM were prepared as well.

Poly(acrylamide) Samples

For the preparation of PAAm FHG, AAm and its crosslinker BIS were dissolved in DI-H_2O at a certain volume fraction calculated by

$$\nu_{pol} = \frac{(m_{AAm} + m_{BIS})/\rho_{PAAm}}{m_{tot}/\rho_{H2O}}, \tag{7.2}$$

from the mass of Aam, m_{AAm}, the mass of BIS, m_{BIS}, the total mass of the mixture, $m_{tot} = m_{AAm} + m_{BIS} + m_{H2O} \approx m_{H2O}$ and the densities of polyacrylamide (AAm + BIS) and water, $\rho_{PAAm} = 1.3\,\text{g/cm}$ and $\rho_{H2O} = 1.0\,\text{g/cm}$, respectively. The respective crosslinker mole fraction, α (cf. eq. (2.76)), was varied to meet a specific range of mesh size entrapping the FFAM (s. below in Section 7.2). FFAM was added to the mixture before the free radical polymerization, i. e. the gelling process of the hydrogel, was initialized by adding APS and TEMED at mass fractions, $\mu_m = 0.2\,\%$ (defined in eq. (7.1)), each. The polymerization occurred at room temperature within $(5 - 10)\,\text{min}$ after initialization under exothermal temperature release. The actual temperature rise during polymerization for selected hydrogels was measured with a fiber-optic thermometer during polymerization, (cf. Appendix A.5.3, Fig. A.17), amounting to

$\Delta T \approx (10 - 15)$ K. Consequently, PAAm hydrogels prepared at room temperature, $T_R \approx 25\,°C$, experienced a maximum temperature of $T \leq 40\,°C$ during polymerization.

Eight PAAm samples with a constant polymer volume fraction of $\nu_{pol} = 8\%$ and various crosslinker mole fractions $\alpha = (0.05, 0.075, 0.1, 0.15, 0.2, 0.3, 0.4, 0.5)\%$ were prepared for this thesis. The FFAM concentration was $c = 0.3\,mg(Fe)/mL$ for the PPAm FHG prepared.

Reference PAAm hydrogels not containing FFAM were prepared as well.

7.1.2. Rheological Characterization of Ferrohydrogels

LM-Agarose and PAAm hydrogels can be characterized by their elasticity (i.e. softness / stiffness) and the mean mesh size of the underlying network structures [141]. The elasticity can be quantified by measuring the storage modulus[1], G', and its complex counterpart, the loss modulus, G'', from oscillatory shear measurements at constant strain [330]. Such rheological characterization was performed using an AR-G2 rheometer (TA Instruments, New Castle, DE, USA) with a plate-to-plate geometry designed for circular samples with diameters of $40\,mm$. A solvent trap sealed with DI-H_2O to keep the water vapor pressure constant and prevent solvent evaporation during measurement. From preliminary strain sweep measurements performed at $1\,Hz$ with strains $\gamma = (0.02 - 2.00)\%$, a constant strain of $\gamma = 0.2\%$ was chosen for all samples in subsequent frequency sweep measurements of as prepared PAAm hydrogels. These measurements were conducted in the linear viscoelastic regime for the a frequency range of $f = (0.1 - 100)\,Hz$.

For rheological measurements, FHG samples were prepared in a circular Teflon mold (diameter $d = 40\,mm$, volume $V = 2.6\,mL$) with a total mass of exactly $m_{tot} = 2\,600\,mg$ (to ensure an exact polymer volume fraction of $\nu_{pol} = 8\%$). The samples were measured 1 h after preparation with a rheometer (s. Section 7.1.2).

Agarose Samples

LM-agarose samples were rheologically characterized for mass fractions of $\mu_{agar} \leq 4.0\%$. Samples with $\mu_{agar} > 4.0\%$ were too stiff for adequate sample preparation and had presumably storage moduli $G' \sim 100\,kPa$ (extrapolated from results below). The rheological measurement results are shown in Fig. 7.1. From the rheological data (Fig. 7.1), one easily confirms that $G' \gg G''$, for all mass fractions, indicating that the LM-agarose gels exhibit predominantly elastic mechanical properties [149]. However, the data of G'' is clearly more scattered than G' data, especially at low mass fractions $\mu_{agar} < 2.0\%$. By averaging the data for $f = (0.1 - 1)\,Hz$, the frequency-independent plateau storage modulus, $G'(f) \approx G'_P$, was estimated from Fig. 7.1 and is listed in detail in Appendix A.5.1. The overall range of $G' \approx (200 - 70\,000)\,Pa$ (s. Fig. 7.1 and in Appendix, Table A.9) corroborates the versitility of agarose hydrogels, able to mimick

---

[1]The shear modulus is actually complex, $G = G' + i \cdot G''$, with the real part, the stroage modulus, G', and the imaginary part, the loss modulus, G''. For small frequencies, $G' \gg G''$ holds.

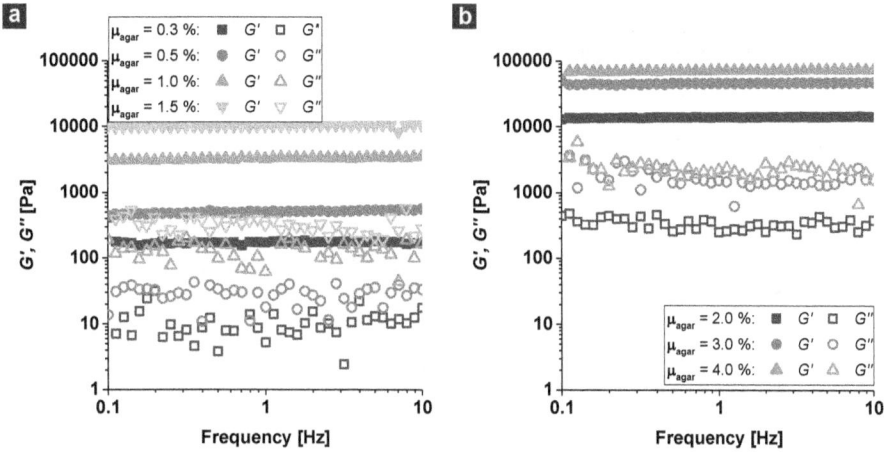

Fig. 7.1.: Rheological measurements showing the storage and loss moduli, G' and G'', measured from oscillatory shear measurements at a constant strain of $\gamma = 0.2\,\%$ and at $T = 25\,^{\circ}$C for LM-agarose samples with mass fractions of (a) $\mu_{\text{agar}} = (0.3 - 1.5)\,\%$ and (b) $\mu_{\text{agar}} = (2.0 - 4.0)\,\%$. Figure provided courtesy of Julian Seifert (Universität zu Köln).

brain tissue ($\approx (500 - 1\,000)\,$Pa), muscle tissue ($\approx (8\,000 - 17\,000)\,$Pa and even collagenous bone ($\approx (25\,000 - 40\,000)\,$Pa) [331].

For agarose hydrogels, the interpretation of the rheological data within the framework of rubber elasticity theory (RET; cf. Section 2.6) was not possible, since the number of active polymer strands, ν_{el}, cannot be estimated. Nevertheless, several methods for the determination of the mean mesh size in agarose gels have been proposed, such as electrophoresis with charged polystyrene particles of defined diameter, estimating the mean mesh size from the largest sized particles still able to penetrate the gel in a fixed time and applied voltage [158], direct mesh size measurement via atomic force microscopy [332], or the indirect determination via light absorbance measurements [333]. In fact, Rhigetti et al. [158] suggested a power law for the mean mesh size in dependence of the agarose mass fraction, μ_{agar} (cf. eq. (7.1)):

$$d_{\text{cross}} = c_1 \cdot \mu_{\text{agar}}{}^{c_2}, \tag{7.3}$$

with the two parameters $c_1 = 140.7\,$nm and $c_2 = -0.7$ determined from fitting to electrophoresis data determined for $\mu_{\text{agar}} = (0.16-1.0)\,\%$ [158]. The authors analyzed highly crosslinked PAAm ($\alpha \geq 10\,\%$) hydrogels as well and fitted the power law from eq. (7.3) to the experimental data, yielding good agreement with different fitting parameters. This fact is confirmed from the PAAm data fitted with a power law above (s. below, Fig. 7.4). Normal agarose and LM-agarose do not have the same elastic properties for the same mass fraction. This directly implies larger pore sizes for LM-agarose, as demonstrated by Narayanan et al. for $\mu_{\text{agar}} = (0.5-3.0)\,\%$ [333]: The experimental data from the authors is listed in Appendix A.5.1, Table A.10 and was used to fit the power law (cf. eq. (7.3)) for the dependency of mean mesh size, d_{approx} on agarose mass fraction, μ_{agar}, as shown in Fig. 7.2. From this it is obvious that LM-agarose has comparatively larger mesh sizes than normal agarose for equal mass fractions.

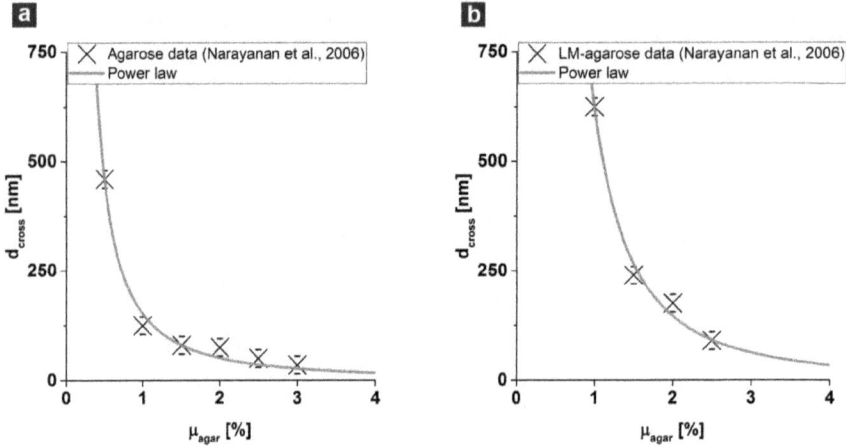

Fig. 7.2.: Mean mesh sizes, d_{approx}, in dependence of agarose mass fraction, μ_{agar}, for (a) normal agarose and (b) low melting (LM-) agarose extracted from Narayanan et al. [333]) and fitted with a power law (eq. (7.3)), yielding $c_1^{agar} = (152 \pm 12)\,$nm and $c_2^{agar} = (-1.58 \pm 0.13)$ for normal agarose and $c_1^{LM\text{-}agar} = (620 \pm 27)\,$nm and $c_2^{LM\text{-}agar} = (-2.09 \pm 0.18)$ for LM-agarose. The fitting quality parameter yields $R_{adj}^2 = 0.984$ for (a) and $R_{adj}^2 = 0.986$ for (b).

For the characterization of LM-agarose hydrogels the power law (eq. (7.3)) is chosen, employing the fitting parameters derived from Narayanan et al. [333] for LM-agarose as mentioned above. From this, approximate mesh sizes in the range of $d_{approx} \approx (5 - 8\,000)\,$nm are estimated for the prepared LM-agarose samples (s. Appendix A.5.1, Table A.9).

Poly(acrylamide) Samples

For PAAm hydrogels, the RET holds [150], and predicts a linear dependence of the shear modulus, G, on the number of elastically active polymer strands per unit volume, ν_{el}, as derived in Section 2.6. For small frequencies, $f \approx (0.1 - 10)\,$Hz in oscillatory shear measurements, the shear modulus can be approximated by the storage modulus, G', which forms a constant plateau value, $G \approx G' \approx G'_P$ [334]. Assuming an ideal network, ν_{el} is proportional to the number of elastically active crosslinks per unit volume within the network, μ_{el}, via the formula [149]:

$$\mu_{el} = \frac{2}{\beta} \cdot \nu_{el}, \tag{7.4}$$

with the crosslinker functionality constant, β ($= 4$ for BIS). By solving eq. (2.75) for ν_{el}, inserting in eq. (7.4), using $G \approx G'_P$ and assuming that the distance between two elastically active crosslinks, d_{cross}, can be estimated by $\mu_{el} \approx d_{cross}^{-3}$, an expression for d_{cross} reads:

$$d_{cross} = \left(\frac{k_B \cdot T \cdot \beta}{2 \cdot G'_P} \right)^{\frac{1}{3}}. \tag{7.5}$$

For PAAm networks, d_{cross} can be interpreted as the mean mesh size of the respective PAAm hydrogel.

Results from rheological measurements for PAAm hydrogels with $\nu_{pol} = 8\%$ and $\alpha = (0.05 - 0.5)\%$ are shown in Fig. 7.3. From Fig. 7.3 it is obvious, that the storage modulus is much

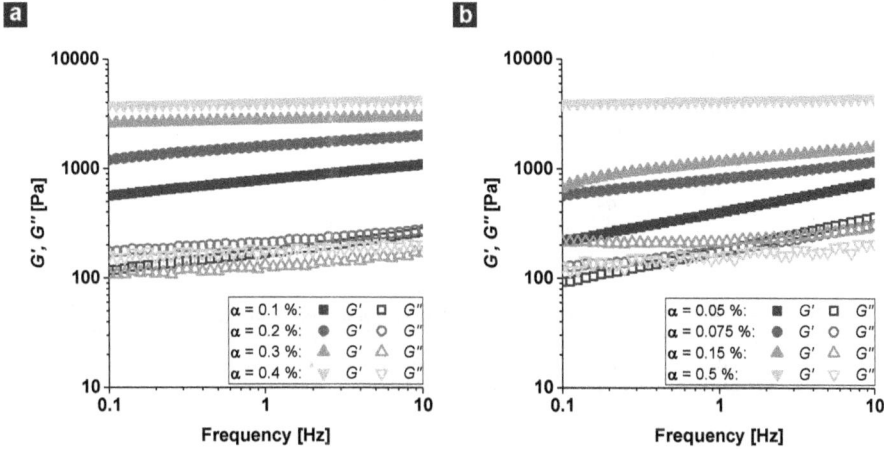

Fig. 7.3.: Rheological measurements showing the storage and loss moduli, G' and G'', measured from oscillatory shear measurements for a constant strain of $\gamma = 0.2\%$ and at $T = 25\,^{\circ}$C for $\nu_{pol} = 8\%$ and (a) $\alpha = (0.1 - 0.4)\%$ and (b) $\alpha = (0.05, 0.075, 0.15, 0.5)\%$. (a) adapted from [264] and (b) provided courtesy of Julian Seifert (Universität zu Köln).

larger than the loss modulus, $G' \gg G''$, for all crosslinker mole fractions, which indicates that the gels exhibit predominantly elastic mechanical properties. Furthermore, the storage moduli amount to $G' \approx (200 - 4\,000)$ Pa, which is equivalent to brain tissue ($\approx (500 - 1\,000)$ Pa) and below muscle tissue ($\approx (8\,000 - 17\,000)$ Pa) [331]. By averaging the data for $f = (0.1 - 5)$ Hz, the frequency-independent plateau storage modulus, $G'(f) \approx G'_P$, was estimated from Fig. 7.3. From using this in eqs. (2.75) and (7.5), the mean mesh size in dependence of the crosslinker mole fraction was subsequently calculated. A detailed list of measured storage and loss moduli and calculated mesh sizes can be found in the Appendix A.5.2, Table A.11. The corresponding data is plotted in Fig. 7.4. As depicted in Fig. 7.4, the data can be well fitted with a power law, cf. eq. (7.3), yielding a fitting quality parameter of $R^2_{adj} = 0.918$. The probed PAAm hydrogels exhibit mean mesh sizes in the range of (12.5 ± 0.1) nm$\leq d_{cross} \leq (26.7\pm3.2)$ nm in dependence of the crosslinker mole fraction $\alpha = (0.05 - 0.5)\%$. Such an ability of precisely tunable mesh sizes is confirmed by literature [159]. A similar decrease in mean mesh size was demonstrated by Wang et al., e. g. reporting $d_{cross} = 6.4$ nm for a similar polymer volume fraction of $\nu_{pol} = 7\%$ and up to five times the crosslinker mole fraction than used in this thesis, $\alpha \approx 2.5\%$ [335].

Even though the polymerization of PAAm hydrogels is exothermal, the effective temperature rise during sample preparation was measured to be only slightly above $10\,^{\circ}$C (cf. Appendix A.5.3). Therefore, temperature effects on FFAM during incorporation in PAAm hydrogels can be neglected.

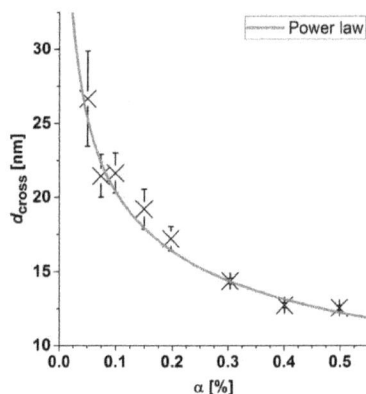

Fig. 7.4.: Mean mesh size, d_{cross}, determined from rheological measurements in dependence of the crosslinker mole fraction, α, for PAAm hydrogels with $\nu_{\text{pol}} = 8\%$. The data was fitted with a power law, $d_{\text{cross}} = c_1 \cdot \alpha^{c_2}$ (cf. eq. (7.3)), yielding $c_1 = (9.8 \pm 0.3)$ nm, $c_2 = (0.32 \pm 0.03)$ with $R^2_{\text{adj}} = 0.918$. The error bars result from error propagation. Adapted from [264].

7.1.3. Dynamic Light Scattering Characterization

As FFAM were subjected to LM-agarose at $T = 70\,°C$, complementary characterization was necessary to check for possible temperature-induced changes of the FFAM composition at the nanoscale (e. g. damaging the citric acid shell and inducing MNP agglomeration). For this, dynamic light scattering (DLS) measurements were performed with FFAM diluted to $c = 300\,\mu g(\text{Fe})/\text{mL}$ in DI-H_2O with $T = 70\,°C$. DLS experiments were carried out and analyzed as described in Section 4.4. Both the intensity-weighted and the volume-weighted intensity signal data of pristine FFAM and FFAM treated with $T = 70\,°C$ are compared. The results including fitting with the probability density function (PDF) of the log-normal distribution are shown in Fig. 7.5. Results from fitting are listed in Table 7.2, while fitting parameters are found in Appendix A.5.6, Table A.12. The hydrodynamic particle size, d_H, does not change for FFAM treated at $T = 70\,°C$, independent of whether intensity-weighted data, volume-weighted data or z_{avg} are compared. Therefore, it is expected that the sodium citrate coating of FFAM is not affected at temperatures $T \leq 70\,°C$. Consequently, it is furthermore expected that FFAM are incorporated in the LM-agarose hydrogels with their coating intact, preserving FFAM monodispersity upon immobilization in agarose hydrogels. Consequently, the DLS data of pristine FFAM is used for the description of FFAM incorporated in both FHG.

When discussing the effects of FFAM immobilization on particle heating in Section 7.2.2, the hydrogel's mesh size is compared to the mean hydrodynamic particle size to estimate the degree of MNP immobilization in the respective hydrogel. Since the primary data in DLS is obtained from the intensity of the backscattered laserlight in DLS measurements, d_H derived from intenstiy-

Table 7.2.: Mean particle hydrodynamic size, d_H, calculated from fitting the log-normal PDF to the intensity-weighted and the volume-weighted DLS signal. R^2_{adj} denotes the fitting quality parameter, z_{avg} and PDI are the respective average hydrodynamic particle size and distribution width, provided by the Malvern software for comparison.

	Intensity-weighted			
	z_{avg} [nm]	PDI	R^2_{adj}	d_H [nm]
FFAM	20.6	0.189	0.981	18.9 ± 6.1
FFAM 70 °C	21.2	0.230	0.993	19.5 ± 6.0

	Volume-weighted			
	z_{avg} [nm]	PDI	R^2_{adj}	d_H [nm]
FFAM	20.6	0.189	0.987	14.4 ± 3.7
FFAM 70 °C	21.2	0.230	0.990	14.9 ± 3.9

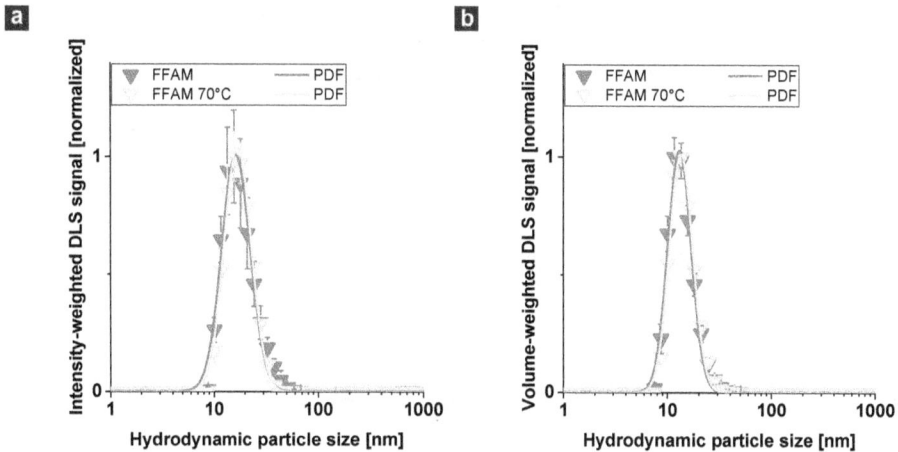

Fig. 7.5.: DLS measurement data for pristine FFAM and FFAM treated at $T = 70\,°C$ showing (a) intensity-weighted and (b) volume-weighted DLS signal. Data was fitted with probabilty density function (PDF) of the log-normal distribution (cf. eq. (3.22)). (a) partially adapted from [264].

weighted data can be considered to be most accurate [336, 337]. Therefore, intensity-weighted DLS data is used to estimate hydrodynamic particle sizes the following analysis.

7.1.4. Transmission Electron Microscopy Characterization

PAAm FHG were analyzed via freeze-fractured transmission electron microscopy (TEM) with a Zeiss LEO 912 Omega device (Carl Zeiss GmbH; s. Section 4.3 for a general description of TEM). Therefore, 55 µL PAAm with $\nu_{pol} = 8\,\%$ and $\alpha = 0.075\,\%$ and an FFAM concentration of $c = 0.3\,mg(Fe)/mL$ was plunge-frozen in liquid nitrogen at $T = 188\,K$. Subsequently, the frozen sample was finely ground and placed on a Cu-grid with a carbon-hole film (Quantifoil Multi A, Quantifoil Micro Tools GmbH, Großlöbichau, Germany) for TEM measurements. Finally, the sample was freeze-dried in a VTR 5036 vacuum drying oven (Heraeus Holding GmbH, Hanau, Germany) overnight[‡]. An exemplary TEM image of FFAM in freeze-dried PAAm hydrogel is shown in Fig. 7.6. As can be seen from Fig. 7.6, single FFAM particles are homogeneously

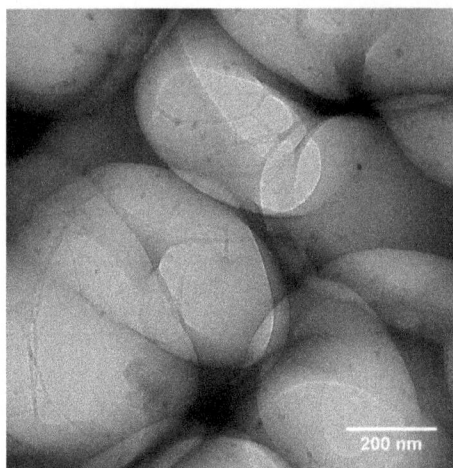

Fig. 7.6.: Exemplary TEM image of freeze-dried PAAm hydrogel ($\nu_{pol} = 8\,\%$ and $\alpha = 0.075\,\%$), containing FFAM with $c = 0.3\,mg(Fe)/mL$ FFAM. Single particles are visible as dark gray spots, homogeneously distributed inside the hydrogel matrix. Taken from [264].

distributed in the PAAm hydrogel, while particle clustering, agglomeration or deformations are not observed. This was also observed for a multitude (> 20) of TEM images on particle-filled PAAm hydrogels (s. Appendix A.18 for additional TEM images of different magnifications) and no particle agglomerates were found. Therefore, it is assumed from freeze-fractured TEM analysis that for FFAM incorporated in PAAm hydrogels agglomeration effects can be neglected.

7.1.5. Vibrating Sample Magnetometry Characterization

In order to validate the monodisperse character of the FFAM inside PAAm hydrogels, vibrating sample magnetometry (VSM) measurements were conducted on all FHG samples (s. Section 4.7 for a general description of VSM). For this, an EV 7 vibrating sample magnetometer (MicroSense, Lowell, MA, USA) operated at $T = 300\,K$ was used to measure the sample virgin magnetization curves for magnetic fields up to $1\,430\,kA/m$. FHG PAAm samples were prepared as described in Section 7.1.1 in cylindrical Teflon moulds with a volume of approximately $55\,\mu L$ with $\nu_{pol} = 8$ and various crosslinker mole fractions ranging from $\alpha = 0.05\,\%$ to $\alpha = 0.5\,\%$[‡]. FFAM with $c = 0.3\,mg(Fe)/mL$ dispersed in water was measured as a reference.

The normalized virgin magnetization curves of FFAM inside PAAm hydrogels are shown in Fig. 7.7. As can be seen from Fig. 7.7, the normalized virgin magnetization curves of the PAAm FHG only differ from the one of FFAM dispersed in water for low fields. Here, magnetization values gradually decrease with increasing crosslinker mole fraction α[‡]. This decrease is attributed to the gradual immobilization of FFAM inside the hydrogels: FFAM are inhibited in their Brownian

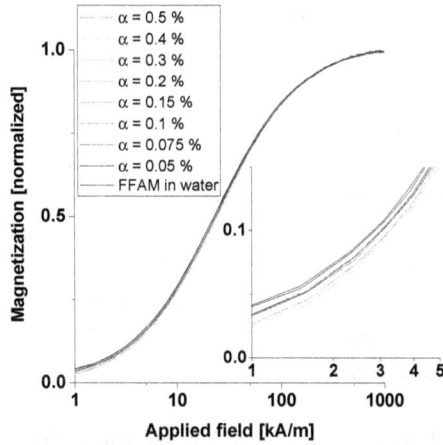

Fig. 7.7.: Normalized virgin magnetization curves of FFAM immobilized in PAAm hydrogels ($c = 0.3\,\text{mg(Fe)/mL}$) for various cross linker mole fractions α and for FFAM dispersed in water for reference. Inset shows a zoom for low fields. Adapted from [264].

(physical) rotation with increasing α, which at low fields significantly contributes to the full alignment of the FFAM magnetic moments in the direction of the applied field [338][‡]. In other words, at low fields, the intrinsic Néel relaxation of FFAM is dominated by anisotropy, and the FFAM magnetic moment mainly aligns with the applied field via Brownian rotation[‡]. Therefore, FFAM dispersed in water show a higher initial magnetization at low fields compared to MNP immobilized in PAAm FHG.

7.2. Effect of Particle Immobilization on Magnetic Particles Heating

The two types of ferrohydrogels (FHG) characterized in Section 7.1 — LM-agarose and poly(acrylamide) (PAAm) — are employed as in situ model systems to study the effects of immobilization of FFAM particles on magnetic particle heating in this section. These FHG possess tunable mechanical properties and mesh sizes of $\sim (10-100)$ nm matching the size of FFAM particles (cf. Section 4.8 for FFAM properties and Section 7.1.2 for hydrogel properties), allowing to gradually immobilized the FFAM particles. Section 7.2.1 covers the experimental details of the heating experiments, while the characteristic effects of gradual immobilization on particle heating are discussed in Section 7.2.2.

7.2.1. Experimental Procedure for Particle Heating Measurements

LM-agarose and PAAm FHG were prepared as described in 2.6 in 4 mL sample vials with a FHG sample volume of $V = 1$ mL. Samples were prepared in triplicate and measured in the Hüttinger setup (s. Section 5.2.1) at $H_0 = 50$ mT$/\mu_0$ and $f = 270$ kHz. These field parameters were chosen to maximize the expected SLP vaulue, while staying in the H_0^2-dependent regime of SLP for FFAM, i.e. keeping $H_0 < H_K/4 \approx 54$ mT$/\mu_0$ (cf. Section 5.2.2, Fig. 5.7). By maximizing the SLP value, the effects of immobilization on particle heating are expected to be most pronounced. From the recorded $T(t)$-curve the mean SLP value was calculated as the average of the triplicate measurement as described in Section 5.2.1, assuming the specific heat capacity of water $c_h = 4.19$ J$/(g \cdot K)$. Such an assumption is supported for LM-agarose by thermal conductivity measurements in agar (mass fraction $\mu_{agar} = 1$ %), reporting in a thermal conductivity of $C_h = 0.55$ W$/(m \cdot K)$ at $T = 15$ °C [339], which is very close to that of water at $T = 15$ °C with $C_h = 0.57$ W$/(m \cdot K)$ [340]). For PAAm hydrogels, the thermal properties are generally assumed to be effectively equal to that of water [341].

Reference LM-agarose and PAAm samples without FFAM were prepared as well for background subtraction.

7.2.2. Effects of Immobilization on Particle Heating

The mean mesh sizes of the PAAm FHG, d_{cross}, were derived from rheological measurements and the mean mesh sizes of LM-agarose FHG, d_{approx}, were estimated from literature data, as described in Section 7.1. The SLP values of gradually immobilized FFAM versus the mean mesh sizes are shown for both hydrogels in Fig. 7.8. For comparison, the SLP values are shown normalized to the value measured in water, SLP$= (343 \pm 7)$ W/g(Fe). For both FHG, the SLP values obviously decrease with decreasing mesh size. More specifically, the SLP values drop significantly for mesh sizes below the mean hydrodynamic particle size, i.e. $d_{mesh} \leq d_H = 18.9$ nm, allowing the conclusion that the degree of immobilization determines the decrease in

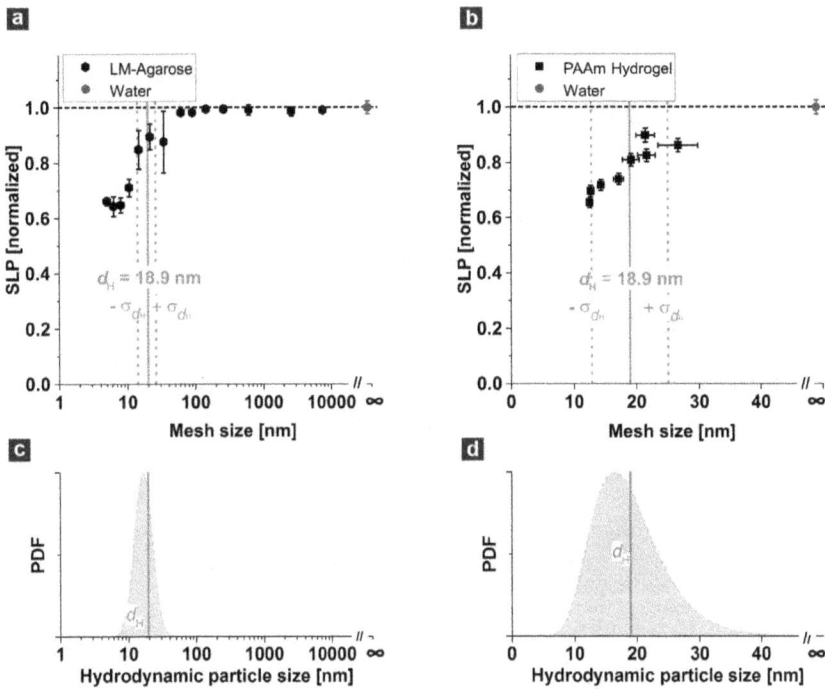

Fig. 7.8.: Normalized SLP values measured for FFAM in (a) LM-agarose and (b) PAAm hydrogels in dependence of the mesh size. Reference SLP for FFAM in water shown in blue. The mean hydrodynamic particle size, d_H, is marked by a solid red line, together with the region of one standard devivaiton ($\pm\sigma_{d_H}$), marked by dashed red lines. The PDF of the log-normal distributed hydrodynamic particle size of pristine FFAM is shown for comparision in (c) and (d) (cf. Fig. 7.5; PDF parameters taken from Table A.12). (b) and (d) adapted from [264].

particle heating efficiency. In the following, the implications of particle immobilization in each hydrogel for particle heating are analyzed individually in detail.

From Fig. 7.8(a,c), it can be deduced that in the case of FFAM immobilization in LM-agarose hydrogels the SLP value is not affected for $d_{approx} \geq 60$ nm (cf. Fig. 7.8a). In other words, the microenvironment of FFAM inside LM-agarose hydrogels with $d_{approx} \geq 60$ nm seemingly does not differ from that of water, when FFAM are probed for their heating efficiency. Accordingly, FFAM has no particles with $d_H > 60$ nm (cf. Fig. 7.8c) For $d_H \geq d_{mesh}$ the onset of physical FFAM immobilization starts, which is responsible for the reduced SLP values. The SLP values drop significantly at $d_{approx} \approx d_H = 18.9$ nm, and saturate at a drop of approx. 35 % compared to the SLP value measured for freely dispersed FFAM in water. This saturation occurs at mesh sizes well below 10 nm. In such narrow meshes, FFAM of all hydrodynamic particle sizes can be expected to be fully immobilized physically.

Please note that two limitations must be kept in mind for the interpretation of the results for LM-agarose: First, the mesh sizes used for Fig. 7.8a are estimations derived from literature (cf. Section 7.1). Consequently, the values may only be interpreted as trends rather than exact numbers. Second, there is evidence reported by Rojas et al. that MNP mixed in agarose hydrogels may form clustered agglomerates [342]. Even though a temperature-induced FFAM agglomeration during LM-agarose sample preparation could be excluded from DLS measurements (s. Section 7.1.3), a formation of FFAM agglomerates in LM-agarose FHG cannot fully be excluded in the present study. This is different for PAAm hydrogels, as TEM (s. Section 7.1.4) and VSM analysis (s. Section 7.1.5) both indicate that there are no FFAM agglomerates inside PAAm FHG. Therefore, PPAm FHG are expected to probe the effects of FFAM immobilization in isolation without superimposed FFAM agglomeration effects:

For PAAm hydrogels, the mesh sizes are precisely determined in the range of $d_{cross} = (12.6 - 26.7)$ nm, corresponding very well to the range of mean hydrodynamic particle sizes within one standard deviation, $d_H = (18.9 \pm 6.1)$ nm. Indeed, the SLP values decrease by approx. 15 % for $d_{cross} = (26.7 \pm 3.2)$ nm to approx. 35 % for $d_{cross} = (12.6 \pm 0.1)$ nm (s. Fig. 7.8b). A maximum of approx. 86 % of the FFAM are expected to be immobilized in the FHG with the smallest mesh size of $d_{cross} = (12.6 \pm 0.1)$ nm, ascribed to the hydrodynamic sizes being higher than the mesh size (estimated from the PDF fit, cf. Fig. 7.8d)[‡]. Therefore, the drop in SLP value is attributed to a strong inhibition of Brownian relaxation inside the hydrogels with smallest mesh sizes.

Interestingly, FFAM immobilized in LM-agarose and PPAm hydrogels show the same maximum decrease in SLP by approx. 35 % for FFAM immobilized in the LM-agarose with lowest mesh sizes (cf. Figs. 7.8(a,b)). This observation hints to the same effect causing the reduction of particle heating, attributed to full physical immobilization of the FFAM. This effect will be analyzed in more detail in the following.

Indeed, a reduction in SLP upon particle immobilization has been be attributed to the inhibition of the Brownian relaxation (rotation) mechanism in literature: A decrease in SLP values by up to approx. 50 % was reported by Ludwig et al. for iron oxide MNP with a particle core size of $d_C = (12 - 15)$ nm, totally immobilized in 10 % polyvinyl alcohol [343]. Furthermore, AC-susceptibility measurements conducted by Soukup et al. confirm such inhibition of Brownian rotation for intracellularly immobilized iron oxide MNP $(d_C = 12 \, nm)$ [305]. Moreover, the SLP value of FFAM (from a different batch but otherwise equally snythesized as the ones used in this thesis) was demonstrated to experience a decrease by up to 70 % upon FFAM internalization inside human adenocarcinoma SKOV-3 cells [245]. As will be discussed in Section 7.4, the combined effects of MNP agglomeration and immobilization — as occurring for intracellular MNP — overall decrease the SLP values even more than expected for the sole inhibition of Brownian relaxation. In the case of PAAm FHG, such agglomeration effects can be neglected as discussed above. Consequently, the reduction in SLP values for FFAM immobilized in PAAm is expected to arise only from inhibition of Brownian rotation.

Assuming an inhibition of Brownain rotation, one can attempt to estimate the contributions of Néel relaxation (independent of immobilization) and Brownian relaxation (immobilization-dependent; inhibited) mechanisms to the overall particle heating with the respecitve relaxation times. Due to the relatively strong field amplitude used for the MFH experiments ($H_0 = 50\,\text{mT}/\mu_0$), the AC field-dependence of Néel and Brownian relaxation times must be taken into account [344]. Neglecting particle interactions, Yoshida and Enpuku derived the following expression from the Fokker-Planck equation the for the AC-field-dependent Brownian relaxation time [345][‡]:

$$\tau_B(H_0) = \frac{\tau_B}{\sqrt{1 + 0.21 \cdot \xi(H_0)^2}}, \tag{7.6}$$

with the zero-field Brownian relaxation time, τ_B (cf. eq. (2.31)), and the reduced field parameter, $\xi(H_0)$ (cf. eq. (2.65)). Eq. (7.6) holds for $\xi(H_0) \leq 20$, which is valid for FFAM with $\xi = 3.1$ (using values for FFAM; cf. Table 4.11). Similarly, the AC-field dependent Néel relaxation time can be estimated for non-interacting particles using [14]:

$$\tau_N(H_0) = \tau_0 \cdot \exp\left(\frac{K_{\text{eff}} \cdot V_M}{k_B \cdot T} \cdot \left(1 - \frac{H_0}{H_K}\right)^2\right), \tag{7.7}$$

with the Néel relaxation time constant $\tau_0 = 1 \cdot 10^{-9}\,\text{s}$ [346], the effective anisotropy constant, K_{eff}, the particle magnetic volume, $V_M = \pi/6 \cdot d_M^3$, and the anisotropy field, H_K (cf. eq. (2.15)). Using eqs. (7.6) and (7.7) and applying the properties of FFAM derived from sample characterization (cf. Table 4.11) as well as setting $T = 313\,\text{K}$ (40 °C; the average saturation temperature achieved during particle heating; cf. Appendix A.5.5) and the viscosity of water $\eta(T = 40\,°\text{C}) = 6.53 \cdot 10^{-4}\,\text{Pa} \cdot \text{s}$, a mean Brownian relaxation time of $\tau_B(H_0 = 40\,\text{kA/m}) = 935.0\,\text{ns}$ and a mean Néel relaxation time of $\tau_N(H_0 = 40\,\text{kA/m}) = 47.9\,\text{ns}$ is calculated[‡]. From this, the effective relaxation time is calculated with eq. (2.70), yielding $\tau_R(H_0 = 40\,\text{kA/m}) = 45.5\,\text{ns}$. From this, FFAM are expected to relax predominantly via Néel relaxation.

As indicated in Section 3.3.1 and comfirmed by literature [42, 163], the qualitative trends of particle heating can be predicted for high anisotropy particles, $K_{\text{eff}} \geq 30\,\text{kJ/m}^3$, with core sizes below $d_C \approx 10\,\text{nm}$ using the linear response theory (LRT). Therefore, using LRT for FFAM, the qualitative dependency of SLP on the particle relaxation time, τ, is fully described by the imaginary component of the complex AC-susceptibility (cf. Section 2.4, eq. (2.69)):

$$SLP(f, \tau) \propto \chi''(f, \tau) = \chi_0 \cdot \frac{2\pi f \cdot \tau}{1 + (2\pi f \cdot \tau)^2}, \tag{7.8}$$

with the initial susceptibility, $\chi_0 = \frac{\mu_0 M_S^2 V_M}{3 k_B T}$, and the frequency of the applied AMF. According to eq. (7.8), the SLP is maximized for the resonance condition $2\pi \cdot f^* \cdot \tau = 1$, with the resonance frequencies $f^*(\tau_R = 45.5\,\text{ns}) \approx 3\,500\,\text{kHz}$, $f^*(\tau_N = 47.9\,\text{ns}) \approx 3\,320\,\text{kHz}$ and $f^*(\tau_B = 935.0\,\text{ns}) \approx 170\,\text{kHz}$ for the case of FFAM. In order to estimate the effective contribution of Néel and Brownian relaxation on SLP for smaller and bigger particles within the size distribution of FFAM, $\chi''(f, \tau)$ is calculated for five exemplary magnetic particle sizes: d_M, $d_M - \sigma_{d_M}$, $d_M - 2\sigma_{d_M}$, $d_M + \sigma_{d_M}$ and $d_M + 2\sigma_{d_M}$. For this, a constant coating thickness of 4.25 nm is assumed for determining the hydrodynamic size, resulting in $d_H = 2 \cdot 4.25\,\text{nm} + d_M$. The coating thickness is deduced from the mean magnetic and hydrodynamic sizes (cf. Table

4.11)[‡]. Furthermore, a size-dependent effective anisotropy constant, $K(d_M)$, is assumed as described by eq. (2.19). With a bulk anisotropy value of $K_B = 11 \, \text{kJ/m}^3$ for magnetite [20], and $K_{\text{eff,FFAM}} = 48 \, \text{kJ/m}^3$ derived from ZFC-FC measurements (cf. Table 4.11), the surface anisotropy constant is calculated from eq. (2.19) as $K_S = 64.3 \, \mu\text{J/m}^2$. Using this K_S in eq. (2.19), $K(d_M)$ is estimated for the above-mentioned exemplary magnetic particle sizes. The FFAM properties and respective relaxation times used for the estimation of $\chi''(f, \tau)$ are listed in Table[‡] 7.3. Even though the values in Table 7.3 are estimates, they allow for the estimat-

Table 7.3.: FFAM properties used for the estimation of $\chi''(f, \tau)$: Magnetic particle size, d_M, hydrodynamic particle size, d_H, effective anisotropy constant, K_{eff}, Néel, $\tau_N(H_0)$, Brownian, $\tau_B(H_0)$ and effective relaxation time, $\tau_R(H_0)$.

Size	d_M [nm]	d_H [nm]	K_{eff} [kJ/m³]	$\tau_N(H_0)$ [ns]	$\tau_B(H_0)$ [ns]	$\tau_R(H_0)$ [ns]
d_M	10.4	18.9	48	47.9	935.0	45.5
$d_M - \sigma_{d_M}$	6.9	15.4	67	6.4	803.3	6.3
$d_M - 2\sigma_{d_M}$	3.4	11.9	124	1.6	400.0	1.6
$d_M + \sigma_{d_M}$	13.9	22.4	39	601.0	768.6	337.3
$d_M + 2\sigma_{d_M}$	17.4	25.9	33	11 296.2	625.2	592.4

ing of $\chi''(f)$ for different particles sizes, which directly translates to the SLP values (cf. eq. (7.8)). For the following interpretation, it is assumed that $\chi''(\tau_R)$ holds for FFAM dispersed in water (allowing both Néel and Brownian relaxation) and $\chi''(\tau_N)$ for FFAM fully immobilized (allowing only Néel relaxation)[‡]. Fig. 7.9c shows that for the AMF frequency of $f = 270 \, \text{kHz}$ and the mean magnetic particles size, d_M, the curves of $\chi''(\tau_R)$ and $\chi''(\tau_N)$ nearly overlap and therefore not much change in SLP is expected upon particle immobilization as the FFAM relaxation is Néel dominated[‡] (i. e. to estimate the change in particle heating expected for inhibited Brownian relaxation upon FFAM immobilization, the difference between $\chi''(\tau_R)$ and $\chi''(\tau_N)$ is compared). Generally, the resonance condition is not met for $f = 270 \, \text{kHz}$, wherefore the general contribution of these particles to the overall particle heating is expected to be low. For smaller-than-average particles (particle sizes $d_M - 2\sigma_{d_M}$ and $d_M - \sigma_{d_M}$; Fig. 7.9(a,b), respectively), $\chi''(\tau_R)$ and $\chi''(\tau_N)$ coincide and the resonance peak shifts to much higher frequencies, even farther away from resonance with the applied frequency of $f = 270 \, \text{kHz}$. For such small particles almost no contribution to the overall SLP is expected; these particles are fully Néel dominated and their immobilization does not have any effect on the SLP[‡].

For larger particles within the size distribution (particle sizes $d_M + \sigma_{d_M}$ and $d_M + 2\sigma_{d_M}$; Fig. 7.9(d,e)), two things are observed: First, for higher particle sizes than d_M, the peak in $\chi''(\tau_N)$ shifts rapidly to small frequencies, while $\chi''(\tau_N)$ shifts much closer to $\chi''(\tau_R)$, indicating an increase of Brownian contributions to particle heating[‡]. In fact, the particles with $d_M + \sigma_{d_M}$ show almost equal contributions from both relaxation mechanisms as $\tau_N \approx \tau_B$ and their $\chi''(\tau_N)$ and $\chi''(\tau_B)$ almost overlap (cf. Fig. 7.9d). Second, the resonance condition for $\chi''(f, \tau_R)$ is met (cf. esp. Fig. 7.9e) and the overall SLP is substantially increased by predominantly Brownian contributions of these larger particles[‡]. If Brownian relaxation contributions are inhibited by particle immobilization in hydrogels, the contribution of these particles to particle heating is strongly reduced[‡]. Even if the absolute number of these larger particles is low in the FFAM suspension (cf. Fig. 7.9f), their Brownian relaxation contributions to the overall SLP value are substantial[‡].

This fits to the observation that for the strongest immobilization states in both LM-agarose and PPAm FHG (i. e. the smallest mesh sizes), the inhibition of Brownian relaxation can cause a decrease of roughly one third in the FFAM heating efficiency (cf. Fig. 7.8). In other words, it may be concluded from these results that Brownian relaxation contributes about on third (approx. 35 %) to the FFAM heating.

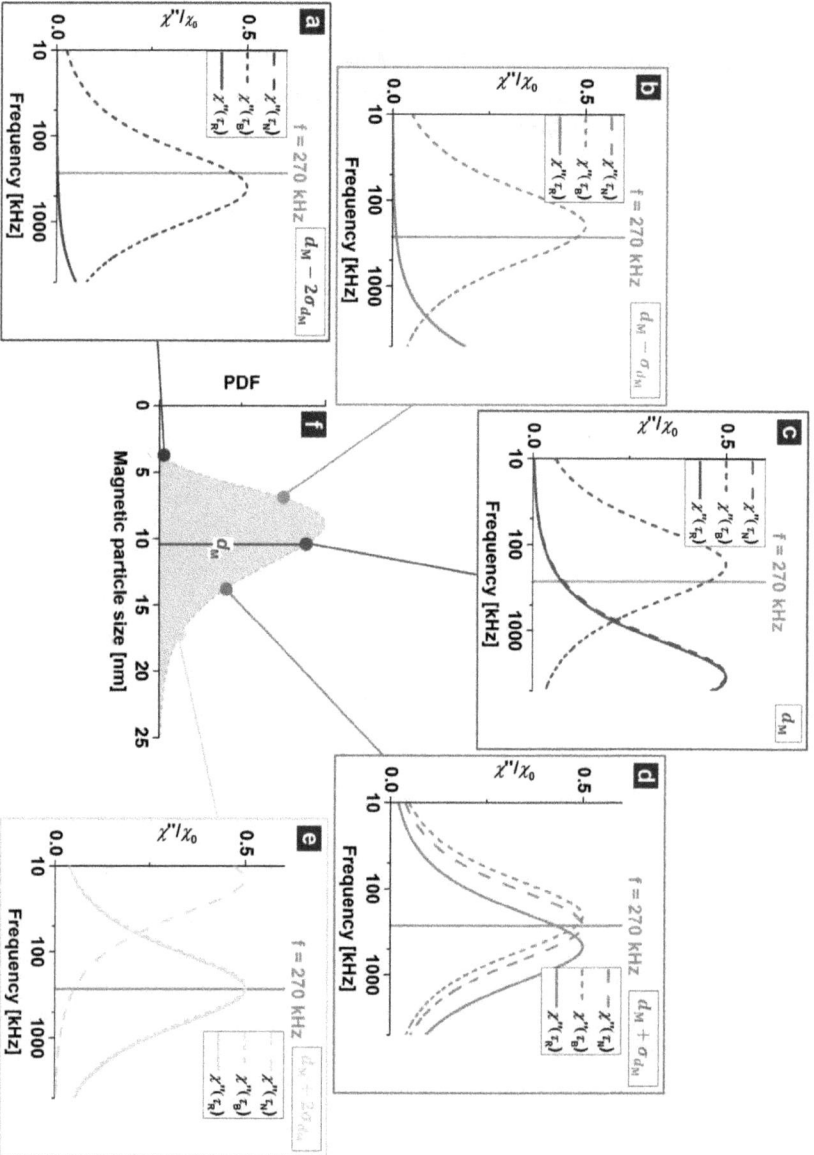

Fig. 7.9.: Imaginary part of the complex AC susceptibility, $\chi''(f)$, cf. eq. (7.8) versus frequency for different Néel (τ_N), Brownian (τ_B) and effective (τ_R) relaxation contributions as well as for five different FFAM particle sizes within the size distribution: (a) particles with $d_M - 2\sigma_{d_M}$, (b) particles with $d_M - \sigma_{d_M}$, (c) particles with mean particle size d_M, (d) particles with $d_M + \sigma_{d_M}$, (e) particles with $d_M + 2\sigma_{d_M}$. (f) shows the PDF of the magnetic size distribution with the mean magnetic particles size, d_M for comparison. The AMF frequency, $f = 270\,\text{kHz}$, used for hyperthermia experiments is marked in red.

7.3. Control and Characterization of Agglomerated Magnetic Particles

Agglomeration of MNP occurs when attractive interparticle forces outbalance repulsive ones, e. g. when van-der-Waals or magnetic attractive forces dominate repulsive electrostatic or steric forces [14, 23]. Genreally, MNP agglomeration is prevented during synthesis by coating MNP with a protective surface coating, stabilizing the MNP in solution [182]. However, upon subjection to physiological media, e. g. for in vivo applications, the ions in the medium attach to the surface of the MNP and change the electrostatic repulsive forces that govern the MNP colloidal stability in solution, potentially leading to MNP agglomeration [99]. Depending on the application, MNP agglomeration could be detrimental or favorable: For example, agglomeration must be prevented in magnetic targeting inside vessels (in order to prevent embolism) [88], while the agglomeration of MNP could enhance particle heating for MFH treatment once MNP have reached the tumor site [287, 347]. However, the effects of agglomerated MNP on particle heating are still under ongoing investigation and it is yet unclear under which conditions MNP agglomeration increases or decreases particle heating effects [12, 27, 276].

In this section, two exemplary agglomerated particles types are studied: One type consists of FFAM particles agglomerated in special unilamellar liposome vesicles, henceforth referred to as *Liposomes Ultra Magnétique* (LUM)[2] throughout this thesis. A second type consists of FFLA particles, with damaged lauric acid protective coating by addition of specific amounts of NaCl, promoting the formation of agglomeration states. The sample preparation for both types is described in Section 7.3.1, followed by the sample characterization with dynamic light scattering (DLS) in Section 7.3.2, transmission electron microscopy (TEM) in Section 7.3.3 and SQUID magnetometry in Section 7.3.4. Finally, Section 7.3.5 summarized the sample properties derived from characterization. These particle types are subsequently used to study and discuss the effects of MNP agglomeration on particle heating in Section 7.4.

7.3.1. Sample Preparation

LUM Sample

LUM particles were provided by the PHENIX Laboratories from UPMC Paris (Paris, France). The preparation steps are described in [195]. These magnetoliposomes consist of a unilamellar liposome shell that contains up to $\nu_{MNP} = 20\,\%$ (volume fraction) of FFAM particles as reported in [195, 245].

LUM were kept in a buffer solution containing $0.11\,M$ sodium chloride, $0.02\,M$ sodium citrate and $0.01\,M$ 2-[4-(2-hydroxyethyl)piperazin-1-yl]ethanesulfonic acid (HEPES), in order to avoid osmotic stress, potentially damaging the liposomes. The iron concentration was $c = 0.3\,mg(Fe)/mL$ for all LUM samples for the following sample characterizations.

---

[2]French naming is used to acknowledge that these particles were provided by the PHENIX Laboratories from UMPC Paris (Paris, France); s. Section 7.3.1.

FFLA Agglomerates

In contrast to LUM, which are basically prepared as agglomerated MNP in a controlled way by encapsulating FFAM particles and thereby inducing dipole-dipole interparticle interactions due to spacial confinement, the FFLA agglomerates are prepared with an arbitrary approach: The lauric acid (LA) shell of FFLA particles can be damaged by addition of sodium chloride (NaCl), increasing attractive interparticle interactions and promoting MNP agglomeration [348]. The addition of NaCl to a FFLA dispersion increases the ionic strength, inducing a decreased solubility for LA and causing the precipitation of LA due to antisolven crystallization [349]. By controlling the concentration of NaCl in MNP solution, the agglomeration state can be controlled.

FFLA particles were taken from the batch described and characterized in Chapter 4 (s. Table 4.11 for a summery of FFLA properties). Sodium chloride (NaCl; analytical grade, Merck KGaA, Darmstadt, Germany) was added to the FFLA suspension obtaining concentrations of $0.0\,M$ NaCl (pristine FFLA), $0.1\,M$ NaCl, $0.25\,M$ NaCl and $0.5\,M$ NaCl, in the final FFLA solution$^\diamond$. The iron concentration was $c = 0.3\,mg(Fe)/mL$ for all samples. FFLA samples treated with NaCl were used as-prepared for the respective measurements (DLS, TEM, SQUID or MFH) the same day they were prepared to assure comparability among the characterization results.

LA crystallization was analyzed visually $1\,h$ after adding NaCl to the FFLA solution: The addition of NaCl cannot be distinguished with the unaided eye for samples with $0.1\,M$ NaCl and $0.25\,M$ NaCl, but for $0.5\,M$ NaCl an increased turbidity was observed, associated with a macroscopic precipitation of LA, as shown in Fig. 7.10. Furthermore, during sample preparation a gel-like consistence due to the formation of a precipitated LA-matrix was noticeable.

Fig. 7.10.: Photography of FFLA samples $1\,h$ after adding (a) $0.0\,M$ NaCl, (b) $0.1\,M$ NaCl, (c) $0.25\,M$ NaCl and (d) $0.5\,M$ NaCl: (a,b,c) show clear transparency, where the bottom of the vials is clearly visible (indicated by light blue arrows), whereas (d) shows a substantial increase in turbidity, which is attributed to macroscopic LA crystallization. (e) shows this crystallization for a lower MNP concentration $c = 0.15\,mg(Fe)/mL$, visible as a precipitated LA matrix in the middle of the sample vial (indicated by yellow arrow). (a)-(d) adapted from [329].

7.3.2. Dynamic Light Scattering Characterization

DLS characterization of all samples was carried out as described in Section 4.4. LUM were diluted in buffer solution to preserve LUM integrity (s. Section 7.3.1). The refractive index was changed accordingly from $n_{H2O}(T = 20\,°C) = 1.330$ to $n_{LUM\text{-}Buffer}(T = 20\,°C) = 1.334$ for the

DLS analysis according to [350] (however, this did not show any significant effect on the DLS results).

Intensity-weighted DLS measurements are chosen for DLS analysis as these are most sensitive to large particles and particle agglomerates (s. discussion in Section 4.4).

LUM Sample

The results from DLS measurements are shown in Fig. 7.11 and listed in Table 7.4, while log-normal PDF fitting parameters are found in Appendix A.5.7, Table A.13. In the intensity-

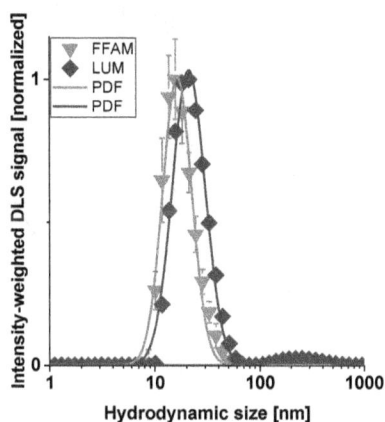

Fig. 7.11.: Hydrodynamic size distribution of LUM particles from intensity-weighted DLS measurements. Data was fitted with the PDF of the log-normal distribution (eq. (3.22)). FFAM DLS data is shown for reference (cf. Fig. 4.9(a)), which nearly coincides with DLS data for LUM, however, LUM also display a second peak at $d_H \approx 325$ nm.

Table 7.4.: Mean hydrodynamic size d_H of LUM particles with fitting quality parameter R^2_{adj} from DLS analysis.

Sample	R^2_{adj}	d_H [nm]
LUM peak 1	0.988	25.4 ± 9.4
LUM peak 2	0.988	326.6 ± 156.0

weighted DLS signal, two peaks are visible for LUM (Fig. 7.11): One distinct peak at $d_H \approx$ 25 nm corresponds to monodisperse FFAM (cf. Fig. 4.9a), while a second, weaker peak at $d_H \approx 325$ nm corresponds to LUM liposomes. This observation of a bimodal distribution of hydrodynamic sizes thus indicates that most FFAM are not clustered inside liposomes. These results will be further interpreted in relation to the results from TEM analysis in Section 7.3.3.

FFLA Agglomerates

The results from DLS measurements are shown in Fig. 7.12 and listed in Table 7.5, while log-normal PDF fitting parameters are found in Appendix A.5.7, Table A.14. For FFLA treated

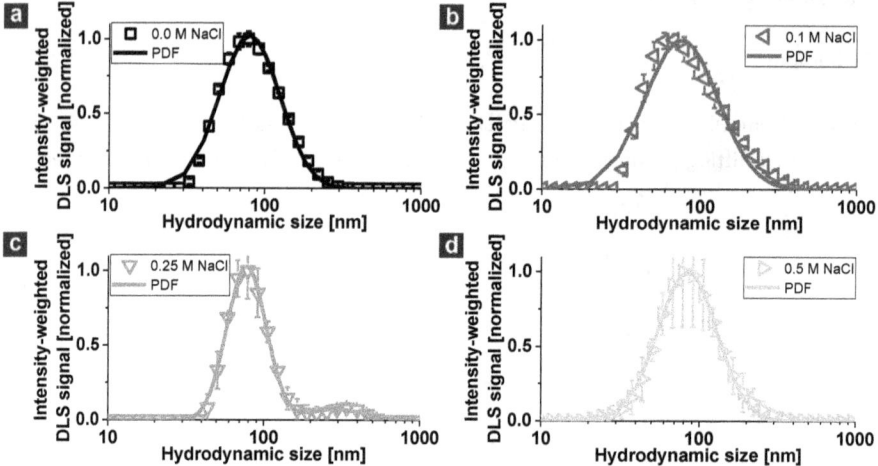

Fig. 7.12.: Hydrodynamic size distribution from intensity-weighted DLS measurements for (a) pristine FFLA, (b) FFLA treated with 0.1 M NaCl, (c) FFLA treated with 0.25 M NaCl and (d) FFLA treated with 0.1 M NaCl. Data was fitted with the PDF of the log-normal distribution (eq. (3.22)). Adapted from [329].

with 0.1 M NaCl, a tendency to particle agglomeration is indicated by the increase in the hydrodynamic size distribution width by 13.0 % (s. Table A.14) and the shoulder forming in the hydrodynamic size distribution for $d_H > 200$ nm, when compared to pristine FFLA (s. Fig. 7.12(a,b)). For FFLA treated with 0.25 M NaCl, a bimodal hydrodynamic size distribution is

Table 7.5.: Mean hydrodynamic size d_H and fitting quality parameter R^2_{adj} from DLS size analysis for NaCl-treated FFLA samples.

NaCl-treatment of FFLA sample	R^2_{adj}	d_H [nm]
0.0 M	0.985	110.8 ± 54.8
0.1 M	0.969	115.7 ± 65.8
0.25 M peak 1	0.985	108.7 ± 31.9
0.25 M peak 2	0.985	386 ± 173
0.5 M	0.997	110.8 ± 49.6

observed, which is attributed to the formation of FFLA agglomerates (s. Fig. 7.12c). The mean hydrodynamic size of the first peak, $d_H \approx 109$ nm, corresponds to pristine FFLA with $d_H \approx 111$ nm, while the second peak indicates FFLA agglomerates with $d_H \approx 385$ nm (s. Fig. 7.12c). In contrast to that, the hydrodynamic size distribution of FFLA treated with 0.5 M NaCl is similar to that of pristine FFLA (s. Fig. 7.12(a,d)). The only difference is the higher uncertainty for the measured sizes for FFLA treated with 0.5 M NaCl. This could be explained by the sedimentation of agglomerated FFLA entangled and immobilized in the precipitated LA-matrix to the bottom of the sample cuvette during DLS measurement. In this way, sedimented agglomerated FFLA will not contribute to the DLS measurement and only freely dispersed FFLA are assessed, resulting in the similarity of the DLS data to that for pristine FFLA.

7.3.3. Transmission Electron Microscopy Characterization

TEM characterization of all samples was carried out as described in Section 4.3 . The iron concentration was $c = 0.3\,\text{mg(Fe)}/\text{mL}$ for each sample.

LUM Sample

Exemplary TEM images of LUM are shown together with the respective cumulative size distribution for LUM in Fig. 7.13. The liposomes are visible in TEM images as black spheres (s. Fig. 7.13(a,b)). TEM images with higher magnification confirm that LUM consist of smaller FFAM particles trapped inside the liposomal vesicles (s. Fig. 7.13b). The mean LUM size could be determined from N= 203 undamaged LUM and was fitted with the CDF of the log-normal distribution (Fig. 7.13c; eq. (4.5)). The results from size analysis are listed in Table 7.6 and the CDF fitting parameters are found in Appendix A.5.8, Table A.15. The mean LUM liposome

Table 7.6.: Mean agglomerate size d_C of LUM particles with fitting quality parameter R^2_{adj} derived from TEM analysis (s. Fig. 7.13c).

Sample	R^2_{adj}	d_C [nm]
LUM	0.979	250.5 ± 71.8

size from TEM analysis, $d_C \approx 250\,\text{nm}$, is in line with observations from previous studies on equally synthesized LUM from a different batch reporting $d_C \approx 260\,\text{nm}$ [195]. Moreover, d_C is approximately 75 nm smaller than the mean hydrodynamic particle size, $d_H \approx 325\,\text{nm}$ (cf. Table 7.5). Assuming a liposome bilayer thickness of $d_{\text{bilayer}} \approx 5\,\text{nm}$ at ambient conditions and in liquid [351], this leaves an approximate size difference of $\Delta d = (325 - 250 - 2 \cdot 5)\,\text{nm} = 65\,\text{nm}$ between the liquid (DLS) and dried (TEM) LUM size. This could indicate that the liposomal vesicles swell in the buffer solution, allowing some individual movement of the entrapped FFAM particles (s. discussion in Section 7.4.2).

FFLA Agglomerates

Exemplary TEM images for each NaCl-treated FFLA sample and for pristine FFLA in Fig. 7.14. The increasing crystallization of LA upon adding NaCl is captured in TEM images as gray shades (cf. Fig. 7.14(d,g,j)), indicating the formation of LA filament structures for 0.25 M NaCl. These LA filament structures entangle in solid, web-like structures for 0.5 M NaCl, presumably assembling the macroscopic LA-matrices observed from visual assessment (cf. 7.10(d,e)). For the LA reference sample not treated with NaCl (cf. Fig. 7.14a), no such gray shades are observed, corroborating the assumption that the otherwise observed structures are crystallized LA. Untreated FFLA appear monodisperse, with only isolated, minor agglomerates, presumably

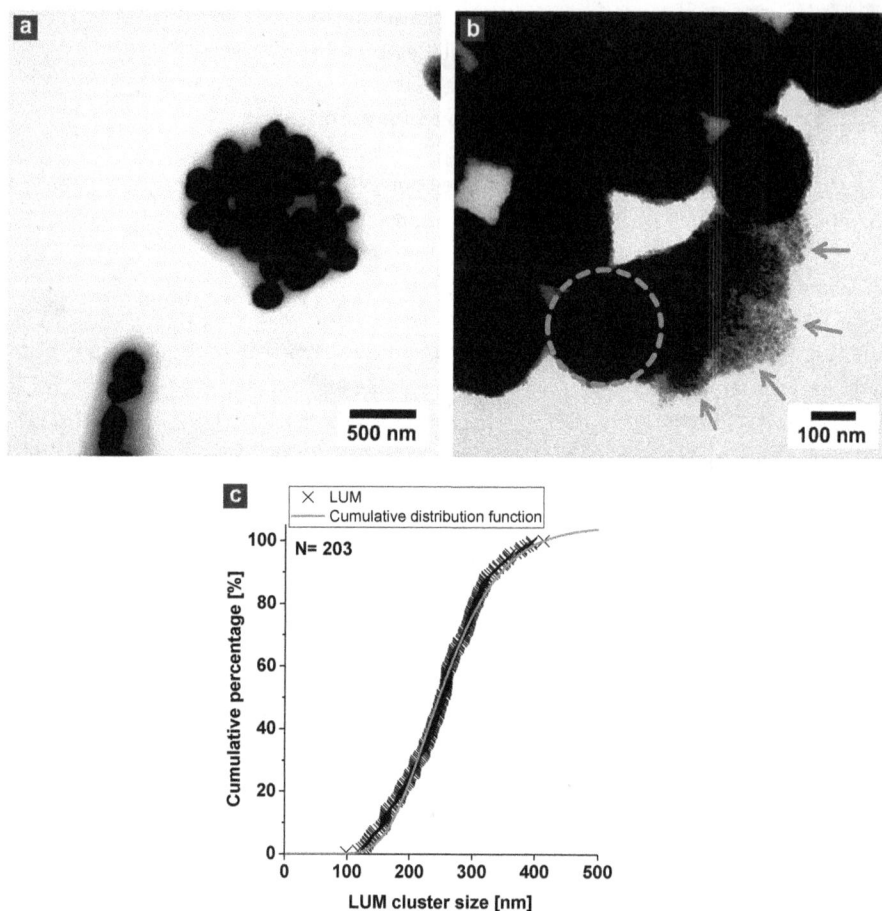

Fig. 7.13.: Exemplary TEM images of LUM with magnifications of (a) $27\,800\times$ and (b) $100\,000\times$. (b) shows a damaged LUM liposome shell (red dashed-line circle); individual FFAM particles are leaking out, as indicated by red arrows. (c) plots the LUM size distribution from TEM image analysis, fitted with the CDF of the log-normal distribution in red.

arising due to sample drying during sample preparation (cf. Fig. 7.14(b,c)). For $0.1\,\mathrm{M}$ NaCl, FFLA agglomerates are observed, which are generally well dispersed and separated from each other (cf. Fig. 7.14(e,f)). Some of these agglomerates form around NaCl crystals, indicating that NaCl crystallization might assist the formation of FFLA agglomerates. Furthermore, FFLA agglomerates are observed clustering along LA-filaments for $0.25\,\mathrm{M}$ NaCl (cf. Fig. 7.14(h,i)). For $0.5\,\mathrm{M}$ NaCl, the FFLA agglomerates appear fully entangled in LA web-like matrix structures, partially packed so densely that they appear black in TEM (i. e. so tightly packed or stacked that they completely block the transmission of the electron beam; cf. Fig. 7.14(k,l)).

From a multitude of TEM images, the size distribution of agglomerated FFLA is estimated as shown in Fig. 7.15. The size and size distribution of pristine FFLA correspond to the data

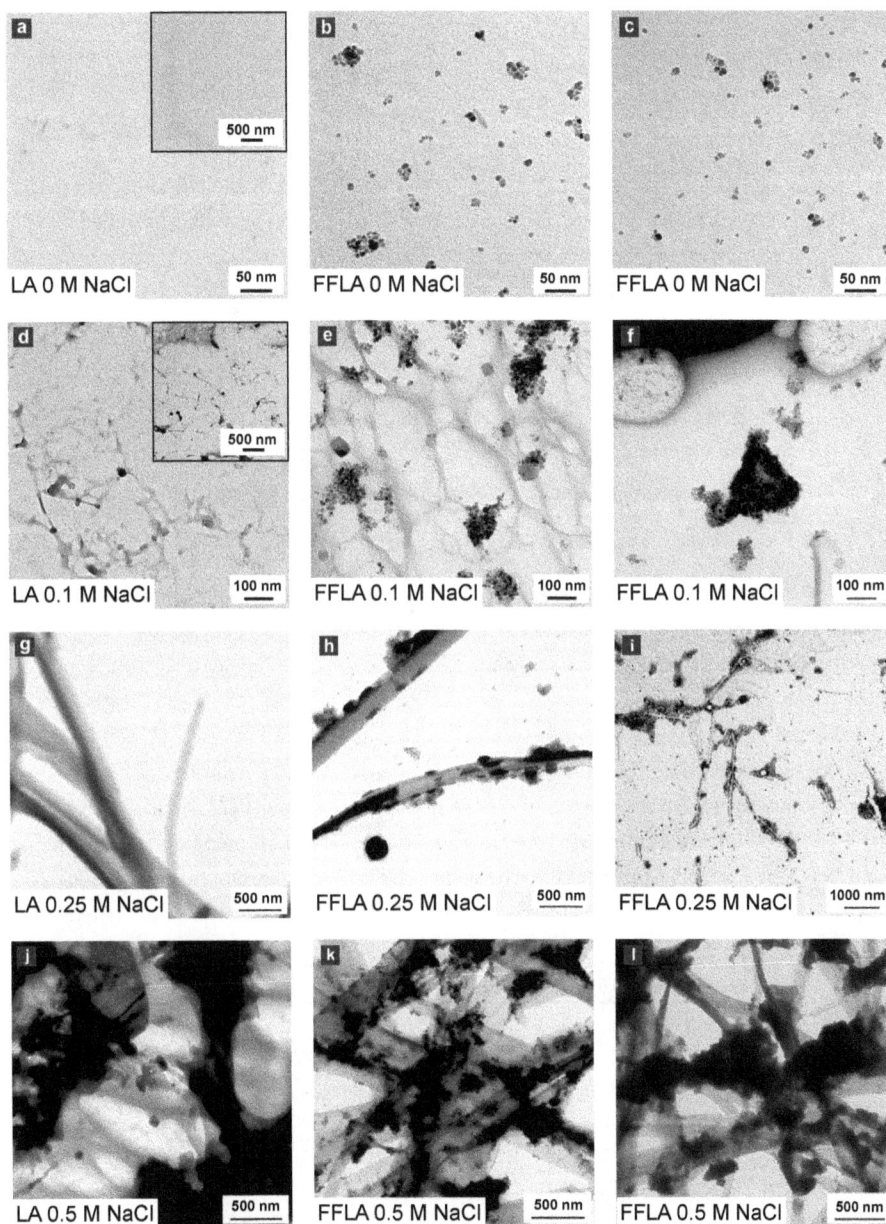

Fig. 7.14.: Exemplary TEM images for samples treated with (a,b,c) 0.0 M NaCl, (d,e,f) 0.1 M NaCl, (g,h,i) 0.25 M NaCl and (j,k,l) 0.5 M NaCl. (a,d,g,j) depict lauric acid (LA) reference samples not containing FFLA: Here the crystallization of LA upon adding NaCl is visible as dark gray shades in the TEM images, forming filaments (g), which finally entangle in web-like structures (j). Insets in (a,d) for comparison. Resolutions: (a,b,c) 200 000×, (d,e,f) 100 000×, (g,h,j,k,l) as well as insets (a,d) 27,800× and (i) 10 000×. Adapted from [328, 329].

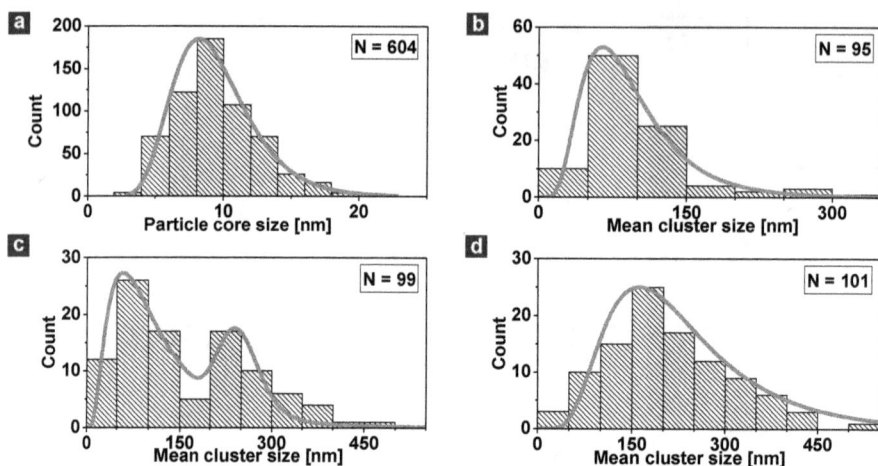

Fig. 7.15.: FFLA agglomerate size distribution estimated from TEM images (cf. Fig. 7.14) for samples treated with (a) 0.0 M NaCl, (b) 0.1 M NaCl, (c) 0.25 M NaCl and (d) 0.5 M NaCl. Fitted with the PDF of the log-normal distribution (eq. (3.22)) shown in red. Adapted from [329].

presented in Chapter 4 (cf. data in Table A.2). As the NaCl-treated FFLA agglomerates showed asymmetrical shape, their size was assessed manually by measuring the maximum concentric length a and then the maximum length b perpendicular to a. Then, the mean cluster size was approximated by averaging[3]: $d_{cluster} = (a + b)/2$. The numbers of FFLA agglomerates counted ($N \approx 100$) is much lower than that for pristine FFLA ($N = 604$), reflecting the difficulty to unambiguously identify both lengths of FFLA agglomerates from TEM images. The results from PDF log-normal size fitting are listed in Table 7.7, while the fitting parameters of PDF are found in Appendix A.5.8, Table A.16). The mean agglomerate size, described by d_C, increases

Table 7.7.: Mean agglomerate size d_C of NaCl-treated FFLA samples with fitting quality parameter R^2_{adj} derived from TEM analysis (s. Fig. 7.15).

NaCl-treatment of FFLA sample	R^2_{adj}	d_C [nm]
0.0 M	0.996	9.5 ± 3.1
0.1 M	0.981	95.4 ± 52.0
0.25 M peak 1	0.852	122.7 ± 95.8
0.25 M peak 2	0.852	248.8 ± 32.5
0.5 M	0.884	238.1 ± 129.5

with increasing the NaCl concentration. Interestingly, a bimodal agglomerate size distribution is observed for 0.25 M NaCl treatment, which is in line with DLS measurements and allows to classify this sample as an "intermediate" state of agglomeration, combining agglomerate sizes observed for 0.1 M NaCl ($d_C \sim 100$ nm) and those observed for 0.5 M NaCl ($d_C \sim 250$ nm).

---

[3]This procedure is identical to the determination of the average lysosome size in Chapter 6, Section 6.2.3. See Fig. 6.6 for an example of the procedure).

Further, TEM measurement artifacts arise from sample preparation: It is expected that upon drying the tendency for agglomeration of MNP is further increased and the values of d_C reported here presumably do not correspond to the ones present in liquid suspension$^\diamond$. An example for this is observed with $d_H < d_C$ for the second peak of the 0.25 M NaCl-treated sample. Nevertheless, TEM provides valuable information by displaying substantial changes in the arrangement of FFLA upon NaCl treatment$^\diamond$ and the size-trends from TEM analysis agree with those for liquid samples from DLS measurements.

7.3.4. Magnetic Properties Characterization

SQUID magnetometry characterization of freeze-dried samples was carried out as described in Section 4.6, measuring the $M(H)$-loop and ZFC-FC curves for each sample. The iron concentration was $c = 0.3\,\mathrm{mg(Fe)/mL}$ for each sample before mixing with mannitol solution (cf. sample preparation in Section 4.6).

LUM Sample

Results from magnetic characterization of LUM are plotted in Fig. 7.16. The saturation

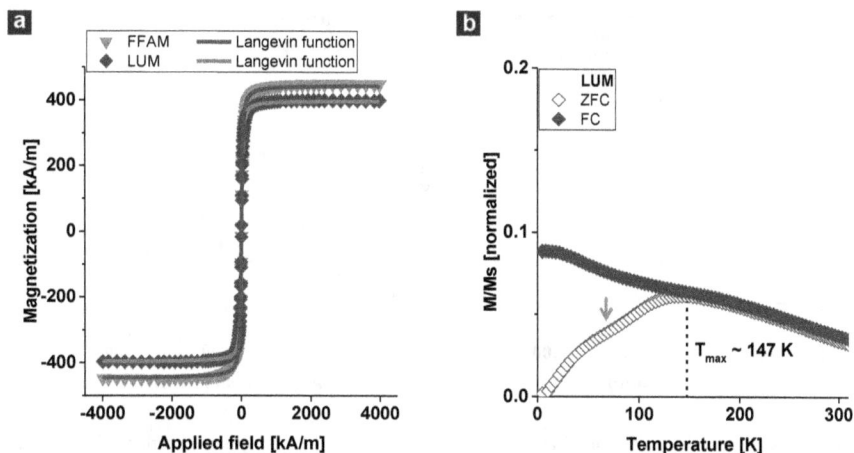

Fig. 7.16.: SQUID measurements for LUM particles: (a) $M(H)$-curve fitted with the Langevin function (s. eq. 4.10; $R^2_{\mathrm{adj}} = 0.999$). $M(H)$-curve of FFAM particles is shown for reference (cf. Fig. 4.13d). (b) ZFC-FC curves measured at $H = 0.8\,\mathrm{kA/m}$ ($\hat{=}1\,\mathrm{mT}/\mu_0$). The red arrow in (b) indicates the position of the peak in ZFC, $T_{\mathrm{max}} = 87\,\mathrm{K}$ measured for FFAM particles (cf. Fig. 4.15d).

magnetization determined from the Langevin fit to the $M(H)$-curve amounts to $M_S = (397.81\pm 0.92)\,\mathrm{kA/m}$. This is a reduction of approx. 11 % when compared to the M_S measured for FFAM (cf. Table 4.7). Such a decrease in magnetization could arise from increased magnetic dipole-dipole interactions due to the small interparticle distance of FFAM clustered inside the liposomes (s. discussion in Section 7.4 for further details). Moreover, the peak in ZFC at $T_{\mathrm{max}} \approx 147\,\mathrm{K}$, is shifted towards higher temperatures for LUM compared to non-agglomerated FFAM partices with $T_{max} \approx 87\,K$ (cf. Fig. 4.15d). Such a shift in the ZFC peak position is in accordance with ZFC measurements for LUM from different batches, reporting $T_{max} \approx 160\,K$ [195] and $T_{max} \approx 170\,K$ [245]. This indicates increased magnetic dipole-dipole interactions as well, as already observed for intracellular MNP (cf. Section 6.4.2). The shoulder observed in the ZFC-curve at $T \approx 87\,K$ most probably corresponds to monodisperse FFAM particles. As individual FFAM are also indicated within LUM dispersions by DLS measurements (cf. Section 7.3.2), this second peak in the ZFC-curve indicates that a part of the FFAM are not clustered inside the liposomal shells, either during synthesis or due to breaking of the shell during sample preparation (as observed by TEM, s. Fig. 7.13b). Note that no blocking temperature and effective anisotropy constant is determined from ZFC measurements, since eq. (4.16) is valid only for non-interacting particles [241, 243].

FFLA Agglomerates

SQUID magnetometry reveals superparamagnetic behavior with negligible hysteresis loops for all FFLA agglomerated samples$^{\diamond}$, as shown in Fig. 7.17a. The saturation magnetization values,

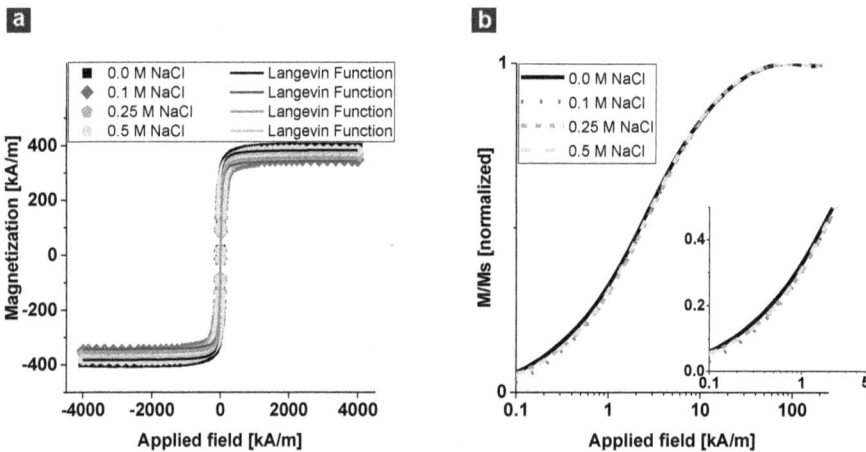

Fig. 7.17.: (a) Magnetization curves for NaCl-treated FFLA samples, (b) virgin magnetization curves for the same samples. Inset shows a zoom at low fields. Adapted from [329].

M_S, are slightly reduced for the agglomerated FFLA (s. Table 7.8). The virgin magnetization curves normalized to M_S also show slightly reduced magnetization for agglomerated FFLA, as shown in Fig. 7.17b. This drop in M_S and especially the reduction in magnetization at low fields (cf. Fig. 7.17) is an indicator for increased magnetic dipole-dipole particle interactions caused by MNP agglomeration [327]. Further evidence on increased magnetic dipole-dipole particle interaction of agglomerated FFLA is provided by ZFC-FC measurements, showing a shift for T_{max} towards higher temperatures, suggesting increased particle interactions caused by agglomeration effects [352], as shown in Fig. 7.18 and Table 7.8. Note that — as well as with LUM agglomerated particles — no blocking temperature and effective anisotropy constant is

Table 7.8.: Magnetic properties of NaCl-treated FFLA samples: M_S saturation magnetization from $M(H)$-curves and T_{max} temperature at maximum ZFC-magnetization from ZFC-FC measurements.

NaCl-treatment of FFLA sample	M_S [kA/m]	T_{max} [K]
0.0 M	385.2 ± 1.8	208 ± 10
0.1 M	350.8 ± 2.2	260 ± 20
0.25 M	363.8 ± 2.1	250 ± 15
0.5 M	381.7 ± 2.0	250 ± 15

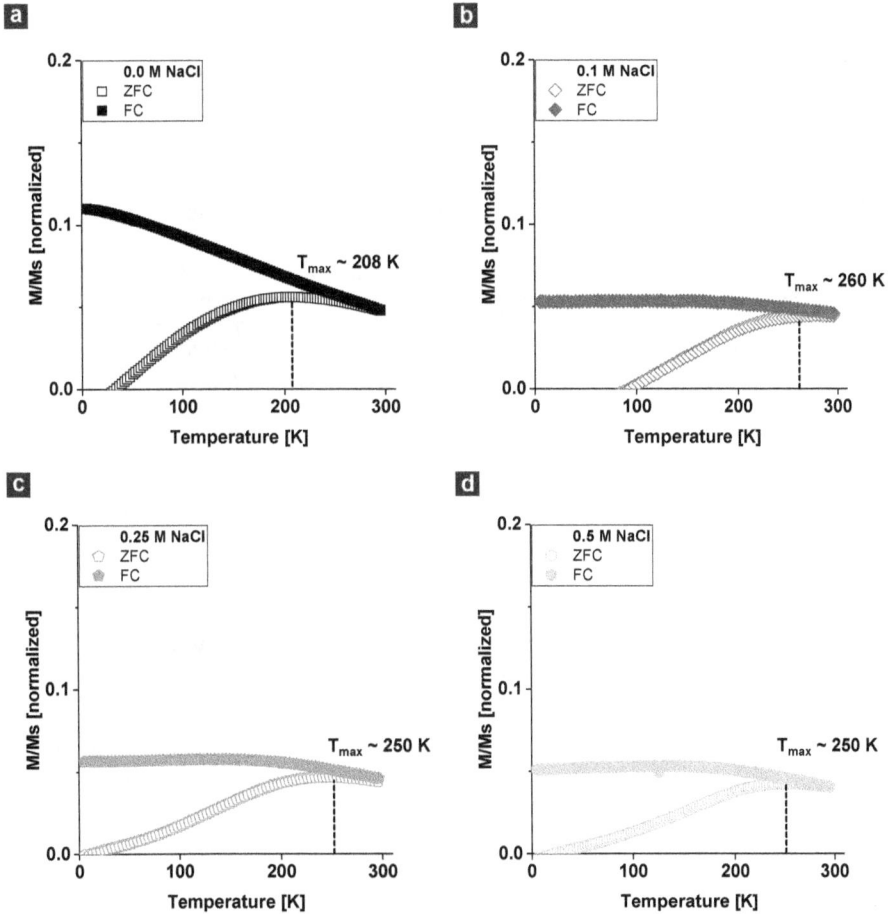

Fig. 7.18.: ZFC-FC measurements performed at $H = 0.8\,\text{kA/m}$ ($\hat{=}1\,\text{mT}/\mu_0$) for FFLA treated with (a) $0.0\,M$ NaCl, (b) $0.1\,M$ NaCl, (c) $0.25\,M$ NaCl and (d) $0.5\,M$ NaCl. Adapted from [329].

calculated from T_{max}, since eq. (4.16) is valid only for non-interacting particles [241,243]. It can, however, be assumed that increased particle interactions arise due to small interparticle distances inside FFLA agglomerates, which can in turn induce higher effective anisotropy constants, $K_{eff}^{\diamond}$ [23].

7.3.5. Sample Characterization Summary

In Table 7.9 the sample properties for LUM and FFLA agglomerates derived from DLS measurement (s. Section 7.3.2), TEM analysis (s. Section 7.3.3) and SQUID magnetometry (s. Section 7.3.4) are summarized. It is noteworthy, that even though samples were prepared differ-

Table 7.9.: Summary of the sample properties for LUM and FFLA agglomerates: Mean hydrodynamic size, d_H, mean agglomerate size, d_C, saturation magnetization, M_S, and temperature at maximum in ZFC-magnetization, T_{max}.

Sample	d_H [nm]	d_C [nm]	M_S [kA/m]	T_{max} [K]
LUM, peak 1	25.4 ± 9.4	250.5 ± 71.8	397.8 ± 0.9	147 ± 10
LUM, peak 2	326.6 ± 156.0			
FFLA (0.0 M)	110.8 ± 54.8	9.5 ± 3.1	385.2 ± 1.8	208 ± 10
FFLA (0.1 M)	115.7 ± 65.8	95.4 ± 52.0	350.8 ± 2.2	260 ± 20
FFLA (0.25 M), peak 1	108.7 ± 31.9	122.7 ± 95.8	363.8 ± 2.1	250 ± 15
FFLA (0.25 M), peak 2	386 ± 173	248.8 ± 32.5		
FFLA (0.5 M)	110.8 ± 49.6	238.1 ± 129.5	381.7 ± 2.0	250 ± 15

ently (liquid for DLS, dried for TEM and freeze-dried for SQUID), the results from independent characterization techniques agree remarkably well with each other, leading to the following conclusion for FFLA samples: FFLA form agglomerate states upon adding NaCl and these states can be adjusted by the amount of NaCl added.

7.4. Effect of Particle Agglomeration on Magnetic Particle Heating

In the present section, the effects of agglomeration and increased magnetic dipole-dipole particle interaction on particle heating are analyzed. First, Section 7.4.1 covers the experimental details. Then, the particle heating of agglomerated samples is assessed and discussed in Section 7.4.2.

7.4.1. Experimental Procedure

LUM and FFLA samples for MFH measurements were prepared as described in Section 7.3.1. Triplicates of each sample were prepared in $4\,\text{mL}$ glass vials containing a sample volume of $V = 1\,\text{mL}$ with an MNP iron concentration of $c = 0.3\,\text{mg(Fe)/mL}$.

Samples were measured in the Hüttinger setup (s. Section 5.2.1) at $H_0 = 50\,\text{mT}/\mu_0$, $f = 270\,\text{kHz}$. From the recorded $T(t)$-curve the mean SLP value was calculated as the average of the triplicate measurement as described in Section 5.2.1, assuming the specific heat capacity of water $c_h = 4.19\,\text{J/(g·K)}$ and furthermore $c_h = 4.11\,\text{J/(g·K)}$ for 0.5 M aqueous NaCl solution [353]. A linear relation was assumed for intermediate NaCl concentrations.

Reference sample solutions (either buffer (LUM) or the nominal NaCl concentrations (FFLA agglomerates)) not containing MNP were prepared for background subtraction.

7.4.2. Effects of Particle Agglomeration on Particle Heating

The averaged SLP values from triplicate MFH measurements ($H_0 = 50\,\text{mT}/\mu_0$, $f = 270\,\text{kHz}$) for both LUM and FFLA agglomerates are shown in Fig. 7.19. For comparison, the SLP value of non-agglomerated MNP is shown as well, i.e. FFAM in Fig. 7.19a and pristine FFLA (0.0 M NaCl) in Fig. 7.19b. Agglomeration increases the SLP value by 25.7% for LUM and by up to 22.2% for 0.25 M NaCl-treated FFLA. Both of these increases are significant compared to non-agglomerated samples (t-testing (two-tailed), $p < 0.05$). This almost identical SLP values correspond to very similar mean agglomerate sizes for LUM ($d_C = 250.5\,\text{nm}$) and 0.25 M NaCl-treated FFLA ($d_C = 248.8\,\text{nm}$; cf. Table 7.9).

In contrast, the SLP value drops by 29.3% for FFLA treated with 0.5 M NaCl, which can be attributed to the simultaneous immobilization of FFLA agglomerates in the precipitated LA matrix: As shown in Section 7.2.2, such immobilization can be associated with an inhibition of Brownian relaxation and a reduction in SLP value.

Overall, these results point towards a trend of increased particle heating for larger ($d_C > 100\,\text{nm}$) MNP agglomerates that are not inhibited in their pyhsical (Brownian) rotation, which is in line with results from several other groups [287, 347, 354, 355]. More specifically, similar LUM samples investigated by other works in literature report an increase in the SLP value by $\approx 60\%$ ($H_0 = 34\,\text{mT}/\mu_0$, $f = 700\,\text{kHz}$) [195] and by $(30 - 40)\%$ ($H_0 = 30\,\text{mT}/\mu_0$, $f = (300 - $

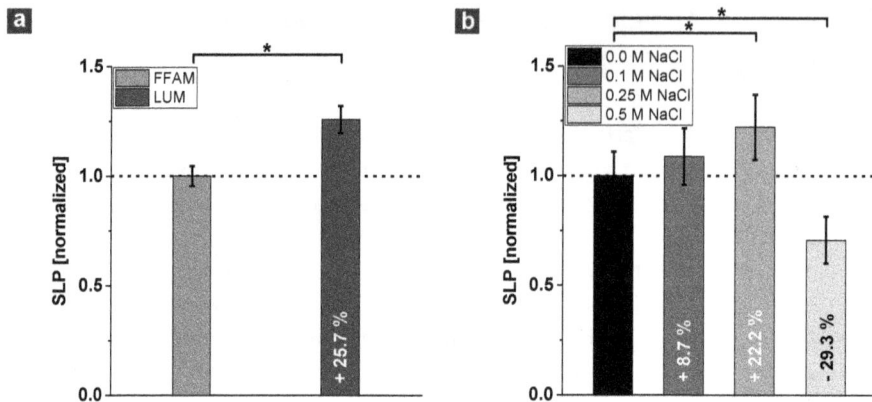

Fig. 7.19.: SLP values from MFH measurements ($H_0 = 50\,\mathrm{mT}/\mu_0$, $f = 270\,\mathrm{kHz}$) for agglomerated MNP: (a) LUM compared to FFAM, (b) FFLA treated with NaCl. A significant increase compared to FFAM (a) or pristine FFLA (0.0 M NaCl) (b) is indicated by the asterisks (t-testing (two-tailed), $p < 0.05$). Partially adapted from [328, 329].

1 100) kHz) [245], respectively. Such an increase in SLP could be explained from increased magnetic dipole-dipole interactions experienced by agglomerated MNP — as evidenced for both LUM and NaCl-treated FFLA from $M(H)$ and ZFC-FC measurements (s. Section 7.3.4) — which can also induce higher effective anisotropy values K_{eff} [23]. Such agglomerated MNP will have an effectively increased anisotropy barrier, $\Delta E = \hat{V}_{\mathrm{M}} \cdot K_{\mathrm{eff}}$ (with $\hat{V}_{\mathrm{M}}$ the effective magnetic volume of a mean MNP agglomerate) [27, 246], when relaxing collectively upon application of an AMF. This collective behavior of interacting MNP has been suggested to increase particle heating at excitation frequencies of $f = (150 - 160)$,kHz and strong field amplitudes $H_0 = (60 - 108)\,\mathrm{mT}/\mu_0$ [287, 347]. At the same time, discussion is still ongoing, whether particle heating is increased or decreased by increased particle interactions upon MNP agglomeration [27, 246, 285]: In fact, the complex interplay between MNP agglomerate intrinsic parameters, such as agglomerate size and size distribution, interparticle interaction-induced effects, like MNP arrangement, and finally the applied field parameters, only allow case-by-case interpretation, considering the characteristics of the specific MNP system under investigation. This complexity was vividly displayed recently by Niculaes et al., who demonstrated an increase in SLP value by approx. 19 % for iron oxide nanocubes clustered in a mixture of dimers and trimers, and contrastingly a decrease by approx. 14 % was reported for larger, centrosymmetric agglomerates (number of nanocubes per cluster $n \geq 4$) [285]. The authors explained the changes in SLP by interparticle interactions, which are enhancing SLP for dimer and trimers due to their chain-like arrangement, but decrease SLP for centrosymmetrically clustered nanocubes. Such an increased heating performance for chained MNP was already predicted theoretically [246] and demonstrated experimentally [356].

However, for the presented case of LUM and NaCl-treated FFLA particles, no such chain-like arrangement could be evidenced. Following the line of argumentation of Dennis et al. [287, 347, 357], the increased heating performance for agglomerated MNP is instead attributed to the interplay between the applied AMF using excitation frequencies of $f \sim 100\,\text{kHz}$ and the increased anisotropy barrier for collectively relaxing agglomerated MNP: As can be easily verified from eqs. (2.32) and (7.7), an increase of the anisotropy barrier $\Delta E = \hat{V}_M \cdot K_{\text{eff}}$ exponentially increases the Néel relaxation time. This increase is further amplified as the effective magnetic volume, $\hat{V}_M$, is higher for agglomerated MNP than for single MNP. Consequently, the Néel relaxation time increases by many orders of magnitude, eventually being blocked at ambient conditions ($T \sim 300\,\text{K}$), so that Brownian relaxation dominates the effective relaxation of the agglomerated MNP [338].

To estimate the size-dependent relaxation behavior of MNP agglomerates, the field-dependent Néel, Brownian and effective relaxation times are calculated using eqs. (2.70), (7.6) and (7.7)[4]. For simplicity, the approximations are based on the properties of pristine FFLA from Table 7.9: a constant coating thickness of 50 nm was added to the varied particle core/agglomerate size, resulting in a hydrodynamic size $d_H[\text{nm}] = d_C[\text{nm}] + 2 \cdot 50\,\text{nm}$. Furthermore, the effective anisotropy constant $K_{\text{eff}} = 78\,\text{kJ/m}^3$ from ZFC-FC measuremets of pristine FFLA (cf. Table 4.8) and the bluk magnetite anisotropy of $K_B = 11\,\text{kJ/m}^3$ [20] were used in eq. (2.19), to calculate a surface anisotropy constant of $K_S = 106.1\,\mu/\text{m}^2$. As discussed in Section 7.2.2, the temperature is assumed as $T = 313\,\text{K}$ ($\hat{=}40\,^\circ\text{C}$) with the respective viscosity $\eta(T = 40\,^\circ\text{C}) = 6.53 \cdot 10^{-4}\,\text{Pa} \cdot \text{s}$. Using these assumptions, the Néel, Brownian and resulting effective relaxation times depend on the particle / agglomerate size, d_C, as shown in Fig. 7.20a. For particle (agglomerate) sizes of $d_C < 10\,\text{nm}$, Néel relaxation dominates and the effective relaxation time, τ_R, follows the Néel relaxation time. However, for $d_C > 15\,\text{nm}$, Brownian relaxation dominates the particle relaxation dynamics. Even though, these numbers are only estimates, as eq. (7.7) is limited to non-interacting particles (which is not the case for agglomerated particles discussed here), the general trends predicted from Fig. 7.20a for particles with $d_C > 100\,\text{nm}$ are in agreement with literature: Here, fully Brownian-dominated relaxation (i. e. Néel relaxation is blocked completely) is shown for magnetite particles with $d_C > (25 - 40)\,\text{nm}^5$ [14, 23, 358]. Consequently, the results from Fig. 7.20a are used to estimate the trends for the AC-susceptibility, χ'', which is proportional to particle heating using eq. (7.8) for the AMF frequency $f = 270\,\text{kHz}$, as shown in Fig. 7.20b. Clearly, χ'' (and with it the contributions to particle heating) is Néel dominated for pristine FFLA with $d_C \approx 10\,\text{nm}$, however, for agglomerated particles with $d_C \approx 100\,\text{nm}$ (such as FFLA treated with 0.1 M NaCl) and $d_C \approx 250\,\text{nm}$ (such as LUM and FFLA treated with 0.25 M NaCl), the particle heating is only constituted of Brownian relaxation contributions (as Néel relaxation is fully blocked). As can be seen from the inset in Fig. 7.20b, the Brownian contributions to particle heating increase up to $d_C \approx 400\,\text{nm}$, where the value saturates. This explains the increased heating for FFLA treated with 0.25 M NaCl

---

[4] As the effective anisotropy constant, K_{eff}, could not be estimated for the agglomerated particles, the individual Néel and Brownian relaxation times cannot be estimated as in Section 7.2.2.

[5] Variation depends on the anisotropy constant, K_{eff}, and the temperature, T, assumed for the approximation.

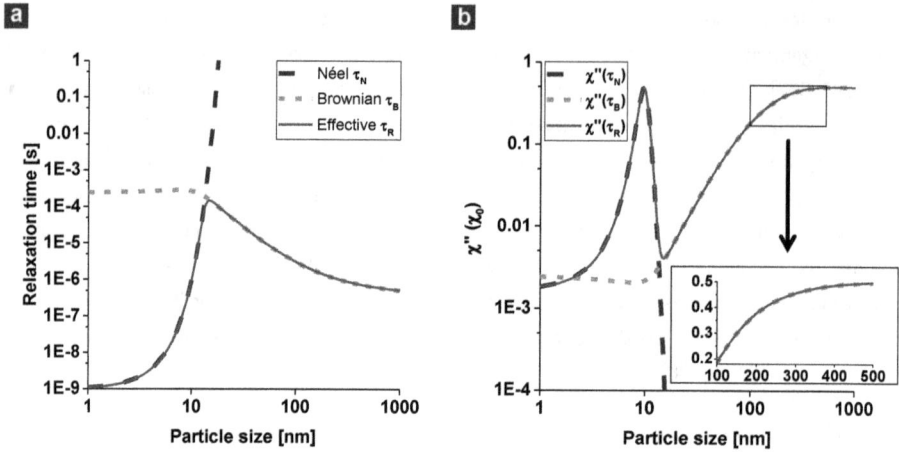

Fig. 7.20.: (a) Relaxation times as a function of particle size, calculated from eqs. (2.70), (7.6) and (7.7), using pristine FFLA parameters and assuming a constant particle coating thickness of 50 nm as well as a size-dependent anisotropy constant (cf. eq. (2.19); s. text). (b) Complex AC-susceptibility, χ'', as a function of particle size, calculated for the Néel (τ_N), Brownian (τ_B) and effective (τ_R) relaxation times shown in (a). Inset in (b) shows a zoom to the marked region. AMF amplitude $H_0 = 50 \, \text{mT}/\mu_0$ and frequency $f = 270 \, \text{kHz}$.

compared to FFLA treated with 0.1 M NaCl with an increased Brownian contribution for the former sample (s. Fig. 7.19b).

An interesting behavior exists for FFLA treated with 0.5 M NaCl, where agglomeration and immobiliztion of FFLA occur simultaneously. Here, particle heating is significantly reduced. Combining the findings of (totally) inhibited Brownian relaxation for immobilized particles (cf. Section 7.2.2) and (partially) blocked Néel relaxation for agglomerated particles, this reduction of the overall particle heating output is expected, since the FFLA agglomerates with $d_C \approx 250 \, \text{nm}$ solely relax via Brownian relaxation, as shown in Fig. 7.20. Immobilization and consequently inhibition of Brownian relaxation consequently cancels the contributions of agglomerated MNP to particle heating. Nevertheless, some smaller FFLA remain in the sample, as evidenced by DLS measurements, cf. Section 7.3.2, which still comtribute to the overall heating via Néel relaxation (cf. Fig. 7.20b). The same superposition effects of agglomeration and immobilization are expected for intracellular MNP [245, 247, 286], as has been discussed in the previous Chapter 6 (cf. Section 6.5).

8. Magnetic Fluid Hyperthermia Efficacy

This chapter demonstrates the feasibility of magnetic fluid hyperthermia (MFH) on pancreatic tumor cells using ML particles. For this, the chapter brings together the findings on the ML uptake kinetics and ML morphology inside cells from Chapter 6 and on the particle heating in cell-mimicking model systems from Chapter 7 and advances towards clinical application of MFH by evaluating MFH efficacy on living cells in vitro. As this study is part of a therapy approach for magnetically targeting endoluminal tumors, in particular the pancreatic ductal adenocarcinoma (PDAC), as outlined in Section 2.5.4, the focus lies on pancreatic tumor cells using ML concentrations that are achievable with current magnetic targeting models. The experimental procedures for in vitro MFH application and assessment of cell survival are covered in Section 8.1. The effects of MFH on cell survival are analyzed in detail in Section 8.2. This chapter is concluded by a remark on the general applicability of MFH, briefly discussing how major challenges arising from interactions of MNP with the (intra)cellular environment in vivo could be met on the basis of the results generated by this thesis (Section 8.3).

This chapter is based on an original publication by the writer [50] (s. Appendix C). Where entire sentences are cited directly from the publication, this is marked with [†] and figures adapted or taken directly from [50] are marked in the respective figure's caption.

8.1. Experimental Procedure

For experiments ML2 particles were used (s. Chapter 4, esp. Section 4.8 for sample characterization).

8.1.1. Cell Culture

The cell lines MiaPaCa-2 and L929 were cultivated as described in Section 6.1: $1 \cdot 10^6$ cells of each cell line were seeded per well in 6-well plates (CELLSTAR) with 1 mL cell-specific medium and stored in the incubator for 24 h. Subsequently, old medium was refreshed with 1 mL cell-specific medium mixed with ML2 particles of final incubation concentrations $c = (0.15, 0.225$ and $0.3)$ mg(Fe)/mL and stored in the incubator ($37\,^\circ$C) for another 24 h. Reference cells for control received 1 mL fresh medium without ML. After incubation, cells were harvested as described in Section 6.1 (quantities: 2.5 mL DPBS, 0.8 mL Trypsin, 3 mL cell-specific medium). Harvested cells were counted with the LUNA automated cell counter and transferred to 15 mL falcon tubes (Eppendorf), which was centrifuged for 5 min (1 200 rpm, $20\,^\circ$C). Supernatant was removed. The cell pellet was processed differently depending on whether MFH or MPS measurements were performed (s. below).

Cell Samples for Magnetic Fluid Hyperthermia

For MFH experiments, three types of samples were prepared: samples not containing ML (*control*), cell samples having internalized ML for 24 h and suspended in fresh medium (*intracellular ML*) and cell samples with internalized ML suspended in their ML-loaded incubation medium (*intra- & extracellular ML*). Each sample was prepared in doublets as described above, one subjected to MFH, the other serving as a MFH-independent control. After centrifugation, the cell pellets were resuspended in 1 mL cell-specific medium (or old ML-loaded incubation medium for intra- & extracellular ML) and transferred to sterile (autoclaved) 4 mL glass vials (Rotilabo). Cell samples for MFH experiments were used in the Hüttinger hyperthermia device directly after preparation as described below (cf. 8.1.2).

Cell Samples for Magnetic Particle Spectroscopy

For MPS experiments, cell samples having internalized ML for 24 h (*intracellular ML*) were prepared in triplicate as described above. After centrifugation, the cell pellets were resuspended in 0.1 mL formalin and otherwise exactly prepared as previously described in Section 6.3.1. The samples were measured in the PTB MPS setup (cf. Section 5.1.1) and averaged over three independent measurements from sample triplicates.

It was observed from earlier experiments that cells might also be removed during the change of the supernatant. Therefore, for an accurate determination of the ML amount per cell, the influence of the processing steps on the residual cell amount used for MPS measurements was estimated using a so-called processing-factor[†]. For this, the cell samples were processed as described above but were finally resuspended in medium instead of agarose and cell counting performed[†] (s. Appendix A.6.1). The average ML-uptake per cell was calculated from the sample iron content derived from the MPS measurements divided by the number of cells with the processing-factor correction[†].

8.1.2. In vitro Magnetic Fluid Hyperthermia

Samples were measured with the Hüttinger hyperthermia setup (cf. 5.2.1) as follows: The AMF with amplitude $H_0 = 50\,mT/\mu_0$ and frequency $f = 270\,kHz$) was applied to in vitro samples for either 30 min or 90 min. The AMF parameters were chosen to maximize the heating effect rather than obey to medical safety constraints (cf. Section 5.3.2). Moreover, the Hüttinger setup cooling settings were adjusted to keep 1 mL of cell medium at 37 °C upon AMF application in order to ensure physiological conditions during MFH. The sample temperature during MFH was recorded simultaneously at the bottom (where the cells settled) and in the middle of the sample vial with two fiber-optic temperature probes connected to the Luxtron 812 thermometer (LumaSense Inc.). Generally, no difference in the temperatures at both position was measured (s. Appendix A.6.3). Therefore, all temperature curves reported in this chapter are those measured at the bottom of the vial (at cell level). Of each of the sample doublets (for control, intracellular ML and intra- & extracellular ML), one sample was subjected to the AMF, to assess the field-dependent effects on cell survival via clonogenic assay (s. below, Section 8.1.3), while the other sample served as a MFH-untreated control. Samples that were not measured were stored in a special iron jacket on a hotplate set to 37 °C to simulate physiological conditions.

For comparison, additional samples not containing ML were prepared as described in Section 8.1.1 and placed on a hotplate set to either 42 or 44 °C for 30 min and 90 min[†]. After treatment, cell survival fraction was assessed via clonogenic assay (s. below)[†]. The sample temperature during heating on the hotplate was measured with the fibre-optic thermometer probe at the bottom of a 4 mL vial containing 1 mL of cell-specific medium.

Additional samples of ML2 suspended in 1 mL cell culture media RPMI and DMEM were prepared in triplicates for each ML concentration $c = (0.15, 0.225$ and $0.3)\,mg(Fe)/mL$ for MFH measurements. These were used as references to compare SLP values for freely dispersed ML and intracellular ML (s. Section 8.2.4 below).

8.1.3. Clonogenic Assay

The assessment of cellular survival after any specific treatment (either with AMF alone, ML alone or AMF and ML) was performed using the method of clonogenic assey (CA). CA is commonly used in oncology and assesses the cell's ability to proliferate and from colonies after any specific treatment (commonly cytotoxic irradiation) [359]. CA is superior to typical colorimetric viability assays, using 3-(4,5-Dimethylthiazol-2-yl)-2,5-diphenyltetrazoliumbromid (MTT) or CellTiter96 AQ$_{ueous}$ (MTS), as it has been shown that the photoabsorption spectra of MTT and MTS assays are distorted in a MNP-concentration-dependent manner [360]. Additionally, CA allows for a long-term evaluation of cell viability over several days.

For CA, cells were counted directly after MFH treatment using the LUNA automated cell counter and seeded with 400 living cells suspended in 2 mL cell-specific medium per well in 6-well plates. Each sample was seeded in triplicate. Seeded cells were placed in the incubator at 37 °C for 12 days. Subsequently, the samples were taken out, the old medium was removed and each well was washed with 2 mL of DPBS (gibco). The washed cells were stained with a mixture of crystal violet (Merck) and 70 % methanol (Sigma Aldrich) in the ratio 1:10 with 1 mL per well. The mixture was left on the cells for 30 min, afterwards each well was washed carefully with 3 mL tap water twice and the well plates were placed under the fume hood for 1 h to dry. The dry-stained well-plates were photographed digitally and further processed with ImageJ [213] and paintNET [212] for counting the number of colonies per well. This colony count was divided by the number of initially seeded cells (400 here) to calculate the plating efficiency (PE). From triplicate CA seeding, the mean PE and standard deviation was determined. From this, the survival fraction (SF) was determined by dividing the mean PE of treated cells by the mean PE of reference samples (controls resp. untreated cells)[†].

8.2. Magnetic Fluid Hyperthermia Effects on Living Cells

8.2.1. Concentration-Dependent MNP Intracellular Uptake

As previously described in detail in Section 6.3, MPS was used to accurately determine the amount of intracellular ML in [pg(Fe)/cell]. The concentration dependent ML-uptake after 24 h of incubation is depicted in Fig. 8.1. The amount of intracellular ML increased with increas-

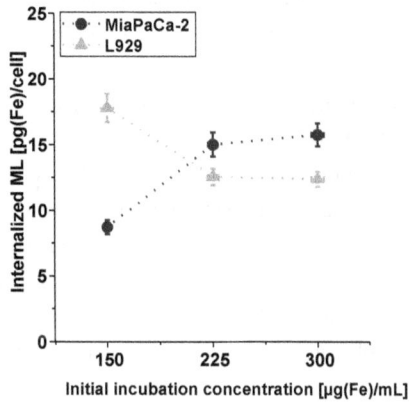

Fig. 8.1.: Initial incubation concentration-dependent uptake of ML internalized after 24 h of incubation. Dotted line to guide the eye. Taken from [50].

ing the incubation concentration for MiaPaCa-2, saturating for $c = 300\,\mu g(Fe)/mL$ at approx. 15 pg(Fe)/cell. Contrastingly, ML-uptake in L929 cells displayed an opposite trend, displaying highest intracellular ML (approx. 18 pg(Fe)/cell) for the lowest incubation concentration of $c = 150\,\mu g(Fe)/mL$ and settling at approx. 12 pg(Fe)/cell at $c = 300\,\mu g(Fe)/mL$. Generally, an increase in intracellular ML with increasing incubation concentration is expected as it was discribed in literature, independent of the cell type [11, 313, 361]. Moreover, previous internalization studies described in Section 6.3 for $c = 150\,\mu g(Fe)/mL$ confirm the ML uptake observed for MiaPaCa-2 (e. g. an internalized amount of approx. 9 pg(Fe)/cell is expected from Fig. 6.10a) and furthermore are in line with the trend of L929 internalizing substantially less ML than MiaPaCa-2, as observed for $c \geq 225\,\mu g(Fe)/mL$. These observation therefore suggest the assumption that some effect significantly altered the ML uptake for L929 at $c = 150\,\mu g(Fe)/mL$. Indeed, it was noticed that ML2 formed large agglomerates in RPMI (L929 cell culture medium) at such low iron concentration with hydrodynamic diameters of up to $d_H \approx 700\,nm$, while no such agglomerates were observed for higher concentrations or for ML2 in DMEM (MiaPaCa-2 cell culture medium) (s. Appendix A.6.2, Fig. A.20). As discussed in Section 6.3, MNP are agglomerated and clustered at the cell membrane prior to internalization. Therefore, agglomerated particles are favored for intracellular uptake due to the reduction of curvature energy, which is necessary for the cell membrane to wrap the (agglomerated) MNP in the initial stage of internalization [304, 319]. Hence, the ML-uptake for L929 at $c = 150\,\mu g(Fe)/mL$ could be increased as observed.

By multiplying the internalized amount of ML from Fig. 8.1 with the individual cell number per sample determined from cell counting (after using the processing factor, s. Appendix A.6.1) the absolute mass of ML internalized per sample, m_{int}, was calculated.

8.2.2. Influence of Bulk Temperature and Treatment Duration on Cell Survival

As described in the experimental procedures above (s. Section 8.1), control, intracellular ML and intra- & extracellular ML samples were prepared. Accordingly, control samples are cells not treated with ML. Intracellular ML describe a situation where ML are only present inside tumor cells but have been in the interstitium (extracellular space) for 24 h prior, while intra- & extracellular ML consist of intracellular ML inside the cells plus ML still present in the extracellular space. This allows to distinguish the cytotoxic effects of intracellular nanoheating (intracellular ML) and those for combined intracellular nanoheating effects and bulk temperature heating (intra- & extracellular ML). All these samples were subjected to the AMF for either 30 min or 90 min and the T(t)-curves were measured. intracellular and intra- & extracellular ML T(t)-curves were fitted with the Box-Lucas function (cf. eq. (5.2)) after a background subtraction with the T(t)-data obtained for control samples was performed (s. Appendix A.6.3 for exemplary T(t)-curves). Furthermore, the T(t)-curves (with background subtraction) were integrated over the Box-Lucas function over time to derive an effective temperature T_{eff} via

$$T_{eff} = \int_0^{t'} \frac{\Delta T \cdot (1 - \exp(-b \cdot t)) + T_0}{t'}\,dt, \tag{8.1}$$

with $dt' = 30$ min or 90 min. T_{eff} describes the average (*effective*) bulk temperature, affecting the cells during the entire MFH treatment and is therefore a more suitable parameter for describing bulk temperature effects on cell survival (compared to e.g. the maximum temperature reached during treatment).

Generally, the AMF had no toxic effect on untreated cells of both cell lines in preliminary testing (s. Appendix A.6.4 for details). Additionally, ML toxicity was assessed independently of MFH treatment, generally good biocompatibility of ML with only mild cytotoxicity esp. at higher ML concentrations (s. Appendix A.6.4 for a detailed discussion). Using eq. (8.1), the effective temperatures reached for all samples during MFH treatment and the respective cellular survival fractions are summarized for MiaPaCa-2 in Fig. 8.2 and for L929 in Fig. 8.3. Generally, a substantial cell damage was observed due to MFH treatment for both cell lines. In order to differentiate between ML-toxicity effects and truely MFH-induced cellular damage, surviving fractions were compared to ML-treated samples via t-testing (2-tailed), $\alpha = 0.05$): Those samples scoring $p < 0.05$ were denoted as statistically significantly damaged by MFH treatement. This revealed significant cellular MFH damage for elevated effective temperatures above $40\,°C$ for both cell lines. In particular, T_{eff} increased with increasing ML concentration of intra- & extracellular ML, reaching a maximum of approx. $44\,°C$ for MiaPaCa-2

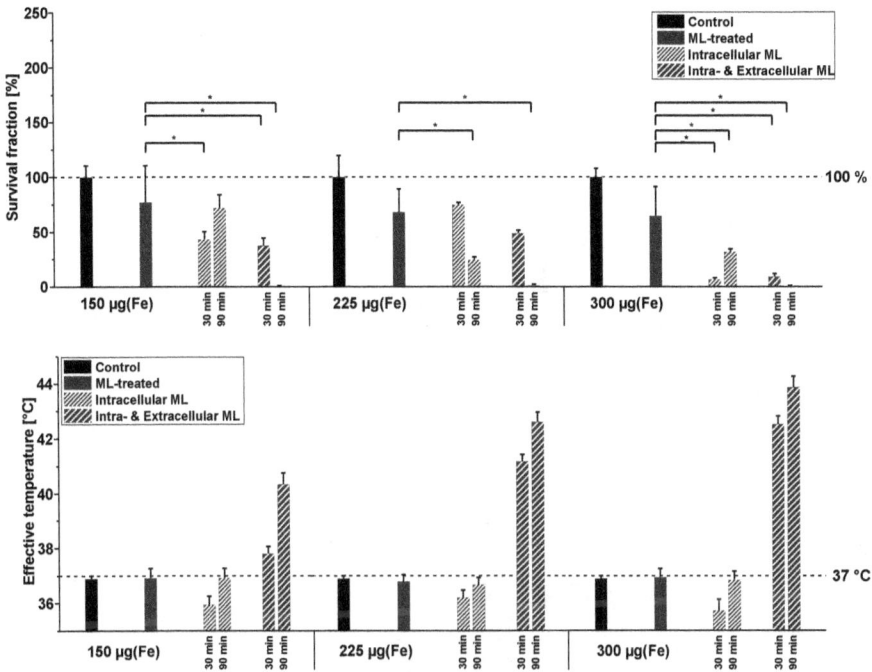

Fig. 8.2.: Survival fractions and effective temperatures reached during MFH treatment for MiaPaCa-2 for either 30 min or 90 min for initial incubation concentrations $c = (150, 225$ and $300)$ µg(Fe)/mL. Controls samples represent cells not treated with either ML or AMF. Statistically significant cell damage due to MFH is marked with an asterix (always compared to ML-treated samples of the respective incubation concentration; t-testing). Taken from [50].

and approx. $42\,°C$ for L929. Highest cell damage was observed for these maximum temperatures and generally for prolonged treatment durations. In fact, a treatment duration of 90 min entirely killed MiaPaCa-2 cells (decreasing survival fraction to $0\,\%$) for $T_{eff} \geq 40\,°C$. On the contrary, L929 with intra- & extracellular ML showed a residual surviving fraction of approx. $7\,\%$ after 90 min of treatment duration and reaching $T_{eff} \approx 42\,°C$, suggesting a higher resistance of L929 to heat treatment in general and to MFH treatment in particular. This observation is in line with the common understanding that tumor cells are more susceptible to heat treatment compared to healthy cells, esp. below $43\,°C$ [362].

Intracellular ML did not experience an effective (bulk) temperature rise above ambient conditions ($37\,°C$) during MFH treatment. However, significant cell damage was still also observed for intracellular ML inside MiaPaCa-2. This suggests that bulk temperature-independent nanoheating effects could significantly cause MFH-induced cell damage without perceptible temperature rise, which is in line with the observations of other research groups (s. discussion in Section 2.5.3). It was observed that the survival fraction was lower for 30 min MFH treatment compared to 90 min of MFH treatment for intracellular ML in both cell lines (cf. Figs. (8.2 and 8.3)). This could arise from the onset of thermotolerance (s. Section 2.5), which is established for cells

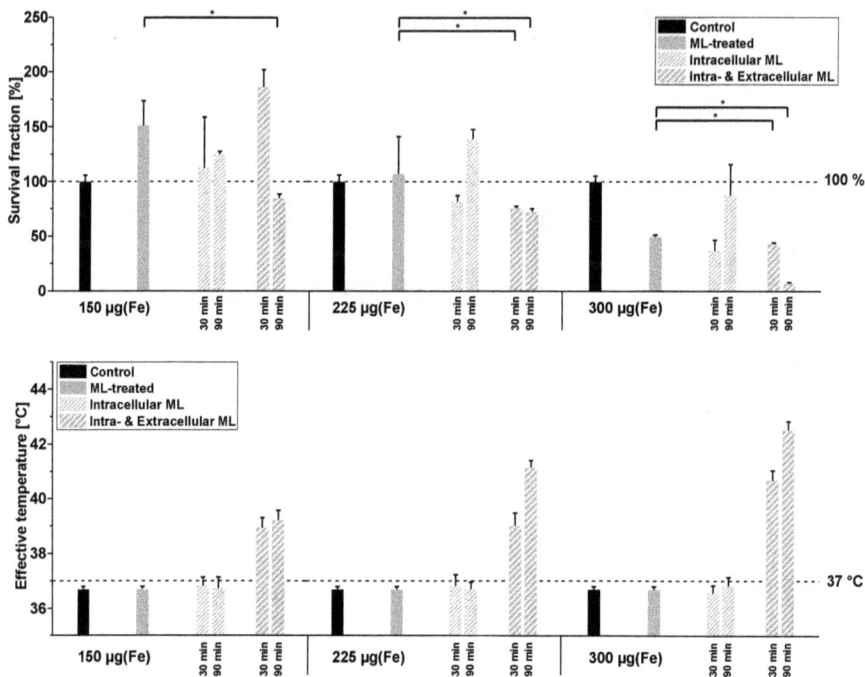

Fig. 8.3.: Survival fractions and effective temperatures reached during MFH treatment for L929 for either 30 min or 90 min for initial incubation concentrations $c = (150, 225$ and $300)$ µg(Fe)/mL. Controls samples represent cells not treated with either ML or AMF. Statistically significant cell damage due to MFH is marked with an asterix (always compared to ML-treated samples of the respective incubation concentration; t-testing). Taken from [50].

treated at $T = (39\text{-}41)\,°C$ (here induced by nanoheating) for treatment durations of at least 30 min [73]. Consequently, the samples treated for 30 min, would not have had enough time to develop such thermotolerance, while contrastingly those cells treated for 90 min could develop thermotolerance. Another hypothesis could be that thermotolerance caused by nanoheating substantially differs from the assumptions above. Recent results suggest a time-dependent apoptotic mode of action for MFH acting over the course of $(6\text{-}18)\,h$ [112]. However, at this time, this hypothesis cannot be followed through completely but requires substantially more evidence from future studies.

8.2.3. Comparison of MFH and Hotplate Heating Effects

In order to distinguish cellular damage caused by bulk temperature heating from that caused by MFH (including nanoheating effects), additional cell samples not containing ML were also heated on a hotplate set to either $42\,°C$ or $44\,°C$. The effective temperatures ranged between $41.5\,°C$ and $44.3\,°C$ for treatment durations of 30 min or 90 min (s. Fig. 8.4a). This partially exceeded the clinically proven thermal damage threshold established at $43\,°C$ [10]. The corresponding SF for both MiaPaCa-2 and L929 decreased significantly for effective temperatures above $42.5\,°C$, therefore confirming the just mentioned thermal damage threshold of $43\,°C$ (s. Fig. 8.4b). Generally, it is observed that cellular damage depends on both, the effective temperature reached and the treatment duration. In detail, however, slight differences occur, as the pancreatic tumor cells MiaPaCa-2 displayed a more prominent cytotoxic effect for prolonged treatment times, whereas healthy L929 cells generally showed increased cytotoxicity in a temperature-dependent manner. These results are in line with the findings for MFH treatment presented in the previous Section 8.2.2, esp. corroborating the afore mentioned fact that tumor cells are more susceptible to heat treatment below $43\,°C$ [362].

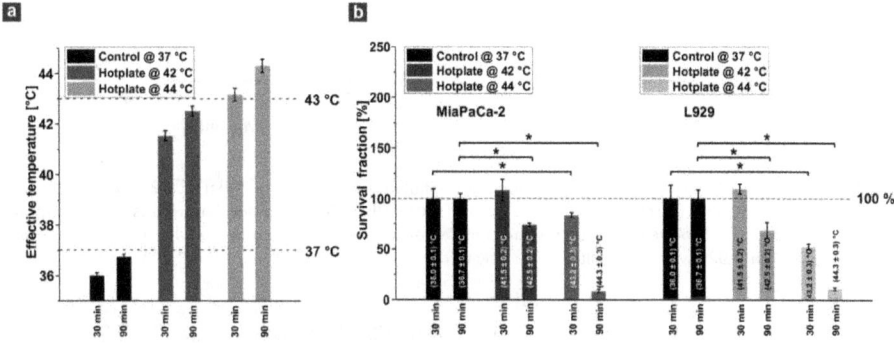

Fig. 8.4.: (a) effective temperatures reached during heating on a hotplate for either 30 min or 90 min. (b) respective surviving fractions for MiaPaCa-2 and L929 cells without ML. Control cells were placed on a second hotplate set to $37\,°C$. Statistically significant cell damage due to heating is marked with an asterix (always compared to control samples of the respective treatment duration; t-testing). Taken from [50].

For convenient comparison of the MFH and hotplate-heated results, the concept of *cumulative equivalent minutes* (CEM) is introduced in the following: In clinical studies, CEM is used to compare cellular damage under varying experimental conditions [10][†]. CEM integrates the temperature-dependent damage rates over the exposure temperatures of the cells accounting for the faster cell damage at higher temperatures[†] and is defined as follows:

$$\text{CEM}_x = \sum_{n=\Delta t_N}^{N} \Delta t_N \cdot R^{x - T'(n)}, \tag{8.2}$$

where $T(t)$ is the temperature data described by the Box-Lucas function (s. above and eq. (5.2)), which is divided in N time-intervals of equal length Δt_N, for which the mean temperature, $T'(n)$, in each interval was calculated according to $T'(n) = (T(t = n) - T(t = n-1))/2$. x [°C] describes the temperature at which cell damage is expected and $R = 0.25$ for $T'(n) < x$ and $R = 0.5$ for $T'(n) \geq x$. Here, $x = 43\,°C$ is chosen, as substantial cellular damage was observed in hotplate experiments above this temperature (s. above). For simplicity, $\Delta t_N = 1\,\text{min}$ was chosen for calculating CEM_{43} (resulting in $N = 30$ or $N = 90$, respectively), which is shown for the samples showing significant cell damage due to MFH (cf. Figs. 8.2 and 8.3) in Fig. 8.5. The survival fraction of MiaPaCa-2 cells dropped below $50\,\%$ in a CEM_{43} time interval

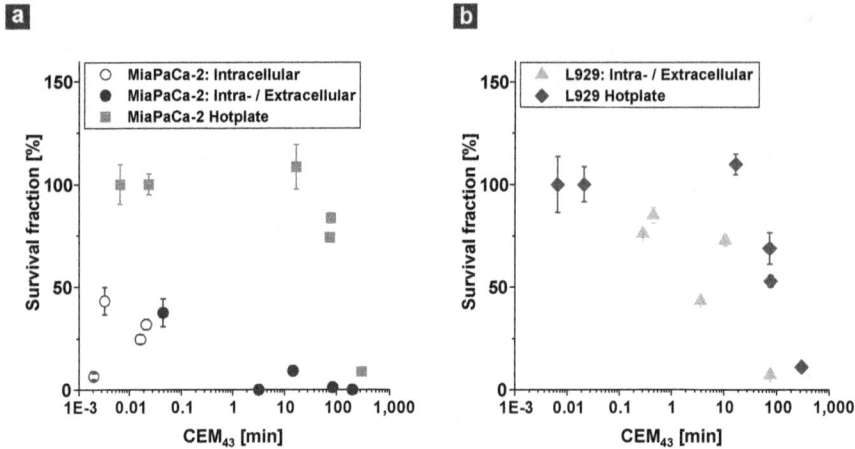

Fig. 8.5.: Survival fraction versus cumulative equivalent minutes (CEM43) for (a) MiaPaCa-2 and (b) L929 cells comparing MFH (intracellular ML and intra- / extracellular ML) and hotplate results. Note that only results from MFH-treatment with significant cell damage are shown, which is why there are no results shown for L929 with intracellular ML. Taken from [50].

of $[0.001, 0.1]$ min for intracellular ML, with no perceptible bulk temperature rise (Fig. 8.5a, cf. Fig. 8.2)[†]. Most importantly, MiaPaCa-2 cells only heated on a hotplate were not damaged in this interval, which suggests an improved cytotoxicity for MFH. At higher CEM_{43} between $[1, 250]$ min, MFH treatment on MiaPaCa-2 samples with intra- & extracellular ML resulted in almost complete cell death, whereas hotplate hyperthermia required approx. 250 min (CEM_{43}) to achieve the same effect[†].

For L929, a similar trend was observed as for MiaPaCa-2: For the intra- & extracellular ML L929 cell samples treated with MFH, the survival fraction dropped from approx. $80\,\%$ to approx. $10\,\%$ for a CEM_{43} time interval of $[0.3, 80]$ min[†]. While for the L929 cells treated with hotplate hyperthermia, the survival fractions experienced the same absolute drop in a CEM_{43} time interval of $[70, 300]$ min (Fig. 8.5b)[†]. For L929, the intracellular ML samples could not be unambiguously identified as significantly damaged by hyperthermia (cf. Fig. 8.3) and were therefore not considered in Fig. 8.5b[†].

All in all, these findings are in line with results of afore mentioned clinical trials on human brain tumors of 14 patients, where an effective damage was achieved in a CEM_{43} time interval of $[5, 500]$ min [64][†]. Furthermore, the results indicate that significant cell damage is induced substantially earlier for cells treated with MFH (i. e. $[1, 10]$ min) compared to the cell treated solely with hotplate hyperthermia (starting at approx. 100 min). Also, the cytotoxic effect of MFH at lower CEM_{43} was more pronounced for MiaPaCa-2 cells than for L929 cells indicating the effective damage-inducing CEM_{43} moreover dependents on the cell type. This increased susceptibility of MiaPaCa-2 cells to MFH treatment compared to healthy L929 cells is an encouraging result hinting at individualized MFH therapy, for which the damaging MFH mechanism can be controlled to reach solely cancer cells[†] (s. discussion below, Section 8.3).

In summary, the results of Sections 8.2.2 and 8.2.3 suggest two factors that contribute to substantial cell damage during MFH-treatment: First, an elevated bulk temperature of above approx. $41.5\,°C$ combined with a prolonged treatment duration of e. g. 90 min as applied in the present study. This is clearly indicated by the decreased survival fraction in the temperature-affected region in the CEM_{43} interval $[1, 100]$ min (cf. Fig. 8.5). Second, an intracellular ML damaging factor, which arises without a perceptible bulk temperature rise (cf. results in Fig. 8.5a for CEM_{43} below 0.1 min)[†]. The latter effect is also detectable in the intra- & extracellular ML samples contributing to the earlier onset of cellular damage for MFH compared to hotplate hyperthermia[†], as described above. This indicates that in MFH other mechanisms of action with a cell damaging effect exist for intracellular MFH, such as nanoheating (s. Section 2.5.3). Even though the hypothesis of nanoheating causing MFH-cytotoxicity is gaining substantial evidence recently (s. discussion in Section 2.5.3), the fact that ML were observed to be packed tightly in intracellular lysosomes in previous internalization studies (s. Chapter 6), also allows the alternative hypothesis of mechanical rupture of cellular membranes (s. Section 2.5): Here, MNP are membrane-bound inside lysosomes, which then mechanically rupture the cell membranes due to physical particle rotation upon AMF-application[†]. However, since in this work AMF frequencies in the order of 100 kHz were applied[†], and the mechanism of membrane rupture was not yet proven at such high frequencies, nanoheating is the most probable effect to explain MFH cytotoxicity throughout this thesis.

8.2.4. Intracellular Particle Heating Efficiency and Thermal Energy per Cell

Another way of analyzing the intracellular damage caused by MFH uses the particle heating efficiency and thermal energy deposited per cell: From T(t)-curves and Box-Lucas fitting, the specific loss power (SLP) was derived as described previously (s. Section 5.2.1 for details on SLP; cf. eq. (5.1)), assuming the specific heat capacity of cell media being equal to that of water. In Fig. 8.6 the mean SLP values for three types of samples (intracellular ML and intra- & extracellular ML as well as ML suspended in cell-specific culture medium containing no cells) are shown in dependency of ML incubation concentration[†]. As the SLP value is generally independent of the duration of treatment after a saturation temperature is achieved, the SLP results shown here are averaged over single results calculated for 30 min and 90 min measurement times[†] (s. Appendix A.6.5 for details). Note further that SLP values of intracellular ML for MiaPaCa-2 cells could not be quantified, as the absolute amount of internalized ML was below the detection limit[†] (s. Appendix A.6.5 again for details). The relative amount of internalized ML per sample was calculated from the absolute number of cells from cell counting multiplied by the mean ML uptake per cell (cf. Fig. 8.1, which was then divided by the absolute amount of ML in the entire sample (intra- + extracellular ML)[†], as shown in Fig. 8.6. The SLP values decrease with increasing the relative amount of internalized ML for MiaPaCa-2 cells, whereas for L929 such a trend could not clearly be identified. Nevertheless, a clear difference between intra- & extracellular ML samples and the samples with ML solely suspended in cell-specific culture medium is observed. The SLP values of the ML suspended freely in cell medium were independent of concentration and higher than those for intra- & extracellular ML samples[†].

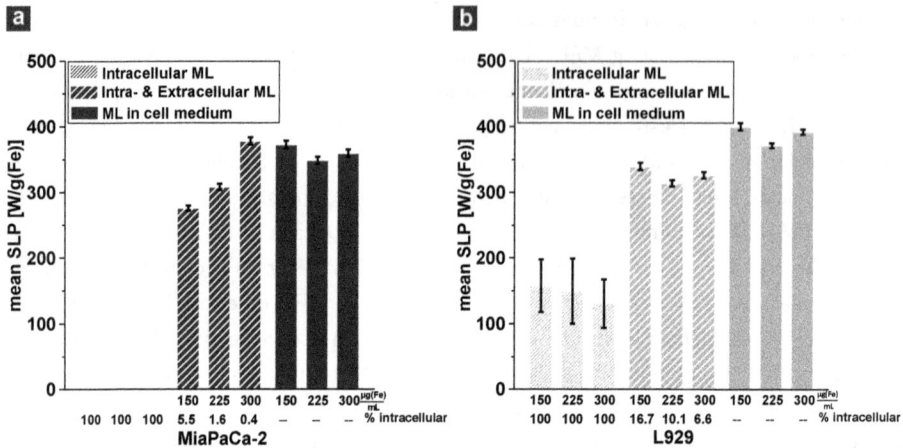

Fig. 8.6.: Particle heating efficiency ($H_0 = 50\,\text{mT}/\mu_0$, $f = 270\,\text{kHz}$), described by the mean specific loss power (SLP) values for intracellular ML, intra- & extracellular ML and ML suspended in cell-specific culture medium for (a) MiaPaCa-2 and (b) L929 cells. The ordinate depicts the ML concentration, below which the relative amount of internalized ML is specified[†]. Note that SLP values of intracellular ML for MiaPaCa-2 cells were below the detection limit and could thus not be quantified[†]. Taken from [50].

This demonstrates the influence of the portion of intracellular ML on the SLP values[†], which are lowered even though[†] the absolute amount of ML was the same for all samples. Moreover, the SLP values of intracellular ML in L929 cells dropped substantially by up to $\approx 65\,\%$[†] compared to the SLP values of ML suspended freely in cell medium (cf. Fig. 8.6b).

The SLP value of intra- & extracellular ML decreased more for MiaPaCa-2 cells, possibly due to the state of the ML aggregates present inside the lysosomes[†]: As observed previously (cf. Sections 6.2.3 and 6.5), MiaPaCa-2 packed intracellular ML more tightly and on average into smaller lysosomes. This would result in ML that are immobilized stronger inside MiaPaCa-2. Such immobilization of clusters of intracellular MNP was shown to reduce SLP values [245, 286] and attributed to the blocking of the Brownian relaxation upon internalization[†], as was also discussed in depth in the previous Chapter 7 (esp. Section 7.2.2). Comparing these findings with the results presented above, it is concluded that the immobilization state of nanoparticles has a strong impact on the heating properties of the whole cell environment[†], which will be discussed in connection to all findings obtained in this thesis in the next Section 8.3.

The SLP values of intracellular ML for MiaPaCa-2 cells could not be quantified as the internalized amount of ML ($(1.5\ldots8.0)\,\mu\text{g(Fe)}$; Fig. 8.6a) was below the setup detection limit[†]. The setup detection limit generally depends on the particle properties and measurement settings (frequency and field amplitude)[†]. Here, no SLP value could be determined below a minimum threshold of $15\,\mu\text{g(Fe)}$ ML per sample (s. Appendix A.6.5)[†]. The reason for the overall lower amount of intracellular ML for MiaPaCa-2 cells—despite its higher uptake (cf. Fig. 8.1)— could be the slower growth rate of the cells compared to L929 cells: As already discussed in Chapter 6, the doubling time for L929 was substantially lower than that for MiaPaCa-2, resulting in approx.

30 % less MiaPaCa-2 cells after 24 h of incubation with ML in previous experiments (cf. Fig. 6.9). Therefore, the number of MiaPaCa-2 cells available for the measurement was much less compared to the cell number of L929 cells[†], resulting in intracellular ML amounts falling below the detection limit.

The mechanism of intracellular nanoheating can further be analyzed in terms of the thermal energy deposited by ML after AMF exposure in its immediate cellular surrounding. Therefore, the *total thermal energy per cell* (TEC) deposited per cell during MFH-treatment was calculated to link the cellular damage to the actual efficiency of MFH-heating, as the total thermal energy is directly proportional to the SLP[†]. In this way, the total thermal energy serves as an indicator for the nanoheating capability of intracellular MNP[†]. TEC was calculated according to:

$$\text{TEC [J/cell]} = m_{\text{int}} \cdot \text{SLP} \cdot t_{\text{HT}}, \tag{8.3}$$

with the internalized amount of ML per cell, m_{int} (taken from mean uptake measurements, cf. Fig. 8.1), the SLP value (cf. Fig. 8.6) and the MFH-treatment time, $t_{\text{HT}} = 30$ min and $t_{\text{HT}} = 90$ min. Fig. 8.7 summarizes the dependency of SF on TEC for the samples which were significantly damaged by MFH for both cell types (cf. Fig. 8.2 and 8.3)[†]. With increasing TEC,

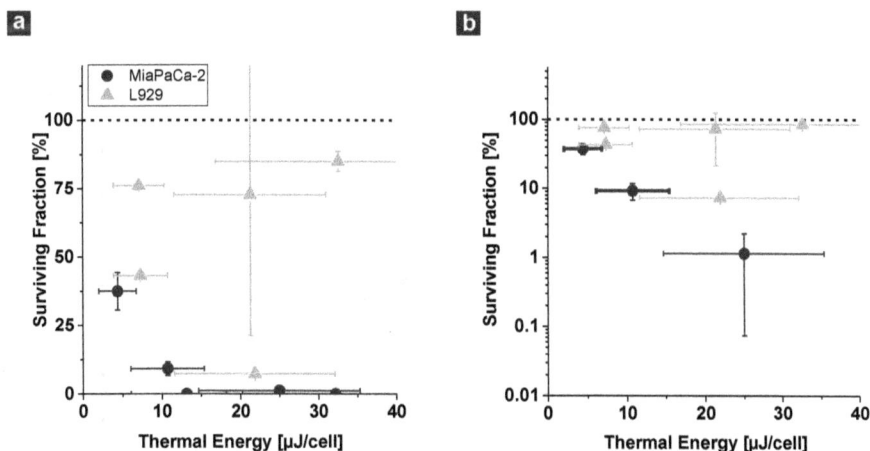

Fig. 8.7.: Survival fraction vs. thermal energy deposited per cell (TEC) of significantly damaged cells. Additional logarithmic plot shown in (b); the dotted line indicates the control's survival fraction with 100 %. Adapted from [50].

a strong decrease in SF was observed, which was more pronounced for MiaPaCa-2 cells (SF < 5 % for TEC ≈ 13 µJ/cell) compared to L929 cells (SF < 5 % for TEC ≈ 22 µJ/cell[†]. These results are in good agreement with findings on MDA-MB-468 adenocarcinoma cells, showing a survival fraction smaller than 5 % for TEC ≈ 13 µJ/cell for similar measurement settings ($H_0 = 47.1$ mT/μ_0, $f = 233$ kHz; $t_{\text{HT}} = 2$ h) [109][†]. Interestingly, L929 cells showed a second branch of healthy cells only slightly below the SF of control measurements even for thermal energies higher than 30 µJ/cell[†]. These findings allow envisioning prospective temperature and time ranges for treatment of pancreatic tumor cells, in which damage could be dealt to cancer cells, while healthy cells remain to a large extent unharmed[†], leaving a 50 % safety margin to damaging healthy cells (derived from comparing critical TEC values necessary for significant cell death: $TEC \approx 13\,\mu J/cell$ for MiaPaCa-2 and $TEC \approx 22\,\mu J/cell$ for L929). TEC can be easily controlled here via the treatment time, which could even compensate for a decrease in SLP value upon internalization, as mentioned above[†] (s. discussion in the next Section 8.3).

8.3. Concluding Remarks on Intracellular Magnetic Fluid Hyperthermia Applicability

This section presents an overview of the facts and statements derived for magnetic particle heating under different conditions throughout this thesis and attempts to forge a comprehensive understanding of what influences particle heating intracellularly. On the basis of this understanding and the results of in vitro MFH treatment of pancreatic tumor cells from the current chapter, some guiding principles for the future application of in vivo MFH will be given in the following.

As confirmed by internalization studies (cf. Sections 6.2 and 6.3), MNP are internalized inside cells, where they are tightly packed inside intracellular lysosomes. It can be assumed that these internalized MNP are mostly non-toxic to cells, if using iron oxide MNP with biocompatible coatings, as proven for the example of ML in this thesis via clonogenic assay analysis. Consequently, in the initial situation such as for an in vivo MFH application, MNP are clustered in agglomerates and immobilized inside lysosomes. As shown in the last Section 8.2.4, this results in an overall decrease in SLP value by approx. 65 % when compared to the SLP of MNP in suspension. Such a decrease in SLP values by approximately two thirds upon MNP internalization was confirmed by several groups [245, 286, 306], and is attributed to the immobilization and agglomeration due to the intracellular arrangement of MNP.

The investigation of isolated particle immobilization on particle heating of FFAM showed a decrease in SLP values by up to one third, attributed to the inhibition of Brownian relaxation (s. Section 7.2.2). Since the particle heating efficiency of FFAM and ML is very simliar (cf. Fig. 5.9), the Brownian contribution to particle heating for ML is assumed to be roughly one third as well. Consequently, immobilization alone cannot account for the reduction in SLP of roughly two thirds observed for intracellular ML. Agglomeration of MNP was also demonstrate to reduce the SLP in Section 7.4, if occuring simultaneously with immobilization. However, this combined agglomeration and immobilization did only decrease SLP values by roughly one third, and the observed redution in SLP value for intracellular MNP cannot be explained satisfacotrily by agglomeration either. Recently, Cabrera et al. demonstrated a decrease in SLP value by approx. 60 % for 21 nm iron oxide MNP inside living cells, which was attributed mainly to demagnetization effects of interacting particles upon intracellular agglomeration [286]: The authors observed the same relative decrease in SLP values for MNP agglomerates with cluster sizes of approximately the same size as intracellular vesicles ($d_H \approx 400$ nm), but only a minor decrease in SLP was reported for gradually Brownian rotation-inhibited MNP suspended in increasing concentrations of glycerol. From this, they concluded that agglomeration effects lead to a smaller hysteresis area (which is proportional to the SLP; cf. Section 2.4) due to intra-aggregate demagnetization effects [162] and are attributed thus as the dominant effect causing the decreased SLP value for intracellular MNP.

It seems however, that also the nanoscale arrangement of agglomerated MNP must be taken into consideration: As discussed in Section 7.4, chained alignment of MNP can increase the SLP value, while centrosymmetrical arrangement of MNP decreases SLP value [162, 285, 356]. Even though it is challenging to control the MNP arrangement on the nanoscale upon internalization, the general arrangement of MNP inside intracellular vesicles was found to vary in agglomerate size and packing density depending on the specific cell line (cf. Section 6.2). It could therefore be favorable to develop specific MNP systems that amplify the SLP values upon internalization in target tumor cell: E.g. large ($d_C > 25$ nm) monodisperse MNP, which are optimized for heating at medically-allowed AMF conditions on the one hand (cf. Section 5.3), could on the other hand be encapsulated in biocompatible shells, which are optimized for high intracellular uptake. If theses core-shell nanocarriers could be fabricated in a way that leaves the MNP freely suspended within its interior even upon intracellular internalization, the Brownian relaxation contributions to particle heating of freely rotating MNP (cf. Sections 7.2.2 and 7.4) could be used, while the intracellular immobilization does only affect the shell, therefore not attenuating particle heating.

A general amplification of SLP value for intracellular MNP is desirable, especially if MFH therapy is combined with magnetically targeted delivery of MNP to the tumor site, as only low amounts of MNP are available at the tumor site (s. Section 2.5.4). Provided that a sufficient amount of MNP is internalized, these intracellular MNP can damage tumor cells via nanoheating effects, even without a perceptible temperature rise (s. Section 2.5.3). This thesis succeeds in identifying three main factors contributing to cell damage by nanoheating during MFH treatment (s. Section 8.2.4): (i) the particle heating efficiency, generally expresses by the SLP value, (ii) the amount of MNP internalized per cell, able to deliver therapeutic heat and (iii) the duration of treatment. All three factors are described by the total thermal energy deposited per cell (TEC; cf. eq. (8.3)). The critical TEC, above which significant cell damage is dealt, has been shown to be cell-type-specific (cf. Section 8.2.4, Fig. 8.7), revealing a substantially higher resistivity of healthy L929 cells to MFH treatment compared to pancreatic tumor cells MiaPaCa-2.

Further investigation beyond this thesis could therefore explore the efficacy of MFH treatment using TEC as a parameter to optimize: The above-mentioned conditions (i) (maximizing SLP) and (ii) (maximizing uptaken MNP) have been shown to be accessible by experiment, and — at least to some extent — controllable: Particle heating and intracellular uptake can be predicted via modeling (cf. Chapters 3 and 6, respectively) and theoretically the biodistribution of MNP inside the body could be imaged with MPI (s. Section 2.3). In fact, combining MFH and MPI holds great potential to form a diagnostic and therapeutic platform from one single MNP system [363–365]. It could furthermore be envisioned to advance MC-simulations to include agglomeration and immobilization effects and predict optimal AMF parameters increasing the intracellular particle heating by taking the individual distribution of MNP within a patient (measured by MPI) into account. In such a treatment scenario, the duration of treatment (factor (iii)) provides a good choice to precisely control and tune the TEC necessary for individualized MFH treatment.

9. Conclusion and Future Directions

Throughout this thesis, magnetic nanoparticles (MNP) have been investigated as therapeutic heating agents for application in innovative cancer therapy via magnetic fluid hyperthermia (MFH). By studying various aspects of MFH — such as the optimization of particle heating considering medical limitation to the applied AMF, the uptake kinetics of MNP inside cells, the changes of particle heating under intracellular conditions and the efficacy of MFH in vitro — this thesis succeeds in answering four key research questions raised in the introduction (Chapter 1) and which are of paramount importance for the advancement of MFH application in tumor therapy:

1. *What MNP (i. e. which MNP properties) maximize the particle heating in MFH?*

2. *How do MNP interact with cells and how can the MNP uptake inside cell be quantified?*

3. *How does particle heating change upon MNP internalization inside cells?*

4. *How efficient is intracellular MFH applied to (tumor) cells; esp. at low MNP concentrations?*

These questions are answered by the main findings of this thesis, as presented in the following. Alongside, future directions on research opportunities resulting from these findings are pointed out.

Question 1 is investigated theoretically using Monte-Carlo (MC-)simulations of stochastic coupled Néel-Brownian magnetic relaxation of MNP to predict particle heating in Chapter 3. From these MC-simulations, the particle core size and effective anisotropy constant, as well as the external field parameters are identified as main contributors to particle heating. It can also be demonstrated that particle heating is indeed described most accurately within the framework of MC-simulations, when compared to established models for particle heating — the linear response theory and the Stoner-Wohlfarth based model theory (s. Section 3.3). Additionally, question 1 is investigated experimentally employing inductive particle heating measurements in Chapter 5, revealing that MNP with large particle core sizes, $d_C > 25$ nm, generate the most heat, expressed as a higher specific loss power (SLP). The trends of field-dependent particle heating predicted by MC-simulations agree well with experimental data, both showing a linear dependence of the SLP on the field-frequency (f) and a square dependency on the field-amplitude (H_0), as long as $H_0 < H_K/4$ holds (where H_K denotes the MNP anisotropy field). As a novelty, MC-simulations are matched to experimental data, showing best fitting for an anisotropy constant of $K_{eff} = 4$ kJ/m^3, lying substantially below magnetite bulk value ($K_{eff} = 11$ kJ/m^3). From these experimentally validated MC-simulations, predictions of SLP are derived in dependence of particle core size and AMF parameters limited to medically-tolerable field strengths (i. e. AMF parameters are low enough to not heat up healthy tissue). Results confirm the experimentally determined highest particle heating for large MNP with $d_C > 25$ nm at frequencies of $f \approx 100\,\text{kHz}$, with the respective field amplitude adjusted to keep within the medically tolerated range. The results suggest an improvement in particle heating by a factor of approx. 12 compared to MNP systems currently used in clinical application of MFH, which is valuable information for the future design of MNP optimized for maximum heat generation and resulted in a publication [251]. In summary, particle heating is maximized for low (lower than bulk) anisotropy MNP with core sizes of $d_C > 25\,\text{nm}$, even under medically-tolerable limitations to the applied field, which precisely answers question 1.

Chapter 6 addresses question 2 by studying how MNP interact with pancreatic tumor cells and healthy control cells in vitro. From morphological analysis via transmission electron microscopy it is observed that MNP first group into agglomerates at the cellular plasma membrane within minutes of incubation with cells and second internalize as agglomerates inside cellular endosomes via endocytosis within tens of minutes (s. Sections 6.2.2 and 6.2.3). Furthermore, MNP are tightly clustered and immobilized inside these cellular endosomes, which merge over time into lysosomes with final average sizes after $24\,\text{h}$ in the range of $d_{lys} \approx (400 - 650)\,\text{nm}$, depending on the cell line. Additionally, some MNP agglomerates are also found attached to the outer cell membrane after $24\,\text{h}$, presumably due to exocytosis. Assuming a three-step process of (1) MNP agglomeration at the outer cell membrane, (2) endocytosis (internalization) of these MNP agglomerates inside the cell and (3) exocytosis (excretion) of MNP agglomerates, the MNP uptake kinetics can be modeled mathematically (s. Sections 6.3.4 and 6.3.5). For modeling, the MNP uptake kinetics is experimentally quantified with magnetic particle spectroscopy (MPS) over an incubation time of up to $24\,\text{h}$. The MNP uptake saturates after $6\,\text{h}$ of incubation in pancreatic tumor cells and after $8\,\text{h}$ of incubation in healthy control cells. This is valuable information for the future planning of MNP targeting duration before applying MFH for localized tumor therapy. In this way, question 2 can be precisely answered using the MNP uptake kinetics model, allowing to predict the MNP uptake inside cells for an arbitrary incubation time.

Chapters 7 and 8 address the change in particle heating upon intracellular agglomeration and immobilization of MNP (question 3). Generally, the SLP decreases by nearly two thirds for MNP inside cells compared to freely dispersed MNP (s. Section 8.2.4). This is in agreement with a decrease in the MPS spectral magnitude of intracellular MNP, attributed to the inhibition of Brownian relaxation upon MNP immobilization inside cells (s. Section 6.3.6), and an increase in the magnetic interparticle interactions for intracellular MNP, leading to demagnetization effects due to MNP agglomeration (s. Section 6.4). This thesis achieves a better understanding of the isolated contributions of either immobilization or agglomeration of MNP to particle heating by unambigously differentiating between both effects: MNP gradually immobilized in agarose and polyacrylamide hydrogels show a decrease in SLP of up to one third, which can be attributed to the inhibition of Brownian relaxation while excluding MNP agglomeration effects (s. Sections 7.1 and 7.2). These results demonstrate the gradual decrease in SLP upon gradual MNP immobilization on the basis of well-characterized hydrogels for the first time [264]. In contrast, MNP agglomerated in liposome shells and MNP agglomerates — formed after damaging their lauric acid shell by adding NaCl — show a significant increase in SLP by up to one forth, as long as the MNP agglomerates are freely dispersed and rotatable [328]. If the MNP agglomerates are additionally immobilized, the SLP decreases by almost one third compared to monodisperse MNP in water [329]. In summary, and to answer question 3, the results reveal a decrease of nearly two thirds in particle heating for intracellular MNP due to a combination of inhibited Brownian relaxation upon MNP immobilization and demagnetization effects arising from the simultaneous MNP agglomeration inside cells.

In Chapter 8, MFH efficacy is evaluated in vitro on pancratic tumor cells MiaPaCa-2 and healthy L929 cells for low MNP concentrations of $c = (0.15...0.3)$ mg(Fe)/mL, specifically addressing question 4. Cytotoxic temperatures ($T \geq 43\,^\circ$C) can be reached even for low MNP concentrations achievable in animal models via magnetic targeting, resulting in significant cell damage. Moreover, significant cell damage is also observed without a perceptible temperature rise, indicating that additional nanoheating effects are present. This assumption is enforced by a higher MFH efficacy, when compared to conventional hyperthermia with the same bulk temperatures but without MNP. Most importantly, the MFH treatment efficacy can be evaluated by a single parameter — the total thermal energy deposited per cell (TEC) — which comprises the amount of intracellular MNP per cell, the SLP generated by the MNP and the duration of MFH treatment. Healthy L929 cells require a 50 % higher TEC to cause significant cell damage compared to pancreatic tumor cells MiaPaCa-2, which leaves a safety margin to only damage tumor cells in in vivo application of MFH [50]. Consequently, question 4 can be answered by evaluating the TEC, confirming that MFH is efficiently even at low MNP concentrations, destroying pancreatic MiaPaCa-2 tumor cells above a certain critical TEC, which is still harmless for healthy L929 cells.

Using TEC as the optimization parameter to evaluate the efficacy of MFH opens the discussion on future research opportunities to advance MFH application for tumor therapy: Further studies should include extensive in vitro analysis of MFH efficacy for many different cell lines (tumor cells and healthy references) in order to assess the individual critical TEC above which significant cell damage is expected for each cell line. Alongside, the MNP uptake kinetics for each cell line could be determined and modeled as presented in Chapter 6 to assess the amount of intracellular MNP per cell. As this work is embedded in a treatment approach for pancreatic tumors using endoscopic magnetic targeting of MNP (s. Section 2.5.4), the following specific suggestions for the application of MFH in this treatment approach can be made: Magnetic targeting of MNP should be performed for approx. 6 h to saturate MNP uptake inside cells prior to applying an AMF for $(60-90)$ min to induce sufficient cell death via MFH. As a future research opportunity, TEC could be used as a control parameter to plan in vivo application of MFH, as it can be tuned by (1) the (intracellular) particle heating, (2) the MNP uptake inside cells and (3) the duration of MFH treatment. TEC can generally be increased by using MNP optimized for particle heating in AMF of medically-tolerated strength (e. g. using low anisotropy MNP with large particle cores, $d_C \leq 25$ nm as suggested by MC-simulations (s. Section 5.3)). Then, the MNP uptake (controlled via the incubation time prior to applying the MFH) and duration of

MFH treatment could be adjusted to match the predetermined critical TEC values necessary to induce tumor-specific cell damage while leaving healthy cells unharmed.

A different research opportunity arises from investigating the suitability of MNP as tracers for the simultaneous application of magnetic particle imaging (MPI) and MFH, as it is striking that MNP with particle core sizes of $d_C = (25 - 28)\,$nm perform best for MPI and MFH alike (s. Chapter 5). From this, a theranostic platform could be envisioned, combining the advantages of imaging and heating in one MNP system and allowing to map the amount and distribution of MNP in real time using MPI, while applying MFH for tumor therapy.

A. Appendix

A.1. Supplementary Information on Chapter 2

A.1.1. MPI Harmonics Expansion

In the following, the generation of higher harmonics in the MPI (or MPS) particle signal is derived from the non-linear MNP magnetization dynamics described by the Langevin function. In particlular, it is shown why the harmonic spectrum contains only odd harmonics; adapted from [32].

As the particle signal in MPI is proportional to the derivative of the MNP magnetization, M with respect to the applied AMF, $H(t)$, i.e. $V_{\text{MNP}}(t) \propto M'(H(t)) = \frac{\mathrm{d}M}{\mathrm{d}H}$ (cf. eq. (2.55)). The MNP magnetization is determined by the Langevin function, $M \propto L(\xi(H(t)))$, with $L := \coth \xi - 1/\xi$ and its argument $\xi(H(t)) = \frac{\mu_0 V_M M_S}{k_B T} \cdot H(t)$, cf. eq. (2.27).

The generation of higher harmonics may be described by expanding $L(\xi)$ into a Taylor series according to:

$$L(\xi) = \frac{1}{3}\xi - \frac{1}{45}\xi^3 + \frac{2}{954}\xi^5 - \frac{1}{4\,725}\xi^7 + \dots \tag{A.1}$$

Without loss of generality, a sinusoidal AMF with $H(t) = -H_0 \cdot \cos(2\pi f \cdot t)$ is chosen, so that the argument of the Langevin function can be rewritten as $\xi(H(t)) = \hat{\xi} \cdot \cos(2\pi f \cdot t)$, with $\hat{\xi} = -\frac{\mu_0 V_M M_S}{k_B T} \cdot H_0$ (cf. eq. (2.28)). With the new argument ξ, eq. (A.1) reads:

$$L(\hat{\xi} \cdot \cos(2\pi f \cdot t)) = \frac{\hat{\xi}}{3} \cdot \cos(2\pi f \cdot t) - \frac{\hat{\xi}^3}{45} \cdot \cos^3(2\pi f \cdot t) + \dots \tag{A.2}$$

With the trigonometric function $\cos^3(x) = \frac{1}{4} \cdot (3\cos(x) + \cos(3x))$, eq. (A.2) can be rewritten as follows

$$L(\hat{\xi} \cdot \cos(2\pi f \cdot t)) = \frac{\hat{\xi}}{3} \cdot \cos(2\pi f \cdot t) - \frac{\hat{\xi}^3}{60} \cdot \cos(2\pi f \cdot t) + \frac{\hat{\xi}^3}{180} \cdot \cos(2\pi \cdot 3f \cdot t) + \dots \tag{A.3}$$

$$= \frac{20\hat{\xi} - \hat{\xi}^3}{60} \cdot \cos(2\pi f \cdot t) + \frac{\hat{\xi}^3}{180} \cdot \cos(2\pi \cdot 3f \cdot t) + \dots \tag{A.4}$$

Eq. (A.3) shows that the first harmonic, $f_1 = f$, and the third harmonic, $f_3 = 3f$, are present in the MPI spectrum. By expanding to the higher order terms $\cos^5$, $\cos^7$,..., it can be shown that all odd harmonics are present in the MPI spectrum. The even harmonics, however, are missing, since the derivatives of the Langevin function are zero for $\xi = 0$, at which the Taylor series is expanded.

A.2. Supplementary Information on Chapter 4

A.2.1. XRD Supplementary Information

This appendix contains the fitted XRD spectra and the ICDD reference data for magnetite and maghemite.

XRD Spectra Peak-Fits

The Pseudo-Voigt peaks fitted according to eq. (4.3) are shown in Fig. A.1(a-d). Furthermore, Fig. A.1e presents the XRD spectrum measured for a blanc glass substrate after lysin coating, showing an increase in refraction intensity between $2\theta = 20°$ and $2\theta = 25°$, followed by a steep decrease up to $2\theta = 40°$. This trend is weakly present in all MNP sample measurements as well.

Fig. A.1f shows the XRD spectrum of pure lauric acid after drying on the substrate, where peaks can be identified between $2\theta = 20°$ and $2\theta = 30°$, as well as for $2\theta \approx 40°$ and $2\theta \approx 50°$. The most prominent peaks were identified between $2\theta = 20°$ and $2\theta = 30°$ [366]. Therefore, these peaks were fitted with the Pseudo-Voigt function. The fitting parameters allowed to estimate the crystal diameter using eq. (4.2), yielding $d = (31.7 \pm 2.6)\,\mathrm{nm}$, which is clearly different from the particle crystalline sizes of $d_{\mathrm{XRD}} \approx 10\,\mathrm{nm}$ (cf. Table A.1).

Table A.1.: Pseudo-Voigt peak fit parameters and particle crystalline size d_{XRD} calculated from these fit parameters using eq. (4.2) for lauric acid sample. The anode wavelength $\lambda_{k_\alpha} = 1,5418367\,\text{Å}$ and the shape factor $K = 0.9$. Particle diameter average shown in bold. Note that the peak width w is converted to radians [rad] for the calculations.

	θ' [°]	$\sigma_{\theta'}$ [°]	w [rad]	σ_w [rad]	R^2_{adj}	d_{XRD} [nm]	$\sigma_{d_{\mathrm{XRD}}}$ [nm]
Lauric Acid	10.105	0.001	0.0044	0.00007	0.996	32.26	0.50
	10.755	0.002	0.0042	0.0001	0.975	33.99	1.12
	11.331	0.004	0.0052	0.0004	0.912	27.19	2.03
	11.921	0.002	0.0044	0.0001	0.977	32.59	0.98
	12.680	0.002	0.0044	0.0002	0.974	32.34	1.45
Average:						**31.67**	**2.60**

ICDD Reference Data

The International Centre for Diffraction Data (ICDD) provides an extensive collection of publications on powder diffractometry spectra of pure element and compounds. From these references, categorized by unique file numbers (PDF = powder diffractometry file), one for magnetite (PDF #19-0629) and one for maghemite (PDF #39-1346) are chosen. Their characteristic angles and respective intensity values are shown in Fig. A.2.

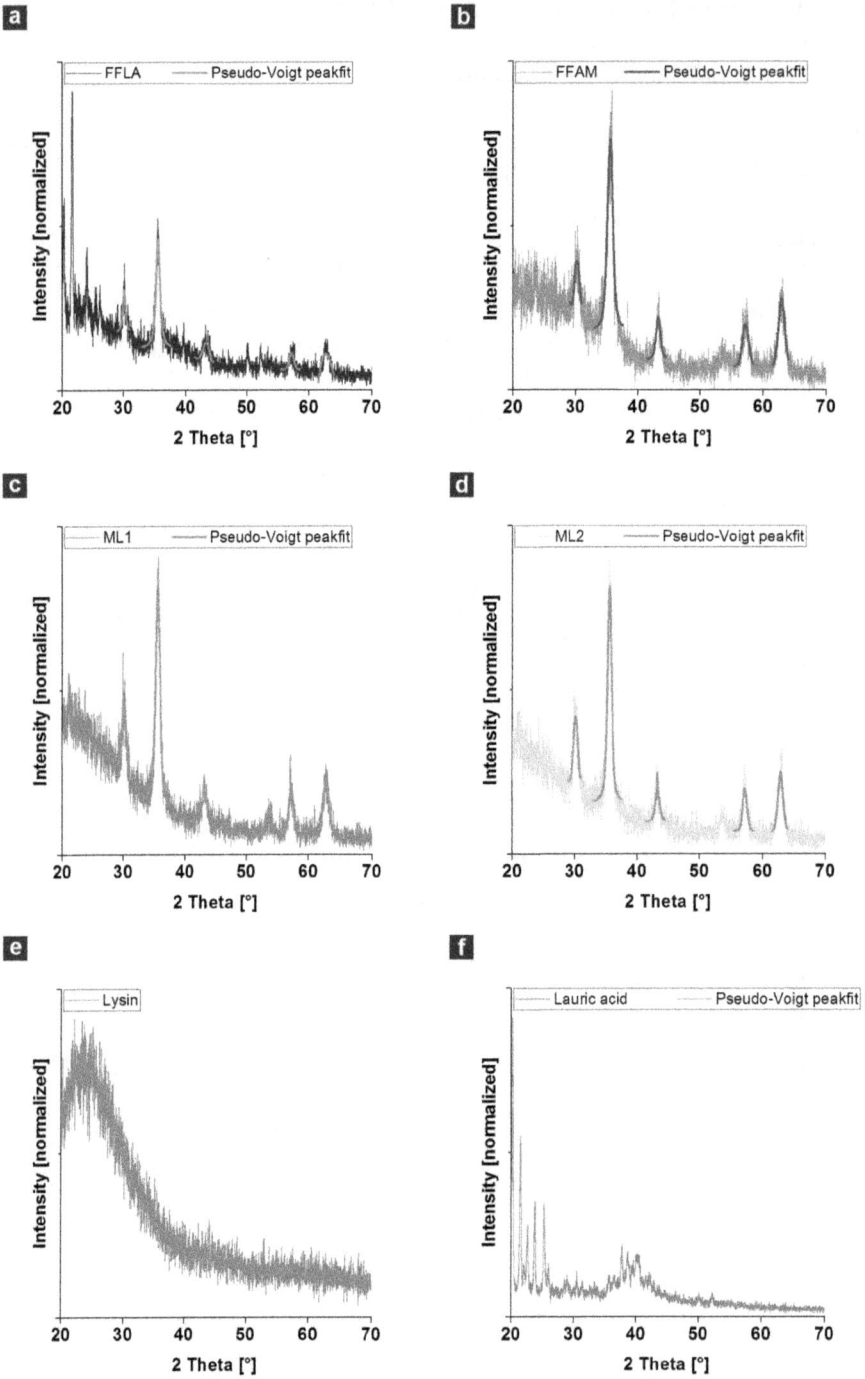

Fig. A.1.: Peakfits to the XRD measurements according to eq. (4.3): (a) FFLA, (b) FFAM, (c) ML1, and (d) ML2. (e) glass substrate with lysin coating and (f) lauric acid.

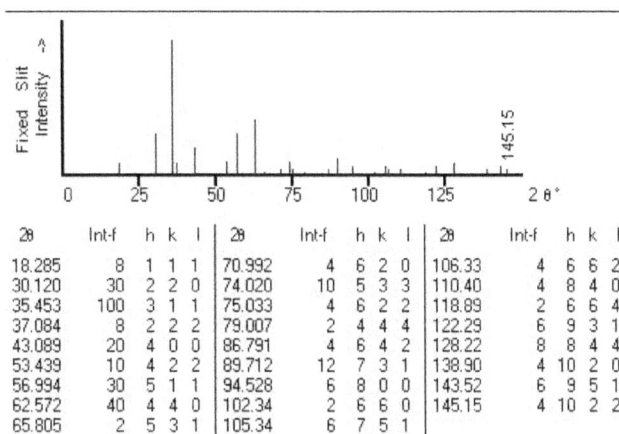

2θ	Int-f	h	k	l	2θ	Int-f	h	k	l	2θ	Int-f	h	k	l
18.285	8	1	1	1	70.992	4	6	2	0	106.33	4	6	6	2
30.120	30	2	2	0	74.020	10	5	3	3	110.40	4	8	4	0
35.453	100	3	1	1	75.033	4	6	2	2	118.89	2	6	6	4
37.084	8	2	2	2	79.007	2	4	4	4	122.29	6	9	3	1
43.089	20	4	0	0	86.791	4	6	4	2	128.22	8	8	4	4
53.439	10	4	2	2	89.712	12	7	3	1	138.90	4	10	2	0
56.994	30	5	1	1	94.528	6	8	0	0	143.52	6	9	5	1
62.572	40	4	4	0	102.34	2	6	6	0	145.15	4	10	2	2
65.805	2	5	3	1	105.34	6	7	5	1					

(a) PDF 19-0629 (magnetite)

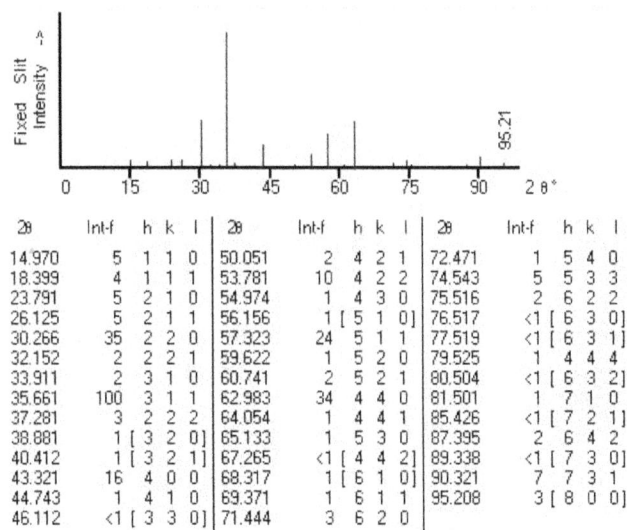

2θ	Int-f	h	k	l	2θ	Int-f	h	k	l	2θ	Int-f	h	k	l
14.970	5	1	1	0	50.051	2	4	2	1	72.471	1	5	4	0
18.399	4	1	1	1	53.781	10	4	2	2	74.543	5	5	3	3
23.791	5	2	1	0	54.974	1	4	3	0	75.516	2	6	2	2
26.125	5	2	1	1	56.156	1	[5	1	0]	76.517	<1	[6	3	0]
30.266	35	2	2	0	57.323	24	5	1	1	77.519	<1	[6	3	1]
32.152	2	2	2	1	59.622	1	5	2	0	79.525	1	4	4	4
33.911	2	3	1	0	60.741	2	5	2	1	80.504	<1	[6	3	2]
35.661	100	3	1	1	62.983	34	4	4	0	81.501	1	7	1	0
37.281	3	2	2	2	64.054	1	4	4	1	85.426	<1	[7	2	1]
38.881	1	[3	2	0]	65.133	1	5	3	0	87.395	2	6	4	2
40.412	1	[3	2	1]	67.265	<1	[4	4	2]	89.338	<1	[7	3	0]
43.321	16	4	0	0	68.317	1	[6	1	0]	90.321	7	7	3	1
44.743	1	4	1	0	69.371	1	6	1	1	95.208	3	[8	0	0]
46.112	<1	[3	3	0]	71.444	3	6	2	0					

(b) PDF 39-1346 (maghemite)

Fig. A.2.: ICDD reference XRD spectra and corresponding intensity values.

209

A.2.2. TEM Supplementary Information

The TEM images and particle core size with the cumulative distribution function (CDF) fit, eq. 4.5, are shown as follows: ML1: Fig. A.3, ML2: Fig. A.4, and FFAM: Fig. A.5.

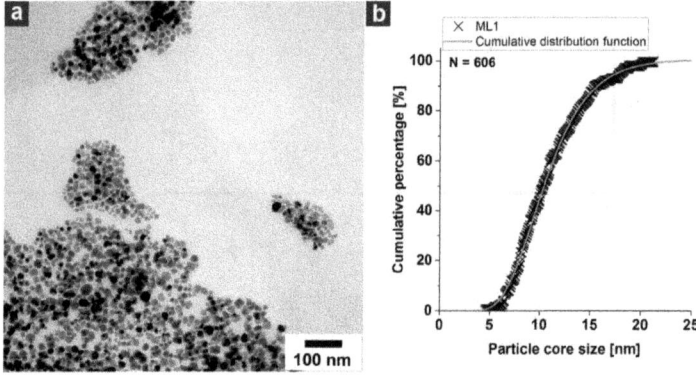

Fig. A.3.: (a) TEM image of ML1 particles with (b) the cumulative size distribution of N particles with the cumulative distribution function fit.

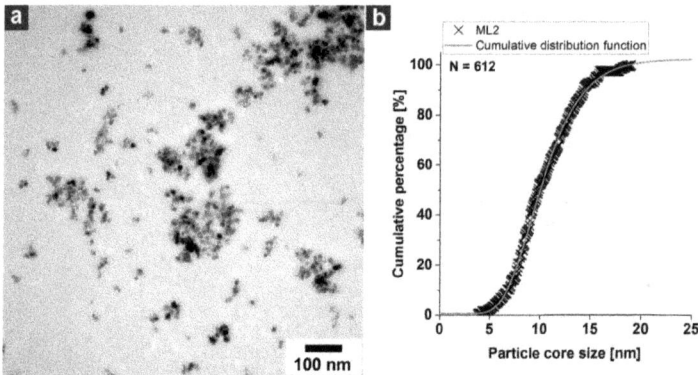

Fig. A.4.: (a) TEM image of ML2 particles with (b) the cumulative size distribution of N particles with the cumulative distribution function fit.

The fit results of the CDF fits are listed in Table A.2 togethter with the calculated particle core size d_C.

The TEM measurements and analysis on samples SEA1 through SEA4 have been conducted in the Krishnan Lab at the Universtiy of Washington (Seattle, WA, USA) and are presented courtesy of Eric Teeman. Representative TEM images together with the CDF calculated from the CDF fitting parameter values for μ and σ are presented in Figs. A.6-A.9.

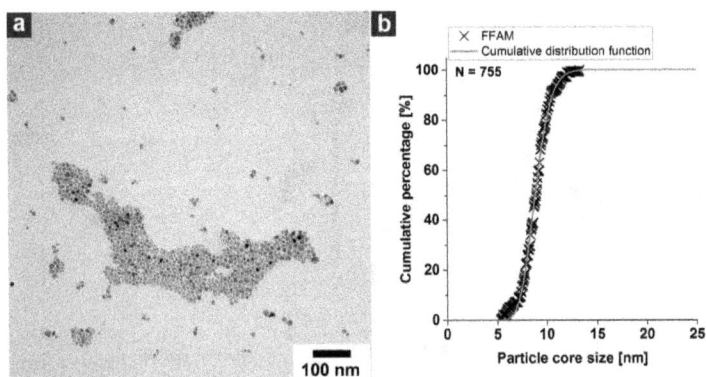

Fig. A.5.: (a) TEM image of FFAM particles with (b) the cumulative size distribution of N particles with the cumulative distribution function fit.

Table A.2.: Cumulative distribution function (CDF) fit parameters and mean particle core size calculated from eqs. (3.23) and (3.24). N denotes the number of particles used for the fitting.

	N	μ	σ	R^2_{adj}	d_C [nm]
FFLA	604	2.222 ± 0.001	0.312 ± 0.002	0.997	9.68 ± 3.10
ML1	606	2.348 ± 0.001	0.324 ± 0.001	0.999	11.03 ± 3.67
ML2	612	2.316 ± 0.001	0.308 ± 0.002	0.998	10.63 ± 3.35
FFAM	755	2.179 ± 0.001	0.146 ± 0.001	0.999	8.93 ± 1.31
SEA1*	>3000	3.0865 ± 0.0009	0.040 ± 0.001	0.999	21.92 ± 0.88
SEA2*	>3000	3.1398 ± 0.0007	0.051 ± 0.001	0.999	23.13 ± 1.16
SEA3*	>3000	3.2308 ± 0.0008	0.079 ± 0.001	0.999	25.38 ± 2.03
SEA4*	>3000	3.3214 ± 0.0009	0.072 ± 0.001	0.999	27.77 ± 1.95

*Data provided courtesy of Eric Teemnan from the Krishnan Labs, University of Washington, Seattle, USA.

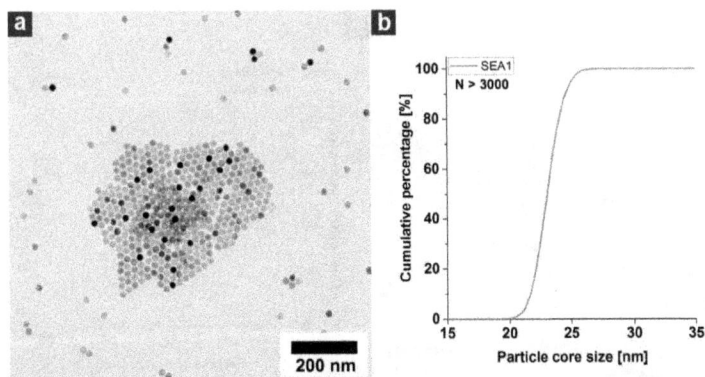

Fig. A.6.: (a) TEM image for SEA1 particles with (b) the respective cumulative distribution function. (a) adapted from [251].

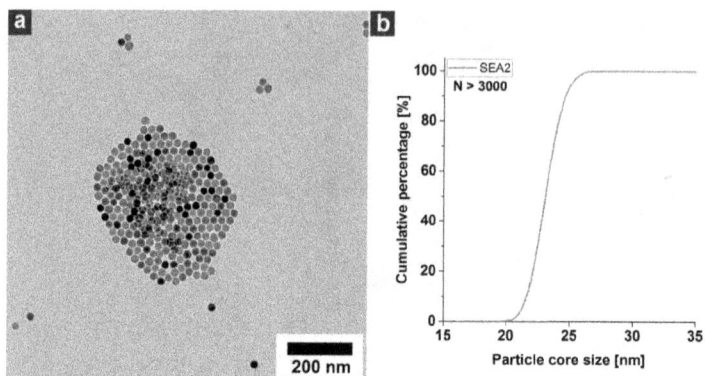

Fig. A.7.: (a) TEM image for SEA2 particles with (b) the respective cumulative distribution function.

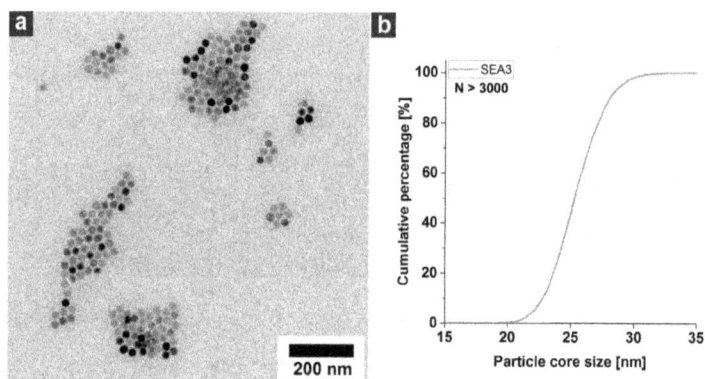

Fig. A.8.: (a) TEM image for SEA3 particles with (b) the respective cumulative distribution function.

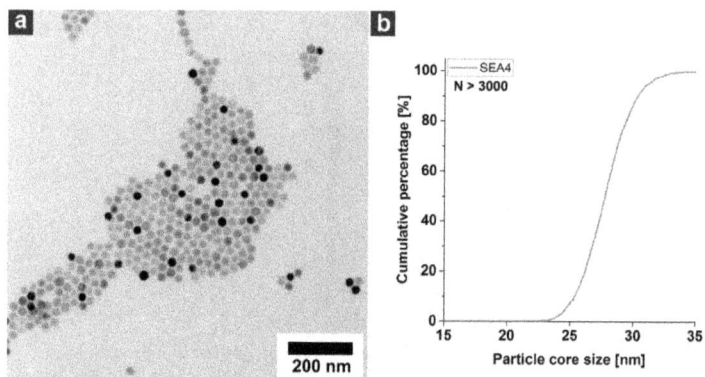

Fig. A.9.: (a) TEM image for SEA4 particles with (b) the respective cumulative distribution function. (a) adapted from [251].

A.2.3. DLS Supplementary Information

The results for fitting the volume-weighted DLS data with the PDF of the log-normal distribution are listed in Table A.3.

Table A.3.: DLS measurement results. Log-normal distribution function fit parameters, μ and σ, and mean particle hydrodynamic size, d_H, calculated from eqs. (3.23) and (3.24). z_{avg} and PdI (=*polydispersity index*) are the respective average particle hydrodynamic size and distribution width, provided by the dedicated Zetasizer software for comparison.

Sample	μ	σ	R^2_{adj}
FFLA	4.000 ± 0.005	0.329 ± 0.008	0.978
ML1	4.163 ± 0.006	0.307 ± 0.006	0.982
ML2	4.465 ± 0.010	0.366 ± 0.012	0.966
FFAM	2.630 ± 0.002	0.253 ± 0.013	0.978
SEA1	4.281 ± 0.004	0.287 ± 0.005	0.979
SEA2	3.634 ± 0.003	0.280 ± 0.005	0.979
SEA3	3.702 ± 0.003	0.219 ± 0.003	0.987
SEA4	4.341 ± 0.004	0.287 ± 0.005	0.978

A.2.4. SQUID Supplementary Information

The $M(H)$ magnetization curves for samples FFLA, ML1, ML2, ML3 and FFAM measured by SQUID were fitted with the Langevin function, cf. eq. (4.10). From this, the particle magnetic size, d_M, was calculated with the Chantrell method, as described in Section 4.6. The resulting fit parameters for Langevin fitting (Fig. 4.13) and log-normal PDF parameters are listed in Table A.4. Note that ML3 is synthesized from FFLA and yields very similar magnetic properties.

Table A.4.: Fitting paramters from Langevin function and Chantrell fitting to SQUID data (cf. Section 4.6): Inverse of the magnetic field at the origin, $\frac{1}{H_0}$ and Langevin function fitting quality, R^2_{adj}, as well as the log-normal PDF parameters, μ and σ.

	$\frac{1}{H_0}$ [Oe^{-1}]	R^2_{adj}	μ	σ
FFLA	0.0026 ± 0.0005	0.999	2.376 ± 0.057	0.437 ± 0.011
ML1	0.0031 ± 0.0009	0.998	2.393 ± 0.093	0.431 ± 0.017
ML2	0.0027 ± 0.0013	0.998	2.357 ± 0.076	0.437 ± 0.014
ML3	0.0026 ± 0.0009	0.998	2.365 ± 0.079	0.446 ± 0.015
FFAM	0.0034 ± 0.0010	0.998	2.292 ± 0.018	0.325 ± 0.003

A.2.5. VSM Supplementary Information

The M(H) magnetization curves for samples SEA1 through SEA4 measured by VSM were fitted with the Langevin function, cf. eq. (4.10). From this, the particle magnetic size, d_M, is calculated with the Chantrell method, as described in Section 4.6. The resulting fit parameters for Langevin fitting (Fig. 4.18) and log-normal PDF parameters are listed in Table A.5.

Table A.5.: Fitting paramters from Langevin function and Chantrell fitting to VSM data (cf. Section 4.7): Inverse of the magnetic field at the origin, $\frac{1}{H_0}$ and Langevin function fitting quality, R^2_{adj}, as well as the log-normal PDF parameters, μ and σ.

	$\frac{1}{H_0}$ [Oe^{-1}	R^2_{adj}	μ	σ
SEA1	0.0353 ± 0.0001	0.999	3.035 ± 0.047	0.181 ± 0.002
SEA2	0.0566 ± 0.0001	0.999	3.135 ± 0.048	0.132 ± 0.001
SEA3	0.0834 ± 0.0002	0.999	3.231 ± 0.050	0.139 ± 0.001
SEA4	0.0656 ± 0.0002	0.999	3.281 ± 0.058	0.250 ± 0.002

A.3. Supplementary Information on Chapter 5

A.3.1. MPS Supplementary Information

The MPS spectra for FFLA, MLs and FFAM are shown in Fig. A.10a and those MPS spectra
for SEA1 throught SEA4 in Fig. A.10. The detection limit for the TUB setup is reached for

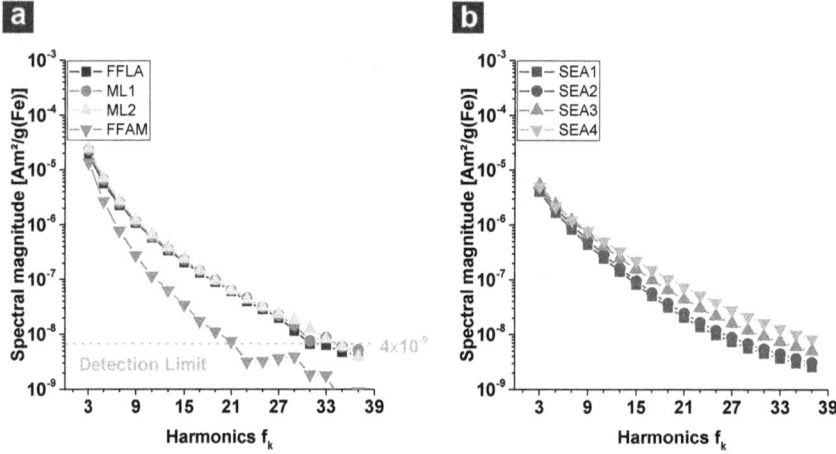

Fig. A.10.: MPS spectra (a) FFLA, MLs and FFAM measured at $f = 10\,\text{kHz}$ and $H_0 = 25\,\text{mT}/\mu_0$ and (b) SEA1
through SEA4 measured at $f = 25\,\text{kHz}$ and $H_0 = 20\,\text{mT}/\mu_0$. Below the detection limit, marked by
a dashed line in (a), the spectral magnitude is dominated by noise.

higher harmonics $(n > 21)$ in Fig. A.10a.

Additional experiments were carried out on sample SEA3, measuring the MPS spectral magnitude in dependence of the applied field amplitude, varied as $H_0 = \{6, 9, 15, 20\}\,\text{mT}/\mu_0$ at
a fixed frequency of $f = 25\,\text{kHz}$. It can be seen from Fig. A.11(a+b), that the spectral
magnitude increases with increasing H_0 for all harmonics n, even for the A_3-normalized MPS
spectrum. The evolution of the parameter A_5/A_3, plotted in Fig. A.11c, also shows an increase
with increasing H_0, as also confirmed by Arami et al. [260]. This confirms that MPS spectra
measured at different field amplitudes, H_0, must not be directly compared quantitatively.

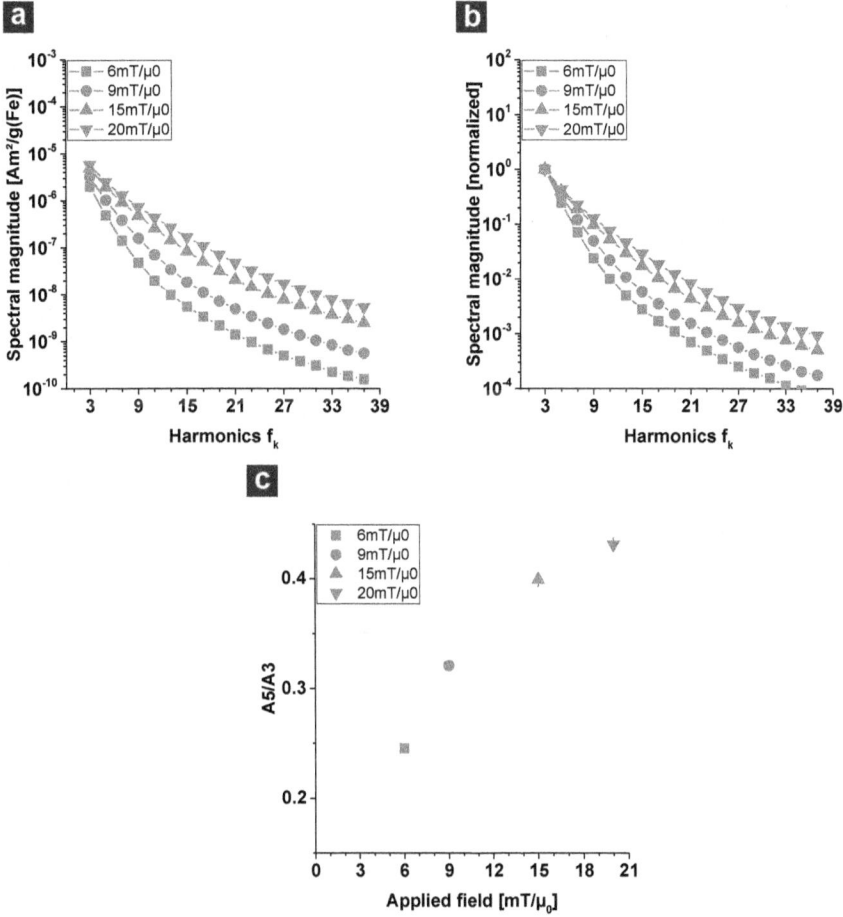

Fig. A.11.: MPS spectra of SEA3 for $f = 25\,\text{kHz}$ and $H_0 = 20\,\text{mT}/\mu_0$: (a) absolute spectral magnitude and (b) A_3-normlaized spectral magnitude. (c) shows the evolution of the A_5/A_3 parameter with increasing H_0.

A.3.2. Magnetic Dipole-Dipole Interaction

Estimation of the Average Interparticle Distance

From the particle properties listed in Table 4.11, the average interparticle distance, $\hat{d}$, can be calculated as follows, first assuming

$$\hat{d} = n^{-\frac{1}{3}}, \tag{A.5}$$

with the particle density in solution,

$$n = \frac{N_{MNP}}{V}, \tag{A.6}$$

including the number of particles, N_{MNP}, suspended in the sample volume, V. The number of particles can be expressed by

$$N_{MNP} = m_{MNP}/m_{core}, \tag{A.7}$$

with the mass of a single MNP

$$m_{core} = N_{UC} \cdot m_{UC}, \tag{A.8}$$

and the mass of all MNP in solution:

$$m_{MNP} = \xi^* \cdot c(Fe) \cdot V. \tag{A.9}$$

N_{UC} and m_{UC} in eq. (A.8) denote the number and mass of unit cells in one magnetite MNP core, respectively. $c(Fe)$ denote the MNP iron concentration and $\xi^* = 1.381$ the mass conversion factor from Fe to Fe_3O_4 in eq. (A.9).

N_{UC} can be calculated from the particle core volume, $V_C = \pi/6 \cdot d_C^3$, and the volume of a single unit cell, $V_{UC} = a_{Fe_3O_4}^3$, with the lattice parameter, $a_{Fe_3O_4} = 0.839$ nm:

$$N_{UC} = V_C/V_{UC}. \tag{A.10}$$

m_{UC} is calculated from the number of 24 Fe-atoms and 32 O-atoms in one Fe_3O_4 unit cell, having atomic weights of $m_{Fe} = 55.845u$ and $m_O = 15.999u$:

$$m_{UC} = 24 \cdot 55.845u + 32 \cdot 15.999u. \tag{A.11}$$

From this, estimates of the average interparticle distances are calculated for each sample, as listed in Table A.6.

Table A.6.: Parameters and final result for calculating the average interparticle distance, $\hat{d}$. The results are calculated for the iron concentrations, c, used for MFH experiements.

Sample	c [mL(Fe)/mL]	N_{UC}	m_{core} [kg]	N_{MNP}	n [1/m³]	$\hat{d}$ [nm]
FFLA	0.300	804	$2.47 \cdot 10^{-21}$	$1.68 \cdot 10^{14}$	$1.68 \cdot 10^{20}$	181.4
ML1	0.300	1 190	$3.66 \cdot 10^{-21}$	$1.13 \cdot 10^{14}$	$1.13 \cdot 10^{20}$	206.7
ML2	0.300	1 065	$3.28 \cdot 10^{-21}$	$1.27 \cdot 10^{14}$	$1.27 \cdot 10^{20}$	199.2
FFAM	0.300	631	$1.94 \cdot 10^{-21}$	$2.13 \cdot 10^{14}$	$2.13 \cdot 10^{20}$	167.4
SEA1	1.498	9 338	$2.87 \cdot 10^{-20}$	$1.80 \cdot 10^{13}$	$7.20 \cdot 10^{19}$	240.3
SEA2	1.315	10 971	$3.37 \cdot 10^{-20}$	$1.35 \cdot 10^{13}$	$5.38 \cdot 10^{19}$	264.9
SEA3	1.513	14 494	$4.46 \cdot 10^{-20}$	$1.17 \cdot 10^{13}$	$4.69 \cdot 10^{19}$	277.4
SEA4	1.499	18 986	$5.84 \cdot 10^{-20}$	$8.86 \cdot 10^{12}$	$3.54 \cdot 10^{19}$	304.4

Estimating the Importance of Magnetic Dipole-Dipole Interactions

The magnetic dipoe-dipole interaction energy, cf. eq. (2.39), can be approximated for two neighboring particles with magnetic moments, $\mu_1 = \mu_2 = \mu = V_M \cdot M_S$, to be within the minimum $\varepsilon_{pp\text{-}IA}^{min}(r) = \frac{\mu^2 \cdot \mu_0}{4\pi \cdot r^3}$ and maximum energy $\varepsilon_{pp\text{-}IA}^{max}(r) = 3 \cdot \frac{\mu^2 \cdot \mu_0}{4\pi \cdot r^3}$. Using the MNP properties from Table 4.11, the magnetic dipole-dipole interaction energies $\varepsilon_{pp\text{-}IA}^{min}(r)$ and $\varepsilon_{pp\text{-}IA}^{max}(r)$ are calculated and plotted against the average interparticle distance, $\hat{d}$, in Figs. A.12 and A.13.

The thermal energy of the system can be approximated by $\varepsilon_{therm} = k_B T \approx 25.8\,\text{meV}$ for $T = 300\,\text{K}$ and is plotted as a reference in Figs. A.12 and A.13. From the interception point of ε_{therm} with ε^{max} the minimum average interparticle distance, $\hat{d}_{min}$, can be estimate, above which thermal energy surely dominates magnetic interaction energy. For $\hat{d} > \hat{d}_{min}$ magnetic interaction effects can be neglected for the MNP system under investigation. In the same way, the maximum average interparticle distance, below which magnetic interaction energy surely dominates thermal energy, $\hat{d}_{max}$, can be estimated from the interception point of ε_{therm} with ε^{min}. For $\hat{d} < \hat{d}_{max}$ magnetic interaction effects must be taken into account for the MNP system under investigation. For $\hat{d}_{min} < \hat{d} < \hat{d}_{max}$ thermal and magnetic interaction energy are expected to be approximately equal (marked as a colored corridor in Figs. A.12 and A.13). The respective values are listed in Table A.7.

Table A.7.: Maximum average interparticle distance, $\hat{d}_{max}$, below which magnetic interaction energy surely dominates thermal energy ; minimum average interparticle distance, $\hat{d}_{min}$, above which thermal energy surely dominates magnetic interparticle energy. Estimated from Figs. A.12 and A.13.

Sample	d_{max} [nm]	d_{min} [nm]
FFLA	14	21
ML1	15	23
ML2	14	22
FFAM	12	18
SEA1	42	61
SEA2	53	77
SEA3	66	96
SEA4	70	103

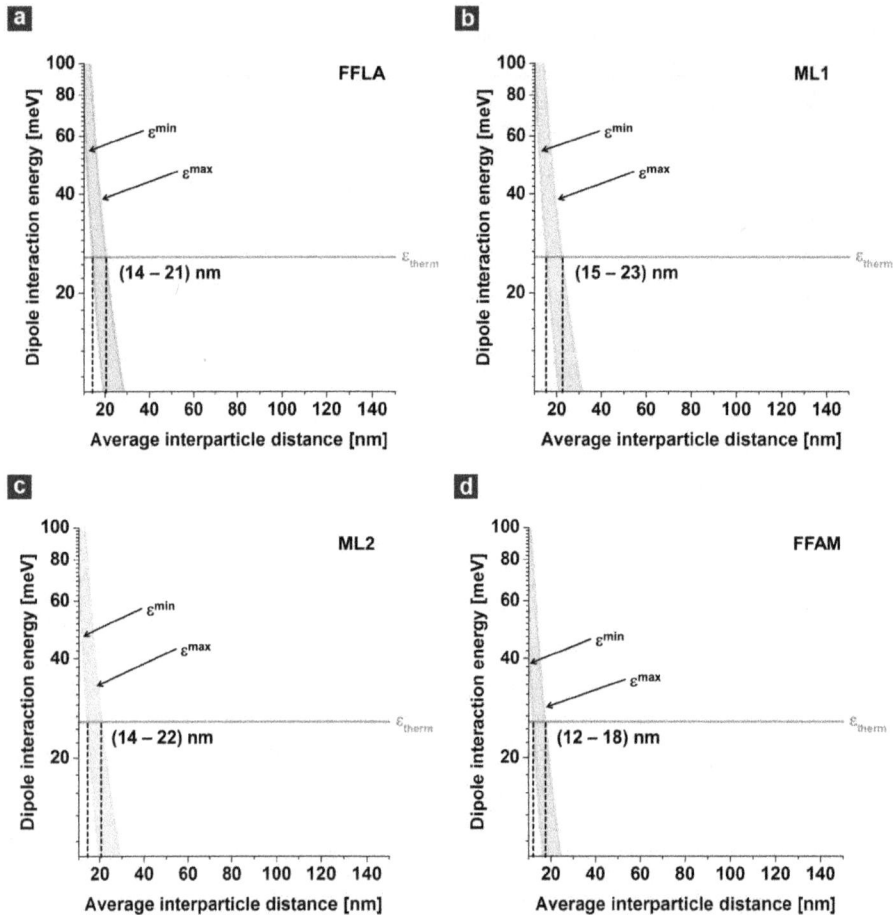

Fig. A.12.: Estimation of the maximum (ε^{max}) and minimum (ε^{min}) magnetic dipole-dipole interaction energy between two neighboring particles in dependence of the average interparticle distance for (a) FFLS, (b) ML1, (c) ML2 and (d) FFAM. The thermal energy at $T = 300$ K, ε_{therm}, is plotted for comparion; the interception point of ε_{therm} with ε^{max} gives the minimum average interparticle distance, above which thermal energy surely dominates magnetic interaction energy (hence magnetic interactions can be neglected); the interception point of ε_{therm} with ε^{min} gives the maximum average interparticle distance, below which magnetic interaction energy surely dominates thermal energy (hence magnetic interactions must be considered). The colored corridor marks the intermediate region, where thermal and magnetic interaction energy are expected to be approximately equal.

$\hat{d}_{min}$ increases generally with the particle volume; therefore the highest value is estimated for SEA4 particles ($d_M = 27.4$ nm) with $\hat{d}_{min} = 103$ nm. Since the average interparticle distances are above $\hat{d} > 180$ nm for all MNP samples studied here (s. Table A.6), magnetic dipole-dipole interactions are neglected for all MFH experiments and MC-simulations.

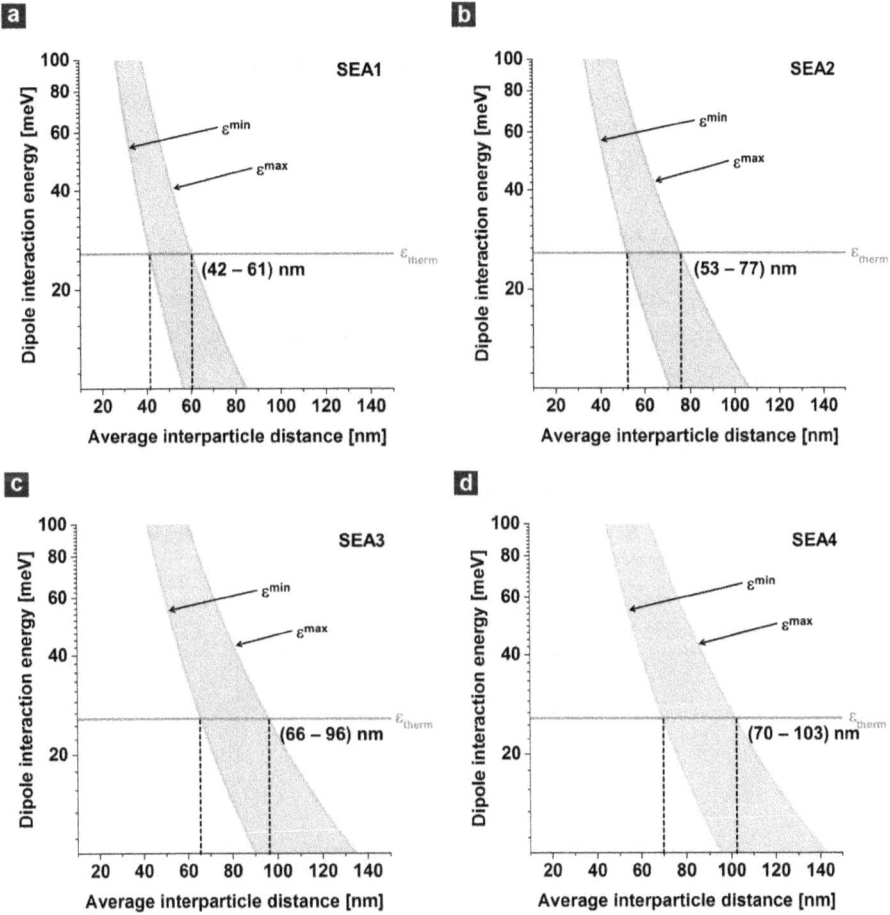

Fig. A.13.: Estimation of the maximum (ε^{max}) and minimum (ε^{min}) magnetic dipole-dipole interaction energy between two neighboring particles in dependence of the average interparticle distance for (a) SEA1, (b) SEA2, (c) SEA3 and (d) SEA4. The thermal energy at $T = 300\,\text{K}$, ε_{therm}, is plotted for comparion; the intercection point of ε_{therm} with ε^{max} gives the minimum average interparticle distance, above which thermal energy surely dominates magnetic interaction energy (hence magnetic interactions can be neglected); the intercection point of ε_{therm} with ε^{min} gives the maximum average interparticle distance, below which magnetic interaction energy surely dominates thermal energy (hence magnetic interactions must be considered). The colored corridor marks the intermediate region, where thermal and magnetic interaction energy are expected to be approximately equal.

A.4. Supplementary Information on Chapter 6

TEM Micrographs for Higher Incubation Concentrations

Additional TEM micrographs for MiaPaCa-2 (Fig. A.14), BxPC-3 (Fig. A.15) and L929 (Fig. A.16) are presented in the following. The experiments are carried out according to the TEM standard procedure as described in seciton 6.2.3 for incubation concentrations $c = 0.225\,\mathrm{mg(Fe)/mL}$ and $c = 0.3\,\mathrm{mg(Fe)/mL}$. Generally, the same trends are observed as for an incubation concnentration of $c = 0.15\,\mathrm{mg(Fe)/mL}$ (cf. Section 6.2.3) and the size of ML-loaded lysosomes is independent of the incubaiton concentration.

Size Distribution Fitting Parameters from Lysosome Size Fitting

The average size of ML-loaded lysosomes, d_{lys}, after 24 incubation was determined from a multitude of TEM images for each cell line. The distribution of d_{lys} was fitted with the CFD (cf. eq. (4.5) for each cell line individually and the fitting parameters of the CFD are listed in Table A.8.

Table A.8.: Cumulative distribution function (CDF) fit parameters and mean lysosome size, d_{lys}, calculated from eqs. (3.23) and (3.24). N denotes the number of lysosomes used for the fitting.

	N	μ [nm]	σ	R^2_{adj}	d_{lys} [nm]
MiaPaCa-2	151	5.939 ± 0.523	0.391 ± 0.024	0.963	410 ± 166
BxPC-3	69	6.390 ± 0.700	0.405 ± 0.016	0.824	646 ± 273
L929	168	6.218 ± 0.539	0.395 ± 0.018	0.881	542 ± 223

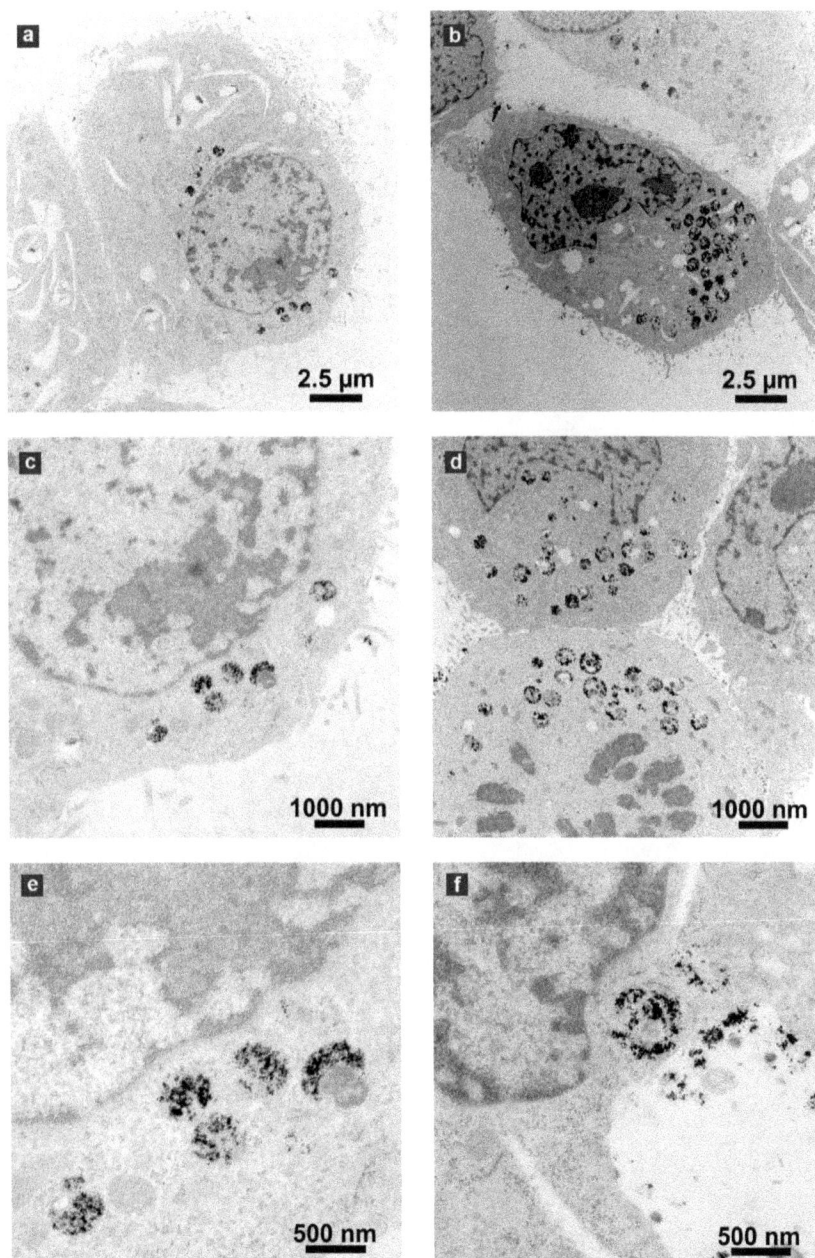

Fig. A.14.: TEM images of MiaPaCa-2 cells after incubation with ML1 particles for 24 h with magnifications (a) 4 646 x, (b) 4 646 x, (c) 10 000 x, (d) 10 000 x, (e) 27 800 x and (f) 27 800 x. ML concentrations: $c = 0.225$ mg(Fe)/mL for (a,c,e) and $c = 0.3$ mg(Fe)/mL for (b,d,f).

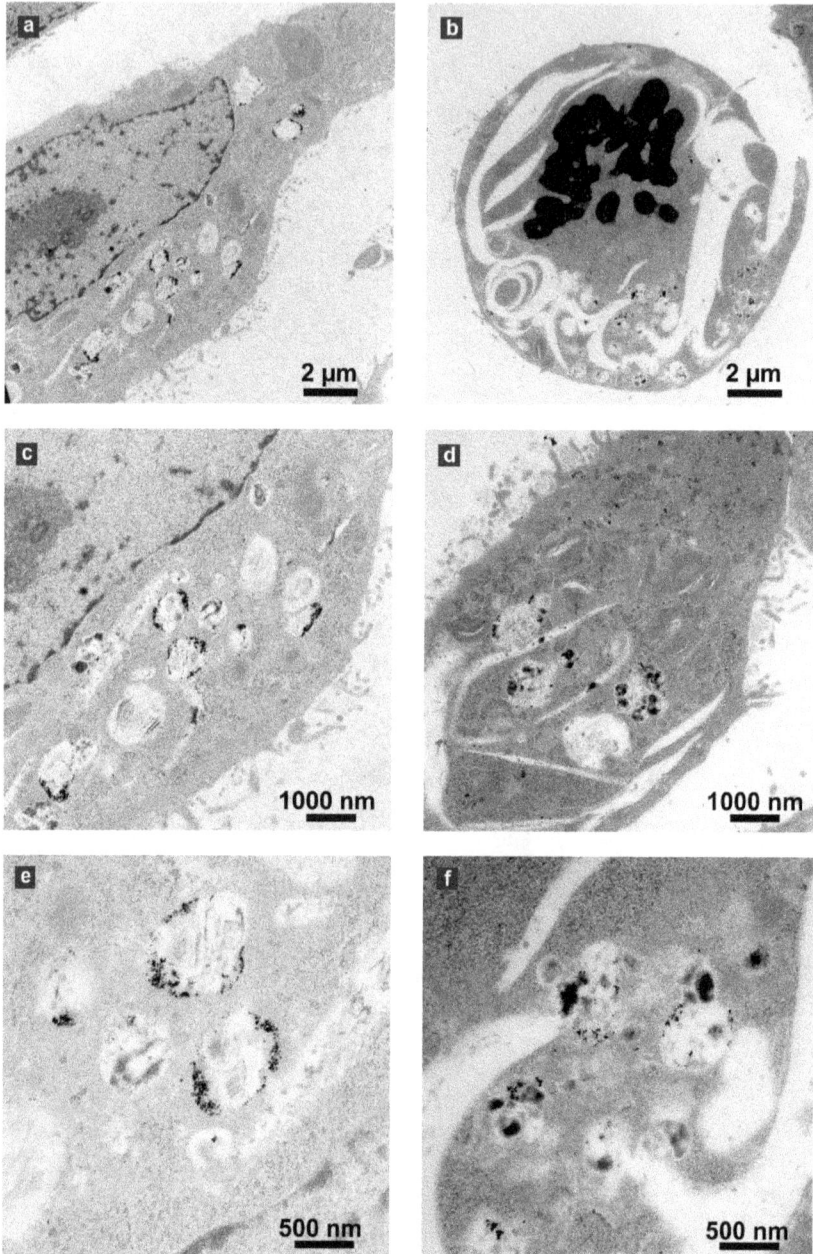

Fig. A.15.: TEM images of BxPC-3 cells after incubation with ML1 particles for 24 h with magnifications (a) 6 000x, (b) 6 000x, (c) 10 000x, (d) 10 000x, (e) 27 800x and (f) 27 800x. ML concentrations: $c = 0.225\,\text{mg(Fe)/mL}$ for (a,c,e) and $c = 0.3\,\text{mg(Fe)/mL}$ for (b,d,f).

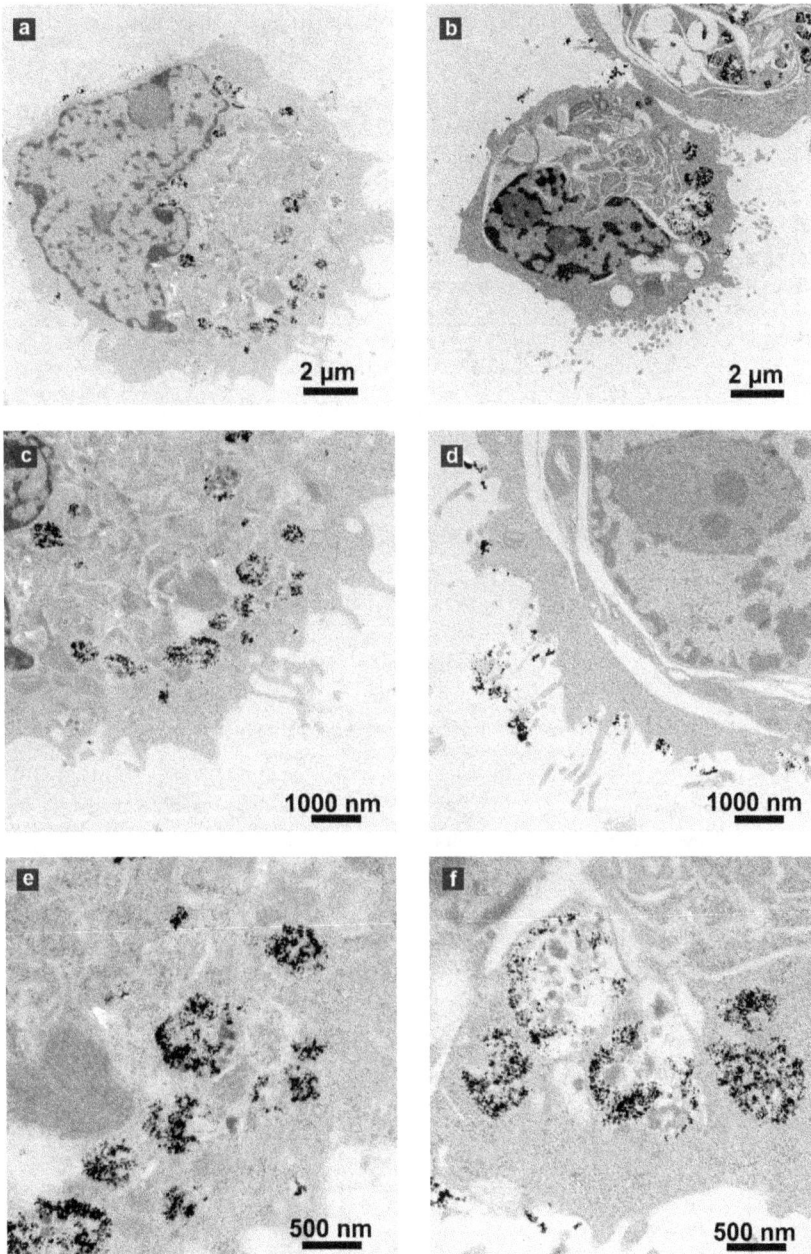

Fig. A.16.: TEM images of L929 cells after incubation with ML1 particles for 24 h with magnifications (a) 6 000x, (b) 6 000x, (c) 10 000x, (d) 10 000x, (e) 27 800x and (f) 27 800x. ML concentrations: $c = 0.225\,\text{mg(Fe)/mL}$ for (a,c,e) and $c = 0.3\,\text{mg(Fe)/mL}$ for (b,d,f).

A.5. Supplementary Information on Chapter 7

A.5.1. Plateau Storage Modulus and Mean Mesh Size of LM-Agarose Hydrogels

The plateau storage modulus, G'_P, extracted from Fig. 7.1 for LM-agarose samples are listed in Table A.9. As described in Section 7.1, the mean mesh size, d_{approx} of low melting (LM-

Table A.9.: Plateau storage moduli, G'_P, determined from rheological measurements at $T = 25\,^\circ\text{C}$ (cf. Fig. 7.1) for LM-hydrogels with the approximate mesh size, d_{approx}, derived from fitting to Fig. 7.2b (s. below).

μ_{agar} [%]	G'_P [Pa]	d_{approx} [nm]
0.3	169 ± 006	7 677.0
0.5	490 ± 020	2 640.0
1.0	$3\,150 \pm 090$	620.0
1.5	$9\,870 \pm 130$	266.0
2.0	$13\,710 \pm 160$	146.0
2.5	$44\,510 \pm 800$	91.3
3.0	$70\,380 \pm 990$	62.4
4.0	—	34.2
5.0	—	21.5
6.0	—	14.7
7.0	—	10.6
8.0	—	8.0
9.0	—	6.3
10.0	—	5.0

)agarose hydrogels is dependent on the agarose mass fraction, μ_{agar} used for preparing the gels but cannot be derived from G'_p as easily as for PAAm hydrogels. The dependencies of d_{approx} on μ_{agar} for agarose and LM-agarose listed in Tbl. A.10 are extracted from the work of Narayanan et al. (Fig. 3 in [333]). This data was used for fitting with a power law (eq. (7.3)) in Fig. 7.2.

Table A.10.: Mean mesh sizes, d_{cross} in dependence of agarose mass fraction, μ_{agar} for agarose and low melting (LM-) agarose samples extracted from Narayanan et al. [333]). Uncertainty on d_{cross} estimated from data extraction.

Agarose		LM-agarose	
μ_{agar} [%]	d_{approx} [nm]	μ_{agar} [%]	d_{approx} [nm]
0.5	460 ± 20		
1.0	125 ± 20	1.0	625 ± 20
1.5	80 ± 20	1.5	240 ± 20
2.0	75 ± 20	2.0	175 ± 20
2.5	50 ± 20	2.5	90 ± 20
3.0	35 ± 20		

A.5.2. Plateau Storage Modulus and Mean Mesh Size of Poly(Acrylamide) Hydrogels

The plateau storage moduli, G'_P, extracted from Fig. 7.3 and the mean mesh sizes calculated from eq. (7.5) for $\nu_{\text{pol}} = 8\,\%$ and $\alpha = (0.05 - 0.5)\,\%$ are listed in Tbl. A.11.

Table A.11.: Plateau storage moduli, G'_P, determined from rheological measurements at $T = 25\,°C$ (cf. Fig. 7.3) for PAAm hydrogels with polymer volume fraction $\nu_{\text{pol}} = 8\,\%$ and the respective crosslinker mole fraction, α. From this, the mean mesh size, d_{cross}, was calculated.

α [%]	G'_P [Pa]	d_{cross} [nm]
0.05	434 ± 156	$26.7 \pm 3.2)$
0.075	833 ± 169	21.5 ± 1.5
0.1	811 ± 153	21.7 ± 1.4
0.15	$1\,161 \pm 244$	19.2 ± 1.3
0.2	$1\,612 \pm 229$	17.2 ± 0.8
0.3	$2\,776 \pm 110$	14.4 ± 0.2
0.4	$3\,985 \pm 147$	12.7 ± 0.2
0.5	$4\,152 \pm 147$	12.6 ± 0.1

A.5.3. Exothermal Temperature Rise during Preparation of Poly(Acrylamide) Hydrogels

Four poly(acrylamide) (PAAm) hydrogels with a high volume fraction of polymer monomers, $\nu_{pol} \approx (20 - 70)\,\%$, (cf. eq. (7.2)) and variable bisacrylamide (BIS) crosslinker mole fraction, $\alpha = (0 - 53)\,\%$, (cf. eq. (2.76)), were prepared according to the procedure desribed in Section 7.1. To estimate the temperature rise during the exotherm polymerization, a fiberoptic thermometer (Lumasense, Santa Barbara, CA, USA) was inserted in the polymer mixture before polymerization was initialized. The resulting temperature rises measured over half an hour are plotted in Fig. A.17, showing a temperature increase by $\Delta T \approx (10 - 15)\,\text{K}$.

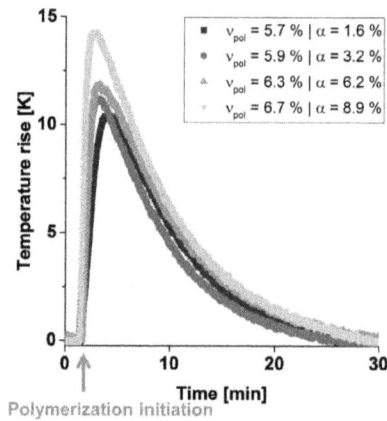

Fig. A.17.: Temperature rise during exothermal polymerization of (poly)acrylamide hydrogels with polymer monomer volume fraction ν_{pol} and crosslinker mole fraction α. The time of polymerization initialization is marked by the red arrow.

A.5.4. Freeze-Fractured TEM Images of Poly(Acrylamide) Ferrohydrogels

Further exemplary freeze-fractured TEM images of PAAm FHG are shown in Fig. A.18 for the representative PAAm hydrogel with $\nu_{pol} = 8\%$ and $\alpha = 0.075\%$.

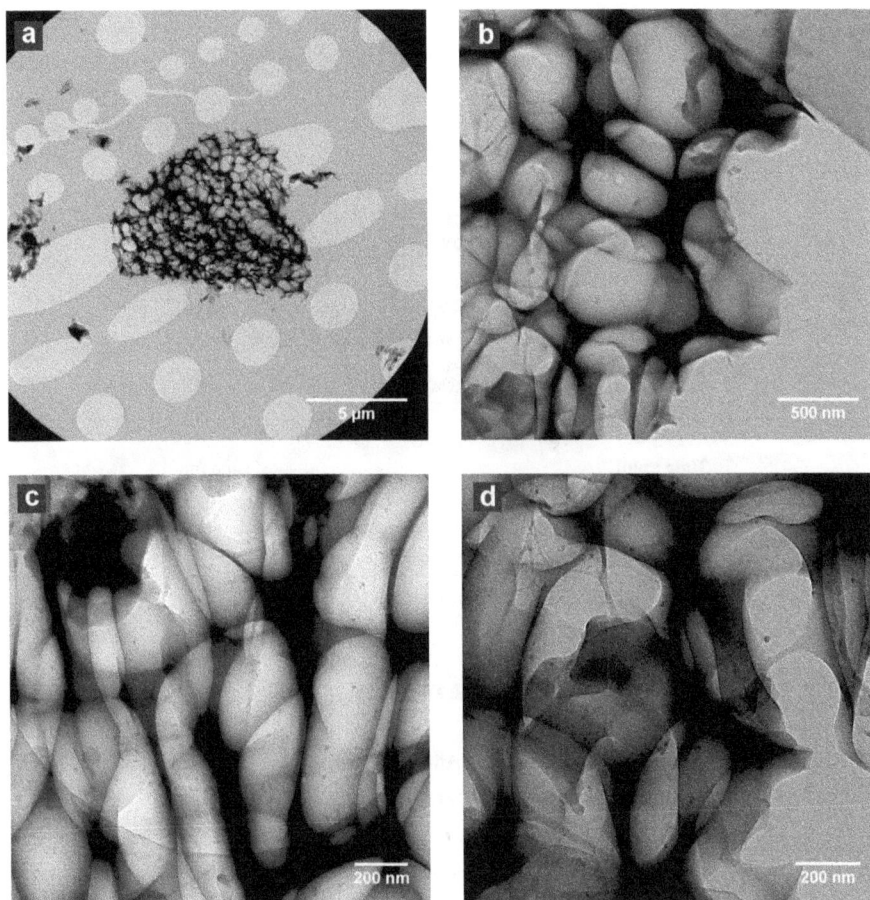

Fig. A.18.: Exemplary TEM images of freeze-fractured PAAm hydrogels ($\nu_{pol} = 8\%$ and $\alpha = 0.075\%$) containing FFAM with $c = 0.3\,mg(Fe)/mL$. Magnifications are (a) $3\,000\times$, (b) $27\,000\times$, (c) $40\,000\times$ and (d) $50\,000\times$. Provided courtesy of Julian Seifert and Stefan Roitsch (Universität zu Köln).

A.5.5. MFH Heating Curves for Ferrohydrogels

Calorimetric hyperthermia heating curves for low melting agarose and poly(acrylamide) ferrohydrogels containing $c = 0.3\,\text{mg(Fe)/mL}$ FFAM particles, measured with AMF parameters $H_0 = 50\,\text{mT}/\mu_0$ and $f = 270\,\text{kHz}$ are shown in Fig. A.19.

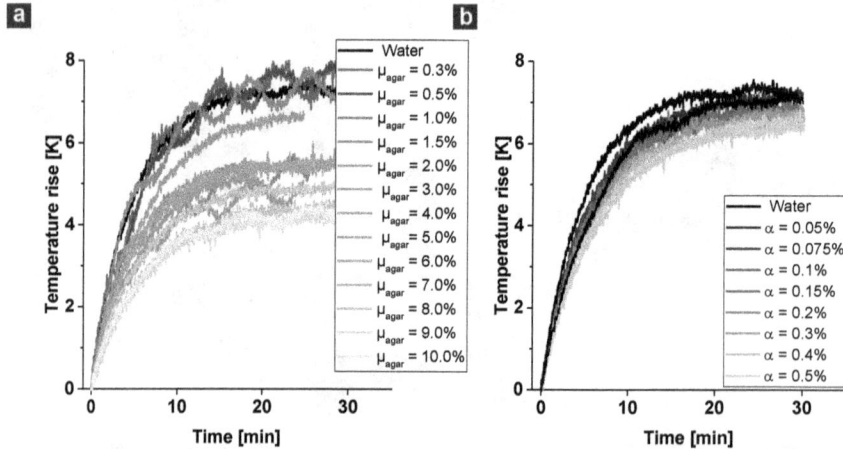

Fig. A.19.: Heating curves of FFAM incorporated in (a) low melting agarose with mass fractions $\mu_{agar} = (0.3 - 10.0)\,\%$ and (b) poly(acrylamide) with $\nu_{pol} = 8\,\%$ for various cross linker fractions $\alpha = (0.05 - 0.5)\,\%$. The heating curve for FFAM in water is shown for reference. The initial temperature for all measurements is $T = 35\,^{\circ}\text{C}$. (b) adapted from [264].

A.5.6. DLS Fitting Parameters for FFAM treated with $T = 70\,^{\circ}\text{C}$

DLS measurements were performed with FFAM in DI-H_2O at $T = 70\,^{\circ}\text{C}$ and at ambient conditions. The results from fitting the DLS data from Fig. 7.5 are listed in Table A.12.

Table A.12.: Parameters from PDF-fitting, μ and σ, fit quality parameter R^2_{adj} and mean particle hydrodynamic size, d_H, calculated with eqs. (3.23) and (3.24). z_{avg} and PDI are the respective average hydrodynamic particle size and distribution width, provided by the Malvern software for comparison.

	z_{avg} [nm]	PDI	μ [nm]	σ	R^2_{adj}	d_H [nm]
			Intensity-weighted			
FFAM	20.6	0.189	2.840 ± 0.008	0.317 ± 0.007	0.981	18.9 ± 6.1
FFAM 70 °C	21.2	0.230	2.881 ± 0.009	0.302 ± 0.004	0.993	19.5 ± 6.0
			Volume-weighted			
FFAM	20.6	0.189	2.634 ± 0.008	0.256 ± 0.005	0.987	14.4 ± 3.7
FFAM 70 °C	21.2	0.230	2.670 ± 0.008	0.255 ± 0.004	0.990	14.9 ± 3.9

A.5.7. DLS Fitting Parameters for Agglomerated Particles

LUM Sample

DLS measurements were performed for LUM and fitted with the log-normal PDF (cf. Fig. 7.11; eq. (3.22)), whose fitting parameters are listed in Table A.13.

Table A.13.: PDF fitting parameters μ and σ, fit quality parameter R^2_{adj} and results for hydrodynamic size d_H for LUM from DLS analysis (cf. Section 7.3.2).

Sample	μ [nm]	σ	R^2_{adj}	d_C [nm]
LUM peak 1	3.170 ± 0.003	0.358 ± 0.007	0.988	25.4 ± 9.4
LUM peak 2	5.686 ± 0.012	0.453 ± 0.007	0.988	326.6 ± 156.0

FFLA Agglomerates

The results and fitting parameters from PDF fitting to the DLS data of FFLA samples treated with NaCl (cf. Fig. 7.12) are summarized in Table A.14.

Table A.14.: DLS measurment PDF fitting parameters μ and σ, fitting quality parameter R^2_{adj} and the mean hydrodynamic size d_H from DLS size analysis for NaCl-treated FFLA samples.

NaCl-treatment	μ	σ	R^2_{adj}	d_H [nm]
0.0 M	4.598 ± 0.012	0.468 ± 0.003	0.985	110.8 ± 54.8
0.1 M	4.611 ± 0.015	0.529 ± 0.002	0.969	115.7 ± 65.8
0.25 M peak 1	4.647 ± 0.003	0.288 ± 0.001	0.985	108.7 ± 31.9
0.25 M peak 2	5.863 ± 0.015	0.430 ± 0.060	0.985	386 ± 173
0.5 M	4.617 ± 0.007	0.427 ± 0.003	0.997	110.8 ± 49.6

A.5.8. TEM Fitting Parameters for Agglomerated Particles

LUM Sample

TEM measurements were performed for LUM and fitted with the log-normal CDF (cf. Fig. 7.13; eq. (4.5)), whose fitting parameters are listed in Table A.15.

Table A.15.: CDF fitting parameters μ and σ, fitting quality parameter R^2_{adj} and results for mean particle (agglomerate) size d_C for LUM from TEM analysis (cf. Section 7.3.3).

Sample	μ [nm]	σ	R^2_{adj}	d_C [nm]
LUM	5.484 ± 0.006	0.281 ± 0.007	0.979	250.5 ± 71.8

FFLA Agglomerates

TEM measurements were performed for NaCl-treated FFLA and the distribution of mean agglomerate sizes was fitted with the log-normal PDF (cf. Figs. 7.14 and 7.15; eq. (3.22)), whose fitting parameters are listed in Table A.15.

Table A.16.: PDF fitting parameters μ and σ, fitting quality parameter R^2_{adj} and the mean (agglomerate) size d_C derived from TEM image analysis for NaCl-treated FFLA samples.

NaCl-treatment	μ	σ	R^2_{adj}	d_C [nm]
0.0 M	2.199 ± 0.002	0.317 ± 0.003	0.996	9.5 ± 3.1
0.1 M	4.428 ± 0.006	0.510 ± 0.025	0.981	95.4 ± 52.0
0.25 M peak 1	4.572 ± 0.016	0.690 ± 0.104	0.852	122.7 ± 95.8
0.25 M peak 2	5.508 ± 0.003	0.131 ± 0.095	0.852	248.8 ± 32.5
0.5 M	5.343 ± 0.008	0.509 ± 0.049	0.884	238.1 ± 129.5

A.6. Supplementary Information on Chapter 8

A.6.1. Cell Processing for MPS

A processing factor, K, was estimated for $1 \cdot 10^6$ cells seeded initially, describing the loss of cells during preparation of cell samples fro MPS measurements due to the removal of the supernatant (where cells were observed to be removed as well). Therefore, cells were counted with the LUNA automated cell counter right after harvest and after the final precessing step (cf. Section 8.1.1). From this, the relative difference in cell count before and after processing, $D[\%]$, is calculated and the processing factor $K = 1 + D[\%]/100$ is derived (cf. Table A.17). The cell count of MPS samples is now multiplied by K to determine the real cell count after sample processing.

Table A.17.: Calculation of the processing factor, K, for MPS cell sample preparation. Adapted from [50].

Cell line	Sample	Cell count before processing	Cell count after processing	Difference	$D[\%]$	K
MiaPaCa-2	1	1 230 000	875 000	−355 000	−28.86	
	2	1 486 500	795 000	−691 500	−46.52	
	3	1 635 000	930 000	−705 000	−43.12	
	Mean				−39.50	0.605 ± 0.077
L929	1	1 515 000	730 000	−785 000	−51.82	
	2	1 434 500	600 000	−834 500	−58.16	
	3	1 380 000	770 000	−610 000	−44.20	
	Mean				−51.59	0.486 ± 0.057

A.6.2. ML Dispersed in Cell Medium

Agglomeration Study

ML2 MNP were suspended in DMEM (MiaPaCa-2 cell culture medium) and RPMI (L929 cell culture medium) at final concentrations of $c = (150, 225$ and $300)\,\mu g(Fe)/mL$. The suspensions were analyzed after 3 h of rest with DLS, as described in Section 4.4, yielding the hydrodynamic diameters (z-averages) shown in Fig. A.20(a,b). When comparing hydrodynamic diameters,

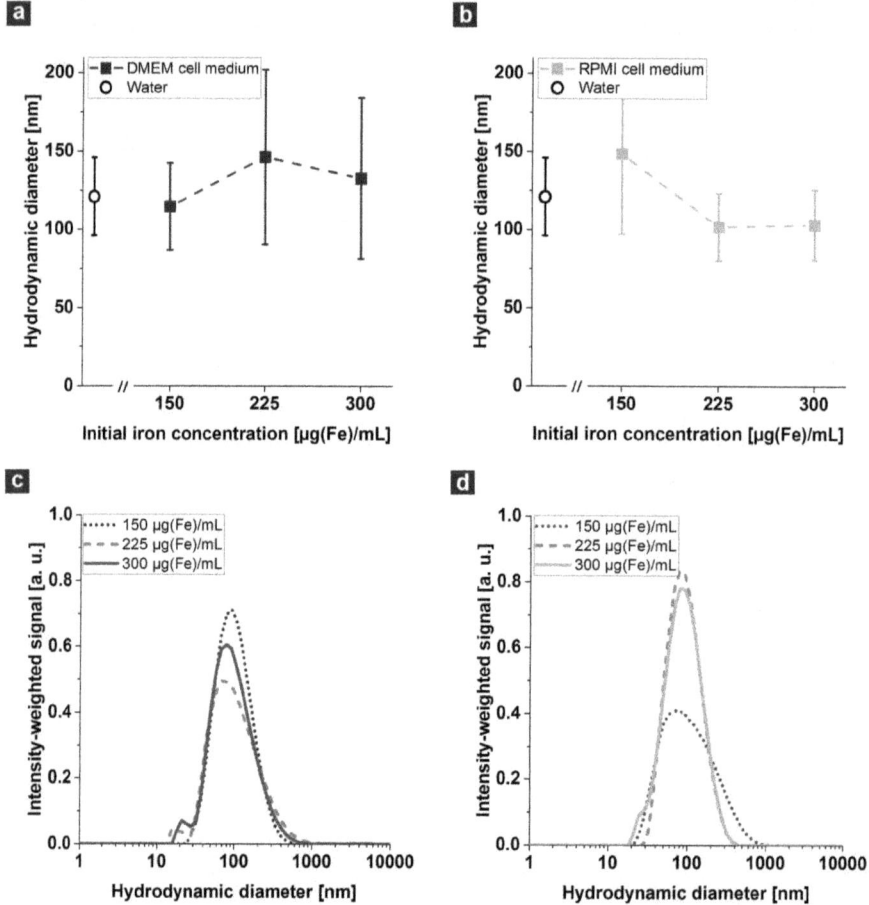

Fig. A.20.: ML2 MNP in cell culture medium at various concentrations: Hydrodynamic diameter (z-average) in (a) DMEM (MiaPaCa-2 cell medium) and (b) RPMI (L929 cell medium), both compared to the z-average measured in water, $z_{avg} = (121.5 \pm 25.3)\,nm$ (cf. Table 4.5). Intensity-weighted DLS signal over hydrodynamic diameter for ML2 dispersed in (c) DMEM and (d) RPMI. Taken from [50].

d_H, to that measured in water, d_H is only slightly increased for ML2 suspended in DMEM at higher concentrations of $c = (225$ and $300)\,\mu g(Fe)/mL$, while it increases substantially to $d_H \approx 150\,nm$ for ML2 suspended in RPMI at $c = 150\,\mu g(Fe)/mL$. Furthermore, the polydisperse character of a sample is reflected in the large error bars. Agglomeration can also be detected from intensity-weighted DLS signal plots (as intensity-weighted signal is most sensitive to large particles, $I \propto d_H^6$, cf. Section 4.4), as shown in Fig. A.20(c,d): Here, the distinct shoulder up to $d_H \approx 700$ nm observed for ML2 at $c = 150\,\mu g(Fe)/mL$ in RPMI also indicates the formation of agglomerates. This tendency for nanostructures of low concentration to form agglomerates is also reported in literature [303, 367]. Contrastingly, the formation of shoulders in the DLS signal for smaller sizes at $d_H \approx 20$ nm for ML2 at higher concentrations in both media indicates a possible breaking of the ML.

Sedimentation Study

In order to determine the sedimentation of ML in cell culture medium, $1\,mL$ of ML2 MNP suspended in DMEM or RPMI was pipetted in 6-plate wells at concentrations of $c = (150$ and $300\,\mu g(Fe)/mL$. The well-plates were incubated for 24 h and afterwards, the iron concentration at the top and the bottom of each well was determined via photometric analysis as described in Section 4.5. As indicated in Fig. A.21, some minor degree of sedimentation was observed for all samples, detectable as an increased iron concentration at the bottom and a decreased one at the top of the well. This is attributed to gravitational sedimentation [368]. However, the sedimentation effect was only significant ($p < 0.05$; t-testing 2-tailed with $\alpha = 0.05$) for ML2 at $c = 150\,\mu g(Fe)/mL$ in RPMI. This is another indicator for a stronger agglomeration of ML in this sample setting, as discussed above.

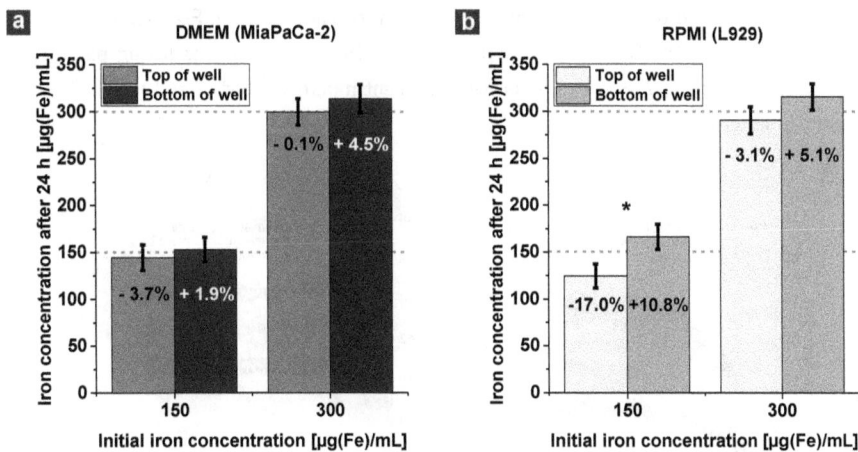

Fig. A.21.: Sedimentation analysis for ML2 after 24 h of incubation in (a) DMEM cell medium and (b) RPMI cell medium. Iron concentration was first measured at the top of the well and subsequently at the bottom (sequence was chosen not to falsify the iron content measured at the top). Initial iron concentrations of $c = (150$ and $300\,\mu g(Fe)/mL$ are indicated by dashed lines. A statistical significant difference in concentration between top and bottom was measured in (b) for $c = 150\,\mu g(Fe)/mL$, indicated with *. Taken from [50].

A.6.3. T(t)-curves Measured in vitro

Exemplary T(t)-curves for intra- & extracellular ML of both cell lines comparing the thermometer position at the bottom of the vial and in the middle of the vial are shown in Fig. A.22. As the temperatures are identical for both positions, further analyses use the T(t)-curves measured at the bottom of the well, where the cells settle. Exemplary T(t)-curves for intra- & extracellular

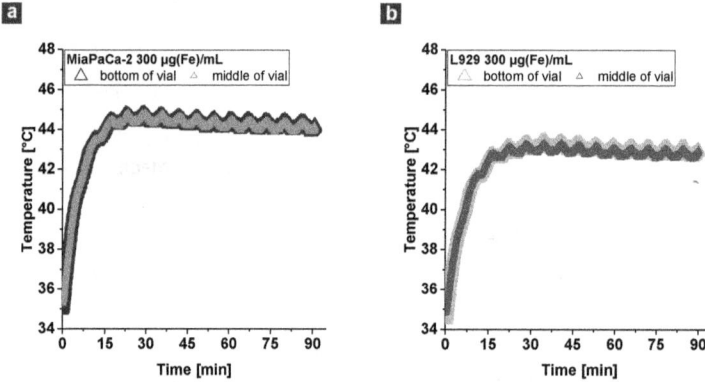

Fig. A.22.: Exemplary T(t)-curves for intra- & extracellular ML with $c = 300\,\mu g(Fe)/mL$ for (a) MiaPaCa-2 and (b) L929, comparing the temperatures measured at the bottom of the vial (cell level) and in the middle of the vial during MFH (AMF applied for 90 min). Taken from [50].

ML of both cell lines and the three incubation concentrations are shown in Fig. A.23. The T(t)-curves were fitted with the Box-Lucas function (cf. eq. (5.2)), from whose fitting parameters the effective temperature, T_{eff}, the cumulative equivalent minutes, CEM_{43}, and the SLP values reported in Chapter 8 were calculated.

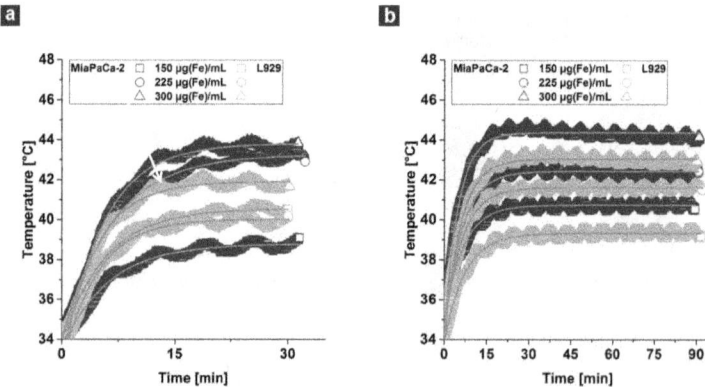

Fig. A.23.: Exemplary T(t)-curves for intra- & extracellular ML with $c = (150, 225$ and $300\,\mu g(Fe)/mL$ for both cell lines MiaPaCa-2 and L929 for (a) 30 min and (b) 90 min of AMF application. Box-Lucas function fitting is shown in red. Fluctuations in the temperature arise from periodic switching of the cooling system (cf. Section 5.2.1). Taken from [50].

A.6.4. Preliminary Cytotoxicity Tests

Before assessing the effects of MFH on cells ($\rightarrow$8.2.2), the effects of ML-treatment and AMF application on cell survival were analyzed isolatedly.

ML Cytotoxicity

The cytotoxic effect of ML presence during $24\,\text{h}$ of incubation on the cells was evaluated via clonogenic assay. Therefore, survival fractions (SF) were calculated by comparing CA results of ML-treated samples (incubated with ML for $24\,\text{h}$ but otherwise untreated) with entirely un-treated control samples. The results depicted in Fig. A.24 show a general moderate decrease in SF for MiaPaCa-2, which is strongest (down to SF$\approx 75\,\%$) for the highest ML concen-tration $c = 300\,\mu\text{g(Fe)}/\text{mL}$. For L929, SF is remarkably higher for cells treated with ML at $c = 150\,\mu\text{g(Fe)}/\text{mL}$ (SF$\approx 150\,\%$), suggesting a stimulating effect of ML on cell growth. Con-trastingly, higher ML concentrations of $c = 300\,\mu\text{g(Fe)}/\text{mL}$ decreased SF down to $\approx 50\,\%$, suggesting a rather toxic effect, while SF was comparable to untreated control samples for $c = 225\,\mu\text{g(Fe)}/\text{mL}$. The toxicity for higher concentrations in both cell lines could either arise

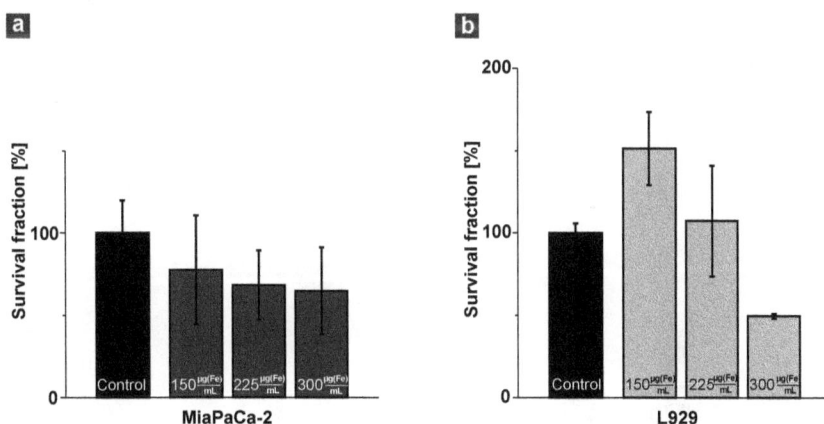

Fig. A.24.: Survival fraction of untreated cells (control) compared to cell incubated with ML2 for $24\,\text{h}$ at different incubation concentrations for (a) MiaPaCa-2 and (b) L929. Taken from [50].

from sedimentation effects [368] (cf. appendix A.6.2) or the breaking of ML into smaller MNP: Sedimented ML could form a dense ML layer on top of adherent cells over incubation time, pos-sibly blocking the nutritious and oxygen supply of the cells, leading to increased toxicity. In this framework, the increased viability of L929 at $c = 150\,\mu\text{g(Fe)}/\text{mL}$ could result from the improved internalization of ML due to the formation of large agglomerates (s. discussion above and in appendix A.6.2): the increased internalization would also increase the nutrient flow into the cells and promote growth. For higher ML concentration, the partial breaking of ML in smaller MNP was observed in both cell culture media ($\rightarrow$A.6.2), which could further promote the formation of a dense and therefore toxic particle layer. In general, small Ag-particles ($d_C \approx 13\,\text{nm}$) showed a remarkably increased toxicity when compared to larger ones ($d_C \approx 80\,\text{nm}$ and $d_C \approx 113\,\text{nm}$) when tested on L929 [369]. This could be another reason for the substantial decrease in SF for L929 at $c = 300\,\mu\text{g(Fe)}/\text{mL}$.

Overall, ML can be classified as mostly biocompatible and only mildly toxic, especially at low concentrations.

AMF Cytotoxicity

Similarly, control samples only containing cells were subjected to the AMF for either 30 min or 90 min and compared to control samples not treated with the AMF. The resulting surviving fractions (SF) are shown in Fig. A.25. No effect of AMF treatment on cell's SF was observed for L929. Interestingly, AMF treatment for 30 min had a growth stimulating effect on MiaPaCa-2, while AMF treatment for 90 min had no effect. Moreover, the variance in SF (represented by the large error bars) was substantially higher for MiaPaCa-2 tumor cell compared to healthy L929 cells, reflecting a higher response of MiaPaCa-2 tumor cells to the AMF. Generally, cell growth can be altered — stimulated or inhibited — by alternating electromagnetic fields [370], but a clear trend is yet unknown: E. g. bone cells have been shown to experience stimulated growth in low frequency AMF ($f \sim 60\,\text{Hz}$) applied for 10 min, which is explained by increased calcium influx [371]. Contrastingly, human cancer cells were reported to show significantly reduced proliferation rates after 24 h application of AMF with $f \sim (100-300)\,\text{kHz}$ [372]. Consequently, the growth simulating effect for MiaPaCa-2 after 30 min of AMF application can not ultimately be explained here.

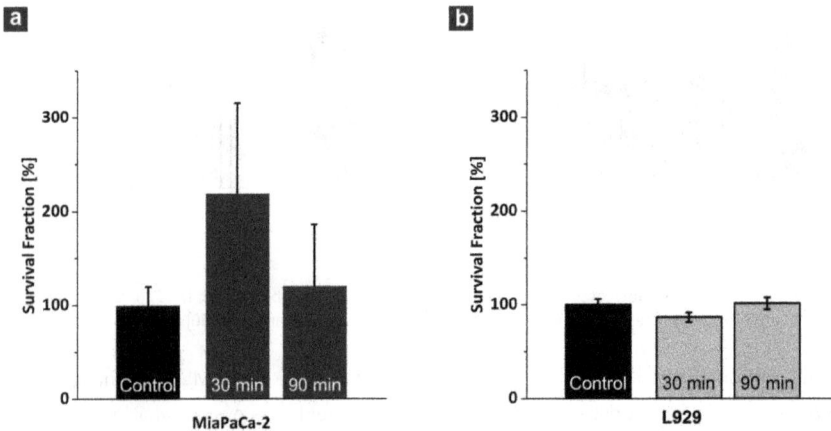

Fig. A.25.: Survival fraction of untreated cells (control) compared to cell subjected to AMF ($H_0 = 50\,\text{mT}/\mu_0$ and $f = 270\,\text{kHz}$) for 30 min or 90 min for (a) MiaPaCa-2 and (b) L929. Taken from [50].

A.6.5. SLP Values of Internalized ML

Detection Limit for Internalized ML in SLP Measurements

The SLP values for each sample are plotted against the internalized amount of ML (m_{int}) in Fig. A.26. As can be seen easily, no SLP values could be determined below the detection limit of $15\,\mu g(Fe)$ due to an insufficient heating (i. e. no heating was measured in T(t)-curves and consequently, no Box-Lucas function fitting could be performed). This results in the fact that SLP values for intracellular ML were only detectable for L929. SLP values were set deliberately to zero for all cases with $m_{int} < 15\,\mu g(Fe)$.

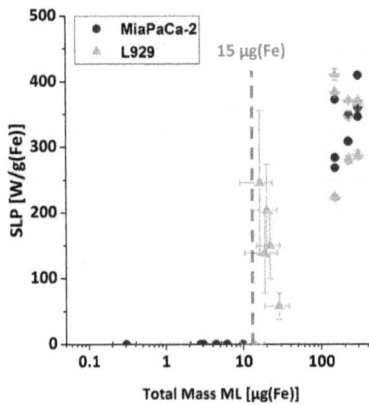

Fig. A.26.: Specific loss power (SLP) versus internalized amount of ML for each individual sample. Below $15\,\mu g(Fe)$ no SLP value could be measured, therefore all values were set to zero here. Adapted from [50].

Mean SLP Values

Table A.18 summarizes the SLP values and internalized amount of ML per samples and its averages used to calculate the thermal energy per cell (TEC) depicted in Fig. 8.7.

Table A.18.: Specific loss power (SLP) values and amount of internalized ML (m_{int}) for each sample individually and their mean value. Adapted from [50].

Initial incubation concentration	Sample	Duration of AMF application	MiaPaCa-2 SLP [W/g(Fe)]	MiaPaCa-2 m_{int} [µg(Fe)]	L929 SLP [W/g(Fe)]	L929 m_{int} [µg(Fe)]
150 µg(Fe)/mL	Intracellular ML	30 min	—	(6.1 ± 2.9)	(232.4 ± 75.5)	(17.2 ± 5.5)
		90 min	—	(9.8 ± 4.6)	(80.6 ± 25.9)	(47.9 ± 15.4)
		Mean	—	(8.0 ± 2.7)	(157.5 ± 39.9)	(32.6 ± 8.2)
	Intra- & extracellular ML	30 min	(283.8 ± 6.0)	(4.3 ± 2.0)	(410.5 ± 8.7)	(11.1 ± 3.6)
		90 min	(268.4 ± 5.7)	(10.5 ± 5.0)	(223.9 ± 4.8)	(39.0 ± 12.5)
		Mean	(276.1 ± 4.1)	(7.4 ± 2.7)	(339.5 ± 5.0)	(25.1 ± 6.5)
	ML in medium	30 min	(372.3 ± 5.7)	—	(399.7 ± 5.5)	—
225 µg(Fe)/mL	Intracellular ML	30 min	—	(3.0 ± 1.0)	—	(13.7 ± 4.5)
		90 min	—	(4.5 ± 1.4)	(149.5 ± 49.6)	(22.0 ± 7.3)
		Mean	—	(3.7 ± 0.9)	(149.5 ± 49.6)	(17.9 ± 4.3)
	Intra- & extracellular ML	30 min	(308.4 ± 6.5)	(3.5 ± 1.1)	(347.2 ± 7.4)	(22.8 ± 7.6)
		90 min	(308.3 ± 6.4)	(3.4 ± 1.1)	(281.0 ± 6.0)	(22.8 ± 7.6)
		Mean	(308.4 ± 4.6)	(3.4 ± 0.8)	(314.1 ± 4.8)	(22.8 ± 5.4)
	ML in medium	30 min	(348.5 ± 5.3)	—	(371.1 ± 3.6)	—
300 µg(Fe)/mL	Intracellular ML	30 min	—	(0.3 ± 0.1)	(57.8 ± 20.1)	(28.5 ± 9.9)
		90 min	—	(2.8 ± 1.0)	(203.6 ± 70.7)	(20.0 ± 7.0)
		Mean	—	(1.5 ± 0.5)	(130.7 ± 36.8)	(24.3 ± 6.1)
	Intra- & extracellular ML	30 min	(346.0 ± 7.3)	(1.2 ± 0.4)	(364.0 ± 7.7)	(19.5 ± 6.8)
		90 min	(409.6 ± 8.7)	(1.7 ± 0.6)	(287.8 ± 6.1)	(20.3 ± 7.1)
		Mean	(377.8 ± 5.7)	(1.5 ± 0.4)	(325.9 ± 4.9)	(19.9 ± 4.9)
	ML in medium	30 min	(359.5 ± 5.4)	—	(391.9 ± 4.0)	—

B. Bibliography

[1] Ferlay J, Soerjomataram I, Ervik M, Dikshit R, Eser S, Mathers C, et al. Cancer Incidence and mortality worldwide: GLOBOCAN 2012 v1.1. IARC CancerBase. 2014;11.

[2] Organization WH, et al.. World Health Report 2018: Monitoring health for the SDGs. WHO; 2018. Available from: http://apps.who.int/iris/bitstream/handle/10665/272596/9789241565585-eng.pdf?ua=1.

[3] Stewart BW, Wild C. World Cancer Report 2014. World Health Organization and International Agency for Research on Cancer. 2014;p. 630.

[4] Statistisches-Bundesamt-Deutschland. Todesursachen in Deutschland. Fachserie. 2017;12(4). Available from: https://www.destatis.de/DE/Publikationen/Thematisch/Gesundheit/Todesursachen/Todesursachen2120400157004.pdf?__blob=publicationFile.

[5] Vaupel P, Kallinowski F, Okunieff P. Blood flow, oxygen and nutrient supply, and metabolic microenvironment of human tumors: a review. Cancer research. 1989;49(23):6449–6465.

[6] EU-Commission. Commission Recommendation of 18 October 2011 on the definition of nanomaterial (2011/696/EU). Official Journal of the European Communities: Legis. 2011;.

[7] Himpsel F, Ortega J, Mankey G, Willis R. Magnetic nanostructures. Advances in physics. 1998;47(4):511–597.

[8] Huang HW, Liauh CT, et al. Therapeutical applications of heat in cancer therapy. Journal of Medical and Biological Engineering. 2011;32(1):1–11.

[9] Mahmoudi K, Bouras A, Bozec D, Ivkov R, Hadjipanayis C. Magnetic hyperthermia therapy for the treatment of glioblastoma: a review of the therapy's history, efficacy and application in humans. International Journal of Hyperthermia. 2018;p. 1–13.

[10] Yarmolenko PS, Moon EJ, Landon C, Manzoor A, Hochman DW, Viglianti BL, et al. Thresholds for thermal damage to normal tissues: an update. International Journal of Hyperthermia. 2011;27(4):320–343.

[11] Wilhelm C, Gazeau F. Universal cell labelling with anionic magnetic nanoparticles. Biomaterials. 2008;29(22):3161–3174.

[12] Blanco-Andujar C, Teran F, Ortega D. Current Outlook and Perspectives on Nanoparticle-Mediated Magnetic Hyperthermia. In: Iron Oxide Nanoparticles for Biomedical Applications. Elsevier; 2018. p. 197–245.

[13] Aharoni A. Introduction to the theory of ferromagnetism. 1st ed. International series of monographs on physics. Oxford; New York: Oxford University Press; 1996.

[14] Krishnan KM. Fundamentals and applications of magnetic materials. 1st ed. Great Clarendon Street, Oxford, OX2, 6DP, UK: Oxford University Press; 2016.

[15] Rosensweig RE. Ferrohydrodynamics. Mineola, N.Y.: Dover Publications; 1997. Originally published: Cambridge ; New York : Cambridge University Press, 1985.

[16] Weiss P. Hypothesis of the molecular field and ferromagnetism. Bulletin de la Societe Francaise de Physique. 1907;(1):95–124.

[17] Barkhausen H. Zwei mit Hilfe der neuen Verstärker entdeckte Erscheinungen. Physik Zeitung. 1919;(20):401.

[18] Slabu I. Synthesis, characterization and application of superparamagnetic iron oxide nanoparticles in medical diagnostics and therapy: MR-visible implants for hernia repair and novel drug targeting models. Shaker Verlag; 2015.

[19] Farle M. Ferromagnetic resonance of ultrathin metallic layers. Reports on Progress in Physics. 1998;61(7):755.

[20] Bickford LR, Brownlow JM, Penoyer RF. Magnetocrystalline anisotropy in cobalt-substituted magnetite single crystals. IEEE Proceedings Part B: Electrical Engineering. 1957;104(5):238–244.

[21] Salazar-Alvarez G, Qin J, Sepelak V, Bergmann I, Vasilakaki M, Trohidou K, et al. Cubic versus spherical magnetic nanoparticles: the role of surface anisotropy. Journal of the American Chemical Society. 2008;130(40):13234–13239.

[22] Bødker F, Mørup S, Linderoth S. Surface effects in metallic iron nanoparticles. Physical Review Letters. 1994;72(2):282.

[23] Krishnan KM. Biomedical nanomagnetics: a spin through possibilities in imaging, diagnostics, and therapy. IEEE transactions on magnetics. 2010;46(7):2523–2558.

[24] Reeves DB. Nonequilibrium dynamics of magnetic nanoparticles in biomedical applications. PhD thesis Dartmouth College. 2015;.

[25] Mamiya H, Jeyadevan B. Hyperthermic effects of dissipative structures of magnetic nanoparticles in large alternating magnetic fields. Scientific reports. 2011;1:157.

[26] Gilbert TL. A phenomenological theory of damping in ferromagnetic materials. IEEE Transactions on Magnetics. 2004;40(6):3443–3449.

[27] Landi GT. Role of dipolar interaction in magnetic hyperthermia. Physical Review B. 2014;89(1):014403.

[28] Usov N, Liubimov BY. Dynamics of magnetic nanoparticle in a viscous liquid: Application to magnetic nanoparticle hyperthermia. Journal of Applied Physics. 2012;112(2):023901.

[29] Reeves DB, Weaver JB. Combined Néel and Brown rotational Langevin dynamics in magnetic particle imaging, sensing, and therapy. Applied physics letters. 2015;107(22):223106.

[30] Weizenecker J, Gleich B, Rahmer J, Dahnke H, Borgert J. Three-dimensional real-time in vivo magnetic particle imaging. Physics in Medicine & Biology. 2009;54(5):L1.

[31] Panagiotopoulos N, Duschka RL, Ahlborg M, Bringout G, Debbeler C, Graeser M, et al. Magnetic particle imaging: current developments and future directions. International journal of nanomedicine. 2015;10:3097.

[32] Knopp T, Buzug TM. Magnetic particle imaging: an introduction to imaging principles and scanner instrumentation. Springer Science & Business Media; 2012.

[33] Saritas EU, Goodwill PW, Croft LR, Konkle JJ, Lu K, Zheng B, et al. Magnetic particle imaging (MPI) for NMR and MRI researchers. Journal of Magnetic Resonance. 2013;229:116–126.

[34] Ferguson RM. Tracer design for magnetic particle imaging: modeling, synthesis, and experimental optimization of biocompatible iron oxide nanoparticles. University of Washington; 2011.

[35] Rahmer J, Weizenecker J, Gleich B, Borgert J. Signal encoding in magnetic particle imaging: properties of the system function. BMC medical imaging. 2009;9(1):4.

[36] Weizenecker J, Borgert J, Gleich B. A simulation study on the resolution and sensitivity of magnetic particle imaging. Physics in Medicine & Biology. 2007;52(21):6363.

[37] Goodwill PW, Conolly SM. The X-space formulation of the magnetic particle imaging process: 1-D signal, resolution, bandwidth, SNR, SAR, and magnetostimulation. IEEE transactions on medical imaging. 2010;29(11):1851–1859.

[38] Goodwill PW, Conolly SM. Multidimensional x-space magnetic particle imaging. IEEE transactions on medical imaging. 2011;30(9):1581–1590.

[39] Ferguson RM, Khandhar AP, Krishnan KM. Tracer design for magnetic particle imaging. Journal of applied physics. 2012;111(7):07B318.

[40] Goodwill P, Tamrazian A, Croft L, Lu C, Johnson E, Pidaparthi R, et al. Ferro-hydrodynamic relaxometry for magnetic particle imaging. Applied Physics Letters. 2011;98(26):262502.

[41] Ferguson RM, Khandhar AP, Arami H, Hua L, Hovorka O, Krishnan KM. Tailoring the magnetic and pharmacokinetic properties of iron oxide magnetic particle imaging tracers. Biomedizinische Technik/Biomedical Engineering. 2013;58(6):493–507.

[42] Carrey J, Mehdaoui B, Respaud M. Simple models for dynamic hysteresis loop calculations of magnetic single-domain nanoparticles: Application to magnetic hyperthermia optimization. Journal of Applied Physics. 2011;109(8).

[43] Hergt R, Dutz S. Magnetic particle hyperthermia—biophysical limitations of a visionary tumour therapy. Journal of Magnetism and Magnetic Materials. 2007;311(1):187–192.

[44] Ibach H, Lüth H. Festkörperphysik: Einführung in die Grundlagen. Springer-Verlag; 2009.

[45] Garcia-Otero J, Garcia-Bastida AJ, Rivas J. Influence of temperature on the coercive field of non-interacting fine magnetic particles. Journal of Magnetism and Magnetic Materials. 1998;189(3):377–383.

[46] Usov N, Grebenshchikov YB. Hysteresis loops of an assembly of superparamagnetic nanoparticles with uniaxial anisotropy. Journal of Applied Physics. 2009;106(2):023917.

[47] Mehdaoui B, Meffre A, Carrey J, Lachaize S, Lacroix LM, Gougeon M, et al. Optimal size of nanoparticles for magnetic hyperthermia: a combined theoretical and experimental study. Advanced Functional Materials. 2011;21(23):4573–4581.

[48] Rosensweig RE. Heating magnetic fluid with alternating magnetic field. Journal of magnetism and magnetic materials. 2002;252:370–374.

[49] Debye PJW. Polar molecules. Chemical Catalog Company, Inc.; 1929.

[50] Engelmann UM, Roeth AA, Eberbeck D, Buhl EM, Neumann UP, Schmitz-Rode T, et al. Combining bulk temperature and nanoheating enables advanced magnetic fluid hyperthermia efficacy on pancreatic tumor cells. Scientific Reports. 2018;8(1):13210.

[51] Christophi C, Winkworth A, Muralihdaran V, Evans P. The treatment of malignancy by hyperthermia. Surgical Oncology-Oxford. 1998;7(1-2):83–90.

[52] Jeyadevan B. Present status and prospects of magnetite nanoparticles-based hyperthermia. Journal of the Ceramic Society of Japan. 2010;118(1378):391–401.

[53] Brace C. Thermal tumor ablation in clinical use. IEEE pulse. 2011;2(5):28–38.

[54] D'Andrea J, Gandhi O, Lords J, Durney C, Johnson CC, Astle L. Physiological and behavioral effects of chronic exposure to 2450-MHz microwaves. Journal of microwave power. 1979;14(4):351–362.

[55] Ritz J, Germer C, Roogan A, Isbert C, Pelz J, Buhr H. Hyperthermic effects of laserinduced thermotherapy-comparison of arterial microembolisation and hepatic blood flow occlusion. In: Gastroenterology. vol. 116. Elsevier; 1999. p. A491–A491.

[56] Brezovich IA, Young JH. Hyperthermia with implanted electrodes. Medical physics. 1981;8(1):79–84.

[57] Robins H, d'Oleire F, Grosen E, Spriggs D. Rationale and clinical status of 41.8 degrees C systemic hyperthermia tumor necrosis factor, and melphalan for neoplastic disease. Anticancer research. 1997;17(4B):2891–2894.

[58] Jordan A, Scholz R, Wust P, Fähling H, Felix R. Magnetic fluid hyperthermia (MFH): Cancer treatment with AC magnetic field induced excitation of biocompatible superparamagnetic nanoparticles. Journal of Magnetism and Magnetic materials. 1999;201(1-3):413–419.

[59] Laurent S, Dutz S, Häfeli UO, Mahmoudi M. Magnetic fluid hyperthermia: focus on superparamagnetic iron oxide nanoparticles. Advances in colloid and interface science. 2011;166(1-2):8–23.

[60] Pankhurst Q, Thanh N, Jones S, Dobson J. Progress in applications of magnetic nanoparticles in biomedicine. Journal of Physics D: Applied Physics. 2009;42(22):224001.

[61] Wust P, Gneveckow U, Wust P, Gneveckow U, Johannsen M, Böhmer D, et al. Magnetic nanoparticles for interstitial thermotherapy–feasibility, tolerance and achieved temperatures. International Journal of Hyperthermia. 2006;22(8):673–685.

[62] Jordan A, Wust P, Fählin H, John W, Hinz A, Felix R. Inductive heating of ferrimagnetic particles and magnetic fluids: physical evaluation of their potential for hyperthermia. International Journal of Hyperthermia. 1993;9(1):51–68.

[63] Atkinson WJ, Brezovich IA, Chakraborty DP. Usable frequencies in hyperthermia with thermal seeds. IEEE Transactions on Biomedical Engineering. 1984;(1):70–75.

[64] Maier-Hauff K, Rothe R, Scholz R, Gneveckow U, Wust P, Thiesen B, et al. Intracranial thermotherapy using magnetic nanoparticles combined with external beam radiotherapy: results of a feasibility study on patients with glioblastoma multiforme. Journal of neuro-oncology. 2007;81(1):53–60.

[65] Johannsen M, Gneveckow U, Thiesen B, Taymoorian K, Cho CH, Waldöfner N, et al. Thermotherapy of prostate cancer using magnetic nanoparticles: feasibility, imaging, and three-dimensional temperature distribution. European urology. 2007;52(6):1653–1662.

[66] Ahlbom A, Bergqvist U, Bernhardt J, Cesarini J, Grandolfo M, Hietanen M, et al. Guidelines for limiting exposure to time-varying electric, magnetic, and electromagnetic fields (up to 300 GHz). Health physics. 1998;74(4):494–521.

[67] on Non-Ionizing Radiation Protection IC, et al. Guidelines for limiting exposure to time-varying electric and magnetic fields (1 Hz to 100 kHz). Health physics. 2010;99(6):818–836.

[68] Henle KJ. Sensitization to hyperthermia below 43 °C induced in Chinese hamster ovary cells by step-down heating. JNCI: Journal of the National Cancer Institute.

1980;64(6):1479–1483.

[69] Li G, Mivechi NF, Weitzel G. Heat shock proteins, thermotolerance, and their relevance to clinical hyperthermia. International journal of hyperthermia. 1995;11(4):459–488.

[70] Li GC, Hahn GM. A proposed operational model of thermotolerance based on effects of nutrients and the initial treatment temperature. Cancer Research. 1980;40(12):4501–4508.

[71] Hahn GM. Comparison of the Malignant Potential of 10T1/2 Cells and Transformants with Their Survival Responses to Hyperthermia and to Amphotericin B. Cancer research. 1980;40(10):3763–3767.

[72] Van Rhoon GC, Samaras T, Yarmolenko PS, Dewhirst MW, Neufeld E, Kuster N. CEM43 C thermal dose thresholds: a potential guide for magnetic resonance radiofrequency exposure levels? European radiology. 2013;23(8):2215–2227.

[73] Bettaieb A, Wrzal PK, Averill-Bates DA. Hyperthermia: Cancer treatment and beyond. In: Cancer Treatment-Conventional and Innovative Approaches. InTech; 2013. .

[74] Gneveckow U, Jordan A, Scholz R, Brüß V, Waldöfner N, Ricke J, et al. Description and characterization of the novel hyperthermia-and thermoablation-system MFH® 300F for clinical magnetic fluid hyperthermia. Medical physics. 2004;31(6):1444–1451.

[75] Gellermann J, Wust P, Stalling D, Seebass M, Nadobny J, Beck R, et al. Clinical evaluation and verification of the hyperthermia treatment planning system hyperplan. International Journal of Radiation Oncology - Biology - Physics. 2000;47(4):1145–1156.

[76] Thiesen B, Jordan A. Clinical applications of magnetic nanoparticles for hyperthermia. International journal of hyperthermia. 2008;24(6):467–474.

[77] Johannsen M, Gneveckow U, Eckelt L, Feussner A, Waldöfner N, Scholz R, et al. Clinical hyperthermia of prostate cancer using magnetic nanoparticles: presentation of a new interstitial technique. International journal of hyperthermia. 2005;21(7):637–647.

[78] Tilly W, Wust P, Rau B, Harder C, Gellermann J, Schlag P, et al. Temperature data and specific absorption rates in pelvic tumours: predictive factors and correlations. International journal of hyperthermia. 2001;17(2):172–188.

[79] Attaluri A, Kandala SK, Wabler M, Zhou H, Cornejo C, Armour M, et al. Magnetic nanoparticle hyperthermia enhances radiation therapy: A study in mouse models of human prostate cancer. International Journal of Hyperthermia. 2015;31(4):359–374.

[80] Torres-Lugo M, Rinaldi C. Thermal potentiation of chemotherapy by magnetic nanoparticles. Nanomedicine. 2013;8(10):1689–1707.

[81] Maier-Hauff K, Ulrich F, Nestler D, Niehoff H, Wust P, Thiesen B, et al. Efficacy and

safety of intratumoral thermotherapy using magnetic iron-oxide nanoparticles combined with external beam radiotherapy on patients with recurrent glioblastoma multiforme. Journal of neuro-oncology. 2011;103(2):317–324.

[82] Hahn GM. Hyperthermia for the engineer: A short biological primer. IEEE Transactions on Biomedical engineering. 1984;(1):3–8.

[83] Miller R, Richards M, Baird C, Martin S, Hall E. Interaction of hyperthermia and chemotherapy agents; cell lethality and oncogenic potential. International journal of hyperthermia. 1994;10(1):89–99.

[84] Urano M. Invited review: for the clinical application of thermochemotherapy given at mild temperatures. International journal of hyperthermia. 1999;15(2):79–107.

[85] Corchero J, Villaverde A. Biomedical applications of distally controlled magnetic nanoparticles. Trends in Biotechnology. 2009;27(8):468–476.

[86] Cole AJ, Yang VC, David AE. Cancer theranostics: the rise of targeted magnetic nanoparticles. Trends in Biotechnology. 2011;29(7):323–332.

[87] J Wang YX, Xuan S, Port M, Idee JM. Recent advances in superparamagnetic iron oxide nanoparticles for cellular imaging and targeted therapy research. Current pharmaceutical design. 2013;19(37):6575–6593.

[88] Estelrich J, Escribano E, Queralt J, Busquets MA. Iron oxide nanoparticles for magnetically-guided and magnetically-responsive drug delivery. International journal of molecular sciences. 2015;16(4):8070–8101.

[89] Kalele S, Narain R, Krishnan KM. Probing temperature-sensitive behavior of pNIPAAm-coated iron oxide nanoparticles using frequency-dependent magnetic measurements. Journal of Magnetism and Magnetic Materials. 2009;321(10):1377–1380.

[90] Herrera AP, Rodriguez M, Torres-Lugo M, Rinaldi C. Multifunctional magnetite nanoparticles coated with fluorescent thermo-responsive polymeric shells. Journal of Materials Chemistry. 2008;18(8):855–858.

[91] Tai LA, Tsai PJ, Wang YC, Wang YJ, Lo LW, Yang CS. Thermosensitive liposomes entrapping iron oxide nanoparticles for controllable drug release. Nanotechnology. 2009;20(13):135101.

[92] Zhang J, Misra R. Magnetic drug-targeting carrier encapsulated with thermosensitive smart polymer: core–shell nanoparticle carrier and drug release response. Acta biomaterialia. 2007;3(6):838–850.

[93] Wilhelm S, Tavares AJ, Dai Q, Ohta S, Audet J, Dvorak HF, et al. Analysis of nanoparticle delivery to tumours. Nature Reviews Materials. 2016;1(5):16014.

[94] Southern P, Pankhurst QA. Commentary on the clinical and preclinical dosage limits of interstitially administered magnetic fluids for therapeutic hyperthermia based on current practice and efficacy models. International Journal of Hyperthermia. 2017;p. 1–16.

[95] Kugler P. Der menschliche Körper: Anatomie Physiologie Pathologie. Elsevier Health Sciences; 2017.

[96] Leonhardt H. Histologie, Zytologie und Mikroanatomie des Menschen. 7th ed. Taschenlehrbuch der gesamten Anatomie. Oxford; New York: Georg Thieme Verlagt Stuttgart - New York; 1985.

[97] Klinke R, Pape HC, Kurtz A, Silbernagl S. Physiologie. Georg Thieme Verlag; 2009.

[98] Doherty GJ, McMahon HT. Mechanisms of endocytosis. Annual review of biochemistry. 2009;78:857–902.

[99] Oh N, Park JH. Endocytosis and exocytosis of nanoparticles in mammalian cells. International journal of nanomedicine. 2014;9(Suppl 1):51.

[100] Mercer J, Schelhaas M, Helenius A. Virus entry by endocytosis. Annual review of biochemistry. 2010;79:803–833.

[101] Iversen TG, Skotland T, Sandvig K. Endocytosis and intracellular transport of nanoparticles: present knowledge and need for future studies. Nano Today. 2011;6(2):176–185.

[102] Parton RG, Simons K. The multiple faces of caveolae. Nature reviews Molecular cell biology. 2007;8(3):185.

[103] Zhang S, Gao H, Bao G. Physical principles of nanoparticle cellular endocytosis. ACS nano. 2015;9(9):8655–8671.

[104] Swanson JA. Shaping cups into phagosomes and macropinosomes. Nature reviews Molecular cell biology. 2008;9(8):639.

[105] Rink J, Ghigo E, Kalaidzidis Y, Zerial M. Rab conversion as a mechanism of progression from early to late endosomes. Cell. 2005;122(5):735–749.

[106] Ganley IG, Carroll K, Bittova L, Pfeffer S. Rab9 GTPase regulates late endosome size and requires effector interaction for its stability. Molecular biology of the cell. 2004;15(12):5420–5430.

[107] Luzio JP, Rous BA, Bright NA, Pryor PR, Mullock BM, Piper RC. Lysosome-endosome fusion and lysosome biogenesis. J Cell Sci. 2000;113(9):1515–1524.

[108] Jordan A, Wust P, Scholz R, Tesche B, Fähling H, Mitrovics T, et al. Cellular uptake of magnetic fluid particles and their effects on human adenocarcinoma cells exposed to AC magnetic fields in vitro. International journal of hyperthermia. 1996;12(6):705–722.

[109] Creixell M, Bohorquez AC, Torres-Lugo M, Rinaldi C. EGFR-targeted magnetic nanoparticle heaters kill cancer cells without a perceptible temperature rise. ACS nano. 2011;5(9):7124–7129.

[110] Schaub NJ, Rende D, Yuan Y, Gilbert RJ, Borca-Tasciuc DA. Reduced astrocyte viability at physiological temperatures from magnetically activated iron oxide nanoparticles. Chemical research in toxicology. 2014;27(12):2023–2035.

[111] Connord V, Clerc P, Hallali N, El Hajj Diab D, Fourmy D, Gigoux V, et al. Real-time analysis of magnetic hyperthermia experiments on living cells under a confocal microscope. Small. 2015;11(20):2437–2445.

[112] Blanco-Andujar C, Ortega D, Southern P, Nesbitt SA, Thanh NTK, Pankhurst QA. Real-time tracking of delayed-onset cellular apoptosis induced by intracellular magnetic hyperthermia. Nanomedicine. 2016;11(2):121–136.

[113] Chiu-Lam A, Rinaldi C. Nanoscale thermal phenomena in the vicinity of magnetic nanoparticles in alternating magnetic fields. Advanced functional materials. 2016;26(22):3933–3941.

[114] Huang H, Delikanli S, Zeng H, Ferkey DM, Pralle A. Remote control of ion channels and neurons through magnetic-field heating of nanoparticles. Nature nanotechnology. 2010;5(8):602.

[115] Polo-Corrales L, Rinaldi C. Monitoring iron oxide nanoparticle surface temperature in an alternating magnetic field using thermoresponsive fluorescent polymers. Journal of Applied Physics. 2012;111(7):07B334.

[116] Dong J, Zink JI. Taking the temperature of the interiors of magnetically heated nanoparticles. ACS nano. 2014;8(5):5199–5207.

[117] Riedinger A, Guardia P, Curcio A, Garcia MA, Cingolani R, Manna L, et al. Sub-nanometer local temperature probing and remotely controlled drug release based on azo-functionalized iron oxide nanoparticles. Nano letters. 2013;13(6):2399–2406.

[118] Taloub S, Hobar F, Astefanoaei I, Dumitru I, Caltun OF. FEM Investigation of Coated Magnetic Nanoparticles for Hyperthermia. Nanoscience and Nanotechnology. 2016;6(1A):55–61.

[119] Pearce J, Giustini A, Stigliano R, Hoopes PJ. Magnetic heating of nanoparticles: The importance of particle clustering to achieve therapeutic temperatures. Journal of nanotechnology in engineering and medicine. 2013;4(1):011005.

[120] Domenech M, Marrero-Berrios I, Torres-Lugo M, Rinaldi C. Lysosomal membrane permeabilization by targeted magnetic nanoparticles in alternating magnetic fields. ACS nano. 2013;7(6):5091–5101.

[121] Tseng P, Judy JW, Di Carlo D. Magnetic nanoparticle–mediated massively parallel mechanical modulation of single-cell behavior. Nature methods. 2012;9(11):1113.

[122] Sadhukha T, Wiedmann TS, Panyam J. Enhancing therapeutic efficacy through designed
aggregation of nanoparticles. Biomaterials. 2014;35(27):7860–7869.

[123] Zhang E, Kircher MF, Koch M, Eliasson L, Goldberg SN, Renstroem E. Dynamic magnetic
fields remote-control apoptosis via nanoparticle rotation. ACS nano. 2014;8(4):3192–
3201.

[124] Mansell R, Vemulkar T, Petit DC, Cheng Y, Murphy J, Lesniak MS, et al. Magnetic
particles with perpendicular anisotropy for mechanical cancer cell destruction. Scientific
reports. 2017;7(1):4257.

[125] Hezel AF, Kimmelman AC, Stanger BZ, Bardeesy N, DePinho RA. Genetics and biology
of pancreatic ductal adenocarcinoma. Genes & development. 2006;20(10):1218–1249.

[126] Rahib L, Smith BD, Aizenberg R, Rosenzweig AB, Fleshman JM, Matrisian LM. Projecting cancer incidence and deaths to 2030: the unexpected burden of thyroid, liver, and
pancreas cancers in the United States. Cancer research. 2014;74:2913–2921.

[127] Brus C, Saif MW. Second line therapy for advanced pancreatic adenocarcinoma: where
are we and where are we going? JOP Journal of the Pancreas. 2010;11(4):321–323.

[128] Kimmelman AC, Hezel AF, Aguirre AJ, Zheng H, Paik Jh, Ying H, et al. Genomic
alterations link Rho family of GTPases to the highly invasive phenotype of pancreas
cancer. Proceedings of the National Academy of Sciences. 2008;105(49):19372–19377.

[129] Ettrich TJ, Perkhofer L, Seufferlein T. Therapie des Pankreaskarzinoms–und sie bewegt
sich doch! DMW-Deutsche Medizinische Wochenschrift. 2015;140(07):508–511.

[130] Glanemann M, Shi B, Liang F, Sun XG, Bahra M, Jacob D, et al. Surgical strategies for
treatment of malignant pancreatic tumors: extended, standard or local surgery? World
journal of surgical oncology. 2008;6(1):123.

[131] Yekebas EF, Bogoevski D, Cataldegirmen G, Kunze C, Marx A, Vashist YK, et al. En bloc
vascular resection for locally advanced pancreatic malignancies infiltrating major blood
vessels: perioperative outcome and long-term survival in 136 patients. Annals of surgery.
2008;247(2):300–309.

[132] Pancreatic Carcinoma: Chemoradiation Compared With Chemotherapy Alone After Induction Chemotherapy (CONKO-007). University-of-Erlangen-Nürnberg-Medical-School.
2018-07-18;Available from: https://clinicaltrials.gov/ct2/show/NCT01827553.

[133] Li WM, Chiang CS, Huang WC, Su CW, Chiang MY, Chen JY, et al.
Amifostine-conjugated pH-sensitive calcium phosphate-covered magnetic-amphiphilic

gelatin nanoparticles for controlled intracellular dual drug release for dual-targeting in HER-2-overexpressing breast cancer. Journal of Controlled Release. 2015;220:107–118.

[134] Hashemi-Moghaddam H, Kazemi-Bagsangani S, Jamili M, Zavareh S. Evaluation of magnetic nanoparticles coated by 5-fluorouracil imprinted polymer for controlled drug delivery in mouse breast cancer model. International journal of pharmaceutics. 2016;497(1-2):228–238.

[135] Roeth AA, Slabu I, Baumann M, Alizai PH, Schmeding M, Guentherodt G, et al. Establishment of a biophysical model to optimize endoscopic targeting of magnetic nanoparticles for cancer treatment. International journal of nanomedicine. 2017;12:5933.

[136] Zohuriaan-Mehr MJ, Kabiri K. Superabsorbent polymer materials: a review. Iranian polymer journal. 2008;17(6):451.

[137] Ahmed EM. Hydrogel: Preparation, characterization, and applications: A review. Journal of advanced research. 2015;6(2):105–121.

[138] Lamouche G, Kennedy BF, Kennedy KM, Bisaillon CE, Curatolo A, Campbell G, et al. Review of tissue simulating phantoms with controllable optical, mechanical and structural properties for use in optical coherence tomography. Biomedical optics express. 2012;3(6):1381–1398.

[139] Pogue BW, Patterson MS. Review of tissue simulating phantoms for optical spectroscopy, imaging and dosimetry. Journal of biomedical optics. 2006;11(4):041102.

[140] Marklein RA, Burdick JA. Controlling stem cell fate with material design. Advanced materials. 2010;22(2):175–189.

[141] Thiele J, Ma Y, Bruekers S, Ma S, Huck WT. 25th Anniversary article: designer hydrogels for cell cultures: a materials selection guide. Advanced materials. 2014;26(1):125–148.

[142] Chrambach A, Rodbard D. Polyacrylamide gel electrophoresis. Science. 1971;172(3982):440–451.

[143] Culjat MO, Goldenberg D, Tewari P, Singh RS. A review of tissue substitutes for ultrasound imaging. Ultrasound in Medicine and Biology. 2010;36(6):861–873.

[144] Menikou G, Yiannakou M, Yiallouras C, Ioannides C, Damianou C. MRI-compatible bone phantom for evaluating ultrasonic thermal exposures. Ultrasonics. 2016;71:12–19.

[145] Frickel N, Messing R, Schmidt AM. Magneto-mechanical coupling in CoFe2O4-linked PAAm ferrohydrogels. Journal of Materials Chemistry. 2011;21(23):8466–8474.

[146] Zhao X, Kim J, Cezar CA, Huebsch N, Lee K, Bouhadir K, et al. Active scaffolds for on-demand drug and cell delivery. Proceedings of the National Academy of Sciences. 2011;108(1):67–72.

[147] Lao L, Ramanujan R. Magnetic and hydrogel composite materials for hyperthermia applications. Journal of materials science: Materials in medicine. 2004;15(10):1061–1064.

[148] Zhang ZQ, Song SC. Thermosensitive/superparamagnetic iron oxide nanoparticle-loaded nanocapsule hydrogels for multiple cancer hyperthermia. Biomaterials. 2016;106:13–23.

[149] Mark JE, et al. Physical properties of polymers handbook. vol. 1076. Springer; 2007.

[150] Treloar L. Physics of Rubber Elasticity (Oxford classic texts in the physical sciences). Oxford University Press; 2005.

[151] Sperling LH. Polymeric multicomponent materials: an introduction. Wiley-Interscience; 1997.

[152] Treloar L. The elasticity and related properties of rubbers. Reports on progress in physics. 1973;36(7):755.

[153] Love A. A treatise on the mathematical theory of elasticity. 4th ed. Cambridge University Press; 1927.

[154] Arnott S, Fulmer A, Scott W, Dea I, Moorhouse R, Rees D. The agarose double helix and its function in agarose gel structure. Journal of molecular biology. 1974;90(2):269–284.

[155] Velasco D, Tumarkin E, Kumacheva E. Microfluidic encapsulation of cells in polymer microgels. Small. 2012;8(11):1633–1642.

[156] Agarose - SERVA Wide Range (molecular biology grade) - Specification Data Sheet. Serva Electrophoresis GmbH. 2016;Cat.No.: 11406.

[157] Agarose Low Melt - Roti®garose - Specification Data Sheet. Carl Roth GmbH + Co KG. 2017;Cat.No.: 6351.

[158] Righetti PG, Brost BC, Snyder RS. On the limiting pore size of hydrophilic gels for electrophoresis and isoelectric focussing. Journal of biochemical and biophysical methods. 1981;4(5-6):347–363.

[159] Tse JR, Engler AJ. Preparation of hydrogel substrates with tunable mechanical properties. Current protocols in cell biology. 2010;p. 10–16.

[160] Horkay F, McKenna GB. Polymer networks and gels. In: Physical properties of polymers handbook. Springer; 2007. p. 497–523.

[161] Tan R, Carrey J, Respaud M. Magnetic hyperthermia properties of nanoparticles inside lysosomes using kinetic Monte Carlo simulations: Influence of key parameters and dipolar interactions, and evidence for strong spatial variation of heating power. Physical Review B. 2014;90(21):214421.

[162] Ovejero JG, Cabrera D, Carrey J, Valdivielso T, Salas G, Teran FJ. Effects of inter-and intra-aggregate magnetic dipolar interactions on the magnetic heating efficiency of iron oxide nanoparticles. Physical Chemistry Chemical Physics. 2016;18(16):10954–10963.

[163] Ruta S, Chantrell R, Hovorka O. Unified model of hyperthermia via hysteresis heating in systems of interacting magnetic nanoparticles. Scientific reports. 2015;5:9090.

[164] Mamiya H, Jeyadevan B. Optimal design of nanomagnets for targeted hyperthermia. Journal of Magnetism and Magnetic Materials. 2011;323(10):1417–1422.

[165] Mamiya H, Jeyadevan B. Formation of Nonequilibrium Magnetic Nanoparticle Structures in a Large Alternating Magnetic Field and Their Influence on Magnetic Hyperthermia Treatment. IEEE Transactions on Magnetics. 2012;48(11):3258–3261.

[166] Van Kampen NG. Stochastic processes in physics and chemistry. vol. 1. Elsevier; 1992.

[167] Garcia-Alvarez D. A comparison of a few numerical schemes for the integration of stochastic differential equations in the Stratonovich interpretation. arXiv preprint arXiv:11024401. 2011;.

[168] Burrage K, Burrage P, Tian T. Numerical methods for strong solutions of stochastic differential equations: an overview. In: Proceedings of The Royal Society of London A: Mathematical, Physical and Engineering Sciences. vol. 460:2041. The Royal Society; 2004. p. 373–402.

[169] Platen E. An introduction to numerical methods for stochastic differential equations. Acta numerica. 1999;8:197–246.

[170] Guidoum AC, Boukhetala K. Itô and Stratonovich stochastic calculus with sSimDiffProc package version 2.9. pdfssemanticscholarorg. 2014;.

[171] Scholz W, Schrefl T, Fidler J. Micromagnetic simulation of thermally activated switching in fine particles. Journal of Magnetism and Magnetic Materials. 2001;233(3):296–304.

[172] Nowak U. Thermally activated reversal in magnetic nanostructures. In: Annual Reviews Of Computational PhysicsIX. World Scientific; 2001. p. 105–151.

[173] Rüemelin W. Numerical treatment of stochastic differential equations. SIAM Journal on Numerical Analysis. 1982;19(3):604–613.

[174] Barlow RJ. Statistics: a guide to the use of statistical methods in the physical sciences. vol. 29. John Wiley & Sons; 1989.

[175] Limpert E, Stahel WA, Abbt M. Log-normal Distributions across the Sciences: Keys and Clues: On the charms of statistics, and how mechanical models resembling gambling machines offer a link to a handy way to characterize log-normal distributions, which can provide deeper insight into variability and probability—normal or log-normal: That is the

question. AIBS Bulletin. 2001;51(5):341–352.

[176] Soto-Aquino D, Rinaldi C. Nonlinear energy dissipation of magnetic nanoparticles in oscillating magnetic fields. Journal of Magnetism and Magnetic Materials. 2015;393:46–55.

[177] Hergt R, Dutz S, Müller R, Zeisberger M. Magnetic particle hyperthermia: nanoparticle magnetism and materials development for cancer therapy. Journal of Physics: Condensed Matter. 2006;18(38):S2919.

[178] Schwertmann U, Cornell RM. Iron oxides in the laboratory. Preparation and characterization. VHC Publishers, Inc.: New York, New York, USA. Illus; 1991.

[179] Thanh NT. Magnetic nanoparticles: from fabrication to clinical application. Boca Raton: CRC press; 2012.

[180] Massart R, Cabuil V. Effect of some parameters on the formation of colloidal magnetite in alkaline-medium-yield and particle-size control. Journal de Chimie Physique et de Physico-Chimie Biologique. 1987;84(7-8):967–973.

[181] Bastús NG, Comenge J, Puntes V. Kinetically controlled seeded growth synthesis of citrate-stabilized gold nanoparticles of up to 200 nm: size focusing versus Ostwald ripening. Langmuir. 2011;27(17):11098–11105.

[182] Lu AH, Salabas E, Schüth F. Magnetische Nanopartikel: Synthese, Stabilisierung, Funktionalisierung und Anwendung. Angewandte Chemie. 2007;119(8):1242–1266.

[183] Jeong U, Teng X, Wang Y, Yang H, Xia Y. Superparamagnetic colloids: controlled synthesis and niche applications. Advanced Materials. 2007;19(1):33–60.

[184] Arami H, Khandhar A, Liggitt D, Krishnan KM. In vivo delivery, pharmacokinetics, biodistribution and toxicity of iron oxide nanoparticles. Chemical Society Reviews. 2015;44(23):8576–8607.

[185] McCormack PL. Ferumoxytol. Drugs. 2012;72(15):2013–2022.

[186] Lu M, Cohen MH, Rieves D, Pazdur R. FDA report: ferumoxytol for intravenous iron therapy in adult patients with chronic kidney disease. American journal of hematology. 2010;85(5):315–319.

[187] Rienso (ferumoxytol): summary of product characteristics. Takeda Global Research and Development Centre (Europe) Ltd. 2012;.

[188] Schiller B, Bhat P, Sharma A. Safety and effectiveness of ferumoxytol in hemodialysis patients at 3 dialysis chains in the United States over a 12-month period. Clinical therapeutics. 2014;36(1):70–83.

[189] ; 2009. Encyclopedia Britannica. Online Library. http://www.britannica.com/.

[190] Khalafalla S, Reimers G. Preparation of dilution-stable aqueous magnetic fluids. IEEE Transactions on Magnetics. 1980;16(2):178–183.

[191] De Cuyper M, Müller P, Lueken H, Hodenius M. Synthesis of magnetic Fe3O4 particles covered with a modifiable phospholipid coat. Journal of Physics: Condensed Matter. 2003;15(15):S1425.

[192] De Cuyper M, Soenen SJH. In: Weissig V, editor. Cationic Magnetoliposomes. vol. 605 of Methods in Molecular Biology. Human Press; 2010. p. 97–111.

[193] Massart R. Preparation of aqueous magnetic liquids in alkaline and acidic media. IEEE transactions on magnetics. 1981;17(2):1247–1248.

[194] Fortin JP, Wilhelm C, Servais J, Menager C, Bacri JC, Gazeau F. Size-sorted anionic iron oxide nanomagnets as colloidal mediators for magnetic hyperthermia. Journal of the American Chemical Society. 2007;129(9):2628–2635.

[195] Bealle G, Di Corato R, Kolosnjaj-Tabi J, Dupuis V, Clement O, Gazeau F, et al. Ultra magnetic liposomes for MR imaging, targeting, and hyperthermia. Langmuir. 2012;28(32):11834–11842.

[196] Kemp SJ, Ferguson RM, Khandhar AP, Krishnan KM. Monodisperse magnetite nanoparticles with nearly ideal saturation magnetization. RSC Advances. 2016;6(81):77452–77464.

[197] Khandhar AP, Ferguson RM, Krishnan KM. Monodispersed magnetite nanoparticles optimized for magnetic fluid hyperthermia: Implications in biological systems. Journal of applied physics. 2011;109(7):07B310.

[198] Park J, An K, Hwang Y, Park JG, Noh HJ, Kim JY, et al. Ultra-large-scale syntheses of monodisperse nanocrystals. Nature materials. 2004;3(12):891.

[199] Khandhar AP, Ferguson RM, Simon JA, Krishnan KM. Tailored magnetic nanoparticles for optimizing magnetic fluid hyperthermia. Journal of Biomedical Materials Research Part A. 2012;100(3):728–737.

[200] Klibanov AL, Maruyama K, Beckerleg AM, Torchilin VP, Huang L. Activity of amphipathic poly (ethylene glycol) 5000 to prolong the circulation time of liposomes depends on the liposome size and is unfavorable for immunoliposome binding to target. Biochimica et Biophysica Acta (BBA)-Biomembranes. 1991;1062(2):142–148.

[201] William WY, Chang E, Sayes CM, Drezek R, Colvin VL. Aqueous dispersion of monodisperse magnetic iron oxide nanocrystals through phase transfer. Nanotechnology. 2006;17(17):4483.

[202] Halliday D, Resnick R, Koch SW, Walker J. Halliday Physik. v. 1, p. 1085. John Wiley

& Sons, Limited; 2003.

[203] Department of Earth & Environmental Sciences RPI. The Electron Microprobe method. http://ees2.geo.rpi.edu/probe/Images/concepts/concept2.html; 22-APR-2018.

[204] Patterson A. The Scherrer formula for X-ray particle size determination. Physical review. 1939;56(10):978.

[205] Swanson HE, Tatge E. Standard X-ray diffraction patterns. Journal of Research of the National Bureau of Standards. 1951;46(4):318.

[206] Schulz D, McCarthy G, Grant-in Aid I. North Dakota State University. Fargo, North Dakota, USA, ICDD Grant-in-Aid. 1987;.

[207] Hufschmid R, Arami H, Ferguson RM, Gonzales M, Teeman E, Brush LN, et al. Synthesis of phase-pure and monodisperse iron oxide nanoparticles by thermal decomposition. Nanoscale. 2015;7(25):11142–11154.

[208] Howard S, Preston K. Profile fitting of powder diffraction patterns. Reviews in Mineralogy and Geochemistry. 1989;20(1):217–275.

[209] Williams DB, Carter CB. The transmission electron microscope. In: Transmission electron microscopy. Springer; 1996. p. 3–17.

[210] Mäntele W. Biophysik. Ulmer Stuttgart; 2012.

[211] Sawada H, Shimura N, Hosokawa F, Shibata N, Ikuhara Y. Resolving 45-pm-separated Si–Si atomic columns with an aberration-corrected STEM. Microscopy. 2015;64(3):213–217.

[212] Brewster R. Paint .NET-Free Software for Digital Photo Editing. Getpaintnet Ultimo acceso. 2017;15.

[213] Schneider CA, Rasband WS, Eliceiri KW. NIH Image to ImageJ: 25 years of image analysis. Nature methods. 2012;9(7):671.

[214] Teeman E. Personal Communication. University of Washington, Krishnan Labs. 2018;.

[215] O'grady K, Bradbury A. Particle size analysis in ferrofluids. Journal of Magnetism and magnetic Materials. 1983;39(1-2):91–94.

[216] Malvern Instruments Ldt. Dynamic Light Scattering: An Introduction in 30 Minutes; Technical Note MRK656-01.

[217] Pecora R. Dynamic light scattering: applications of photon correlation spectroscopy. Springer Science & Business Media; 2013.

[218] Berne B, Pecora R. Dynamic light scattering. USA, New York. 2000;.

[219] ISO-22412:2017. Particle size analysis - dynamic light scattering (DLS); 2018-09-03. Available from: https://www.iso.org/standard/65410.html.

[220] Nobbmann U. FAQ: Peak size or z-average size – which one to pick in DLS?; 2018-09-03. Available from: http://www.materials-talks.com/blog/2014/07/10/faq-peak-size-or-z-average-size-which-one-to-pick-in-dls/.

[221] Khandhar AP. Biomedical imaging and therapy with physically and physiologically tailored magnetic nanoparticles. PhD thesis, University of Washington. 2013;.

[222] De Cuyper M, Hodenius M, Ivanova G, Baumann M, Paciok E, Eckert T, et al. Specific heating power of fatty acid and phospholipid stabilized magnetic fluids in an alternating magnetic field. Journal of Physics: Condensed Matter. 2008;20(20):204131.

[223] Lak A, Niculaes D, Anyfantis GC, Bertoni G, Barthel MJ, Marras S, et al. Facile transformation of FeO/Fe$_3$O$_4$ core-shell nanocubes to Fe$_3$O$_4$ via magnetic stimulation. Scientific reports. 2016;6:33295.

[224] Hollemann A, Wiberg E, Wiberg N. Lehrbuch der Anorganischen Chemie, 91.-100. Auflage. Walter de Gruyter, Berlin. 1985;999:1012–1021.

[225] Olesik JW. Elemental analysis using ICP-OES and ICP/MS. Analytical Chemistry. 1991;63(1):12A–21A.

[226] Bhadani S, Tiwari M, Agrawal A, Kavipurapu CS. Spectrophotometric determination of Fe (III) with tiron in the presence of cationic surfactant and its application for the determination of iron in Al-alloys and Cu-based alloys. Microchimica Acta. 1994;117(1-2):15–22.

[227] Clarke J. SQUIDS. Scientific American. 1994;271(2):46–53.

[228] Script: Superconductivity and SQUID magnetometer. RWTH Aachen University; 2011.

[229] Ashcroft NW, Mermin ND. Solid state physics. Science: Physics. Saunders College; 1976.

[230] Chantrell R, Popplewell J, Charles S. Measurements of particle size distribution parameters in ferrofluids. IEEE Transactions on Magnetics. 1978;14(5):975–977.

[231] Bradbury A, Menear K, O'Grady K, Chantrell R. Magnetic size determination for interacting fine particle systems. IEEE Transactions on Magnetics. 1984;20(5):1846–1848.

[232] Winkler K, Murphy W. Rock physics and phase relations: a handbook of physical constants. American Geophysical Union; 1995.

[233] Hu P, Kang L, Chang T, Yang F, Wang H, Zhang Y, et al. High saturation magnetization Fe3O4 nanoparticles prepared by one-step reduction method in autoclave. Journal of Alloys and Compounds. 2017;728:88–92.

[234] Shete P, Patil R, Tiwale B, Pawar S. Water dispersible oleic acid-coated Fe3O4 nanoparticles for biomedical applications. Journal of Magnetism and Magnetic Materials. 2015;377:406–410.

[235] Pal S, Dutta P, Shah N, Huffman G, Seehra M. Surface Spin Disorder in Fe_{3} O_{4} Nanoparticles Probed by Electron Magnetic Resonance Spectroscopy and Magnetometry. IEEE transactions on magnetics. 2007;43(6):3091–3093.

[236] Schmitz-Antoniak C, Schmitz D, Warland A, Darbandi M, Haldar S, Bhandary S, et al. Suppression of the Verwey Transition by Charge Trapping. Annalen der Physik. 2018;530(3):1700363.

[237] Nedelkoski Z, Kepaptsoglou D, Lari L, Wen T, Booth RA, Oberdick SD, et al. Origin of reduced magnetization and domain formation in small magnetite nanoparticles. Scientific reports. 2017;7:45997.

[238] Hai HT, Yang HT, Kura H, Hasegawa D, Ogata Y, Takahashi M, et al. Size control and characterization of wustite (core)/spinel (shell) nanocubes obtained by decomposition of iron oleate complex. Journal of colloid and interface science. 2010;346(1):37–42.

[239] Pichon BP, Gerber O, Lefevre C, Florea I, Fleutot S, Baaziz W, et al. Microstructural and magnetic investigations of wustite-spinel core-shell cubic-shaped nanoparticles. Chemistry of Materials. 2011;23(11):2886–2900.

[240] Unni M, Uhl AM, Savliwala S, Savitzky BH, Dhavalikar R, Garraud N, et al. Thermal decomposition synthesis of iron oxide nanoparticles with diminished magnetic dead layer by controlled addition of oxygen. ACS nano. 2017;11(2):2284–2303.

[241] Tamion A, Hillenkamp M, Tournus F, Bonet E, Dupuis V. Accurate determination of the magnetic anisotropy in cluster-assembled nanostructures. Applied Physics Letters. 2009;95(6):062503.

[242] Tournus F, Bonet E. Magnetic susceptibility curves of a nanoparticle assembly, I: Theoretical model and analytical expressions for a single magnetic anisotropy energy. Journal of Magnetism and Magnetic Materials. 2011;323(9):1109–1117.

[243] Tournus F, Tamion A. Magnetic susceptibility curves of a nanoparticle assembly II. Simulation and analysis of ZFC/FC curves in the case of a magnetic anisotropy energy distribution. Journal of Magnetism and Magnetic Materials. 2011;323(9):1118–1127.

[244] Tournus F, Blanc N, Tamion A, Hillenkamp M, Dupuis V. Dispersion of magnetic anisotropy in size-selected CoPt clusters. Physical Review B. 2010;81(22):220405.

[245] Di Corato R, Espinosa A, Lartigue L, Tharaud M, Chat S, Pellegrino T, et al. Magnetic hyperthermia efficiency in the cellular environment for different nanoparticle designs. Biomaterials. 2014;35(24):6400–6411.

[246] Branquinho LC, Carrião MS, Costa AS, Zufelato N, Sousa MH, Miotto R, et al. Effect of magnetic dipolar interactions on nanoparticle heating efficiency: Implications for cancer hyperthermia. Scientific reports. 2013;3:2887.

[247] Levy M, Wilhelm C, Luciani N, Deveaux V, Gendron F, Luciani A, et al. Nanomagnetism reveals the intracellular clustering of iron oxide nanoparticles in the organism. Nanoscale. 2011;3(10):4402–4410.

[248] Lévy M, Gazeau F, Bacri JC, Wilhelm C, Devaud M. Modeling magnetic nanoparticle dipole-dipole interactions inside living cells. Physical Review B. 2011;84(7):075480.

[249] Quantum Design Inc . VSM Option User's Manual, 1096-100, Rev. B0; February 2011.

[250] Foner S. Vibrating Sample Magnetometer. Review of Scientific Instruments. 1956;27(7):548–548.

[251] Engelmann UM, Shasha C, Teeman E, Slabu I, Krishnan KM. Predicting size-dependent heating efficiency of magnetic nanoparticles from experiment and stochastic Néel-Brown Langevin simulation. Journal of Magnetism and Magnetic Materials. 2019;471:450–456.

[252] Roca A, Morales M, O'Grady K, Serna C. Structural and magnetic properties of uniform magnetite nanoparticles prepared by high temperature decomposition of organic precursors. Nanotechnology. 2006;17(11):2783.

[253] Chen DX, Sanchez A, Taboada E, Roig A, Sun N, Gu HC. Size determination of superparamagnetic nanoparticles from magnetization curve. Journal of Applied Physics. 2009;105(8):083924.

[254] Löwa N, Radon P, Kosch O, Wiekhorst F. Concentration dependent MPI tracer performance. International Journal on Magnetic Particle Imaging. 2016;2(1).

[255] Friedrich RP, Janko C, Poettler M, Tripal P, Zaloga J, Cicha I, et al. Flow cytometry for intracellular SPION quantification: specificity and sensitivity in comparison with spectroscopic methods. International journal of nanomedicine. 2015;10:4185.

[256] Gräfe C, Slabu I, Wiekhorst F, Bergemann C, von Eggeling F, Hochhaus A, et al. Magnetic particle spectroscopy allows precise quantification of nanoparticles after passage through human brain microvascular endothelial cells. Physics in Medicine & Biology. 2016;61(11):3986.

[257] Ficko BW, NDong C, Giacometti P, Griswold KE, Diamond SG. A feasibility study of nonlinear spectroscopic measurement of magnetic nanoparticles targeted to cancer cells. IEEE Transactions on Biomedical Engineering. 2017;64(5):972–979.

[258] Schilling M, Ludwig F, Kuhlmann C, Wawrzik T. Magnetic particle imaging scanner with 10-kHz drive-field frequency. Biomedizinische Technik/Biomedical Engineering.

2013;58(6):557–563.

[259] Draack S, Viereck T, Kuhlmann C, Schilling M, Ludwig F. Temperature-dependent MPS measurements. International Journal on Magnetic Particle Imaging. 2017;3(1).

[260] Arami H, Ferguson R, Khandhar AP, Krishnan KM. Size-dependent ferrohydrodynamic relaxometry of magnetic particle imaging tracers in different environments. Medical physics. 2013;40(7).

[261] Khandhar AP, Ferguson RM, Arami H, Krishnan KM. Monodisperse magnetite nanoparticle tracers for in vivo magnetic particle imaging. Biomaterials. 2013;34(15):3837–3845.

[262] Ferguson RM, Khandhar AP, Kemp SJ, Arami H, Saritas EU, Croft LR, et al. Magnetic particle imaging with tailored iron oxide nanoparticle tracers. IEEE transactions on medical imaging. 2015;34(5):1077–1084.

[263] Yu EY, Bishop M, Zheng B, Ferguson RM, Khandhar AP, Kemp SJ, et al. Magnetic particle imaging: a novel in vivo imaging platform for cancer detection. Nano letters. 2017;17(3):1648–1654.

[264] Engelmann U, Seifert J, Mues B, Roitsch S, Menager C, Schmidt A, et al. Heating efficiency of magnetic nanoparticles decreases with gradual immobilization in hydrogels. Journal of Magnetism and Magnetic Materials. 2019;471:486–494.

[265] Jones C; 2017. NanoTherics, Ltd., Keele, Staffordshire, UK. Personal communication.

[266] Huang S, Wang S, Gupta A, Borca-Tasciuc D, Salon S. On the measurement technique for specific absorption rate of nanoparticles in an alternating electromagnetic field. Measurement Science and Technology. 2012;23(3):035701.

[267] Gonzales-Weimuller M, Zeisberger M, Krishnan KM. Size-dependant heating rates of iron oxide nanoparticles for magnetic fluid hyperthermia. Journal of magnetism and magnetic materials. 2009;321(13):1947–1950.

[268] Bordelon DE, Cornejo C, Grüttner C, Westphal F, DeWeese TL, Ivkov R. Magnetic nanoparticle heating efficiency reveals magneto-structural differences when characterized with wide ranging and high amplitude alternating magnetic fields. Journal of Applied Physics. 2011;109(12):124904.

[269] Wildeboer R, Southern P, Pankhurst Q. On the reliable measurement of specific absorption rates and intrinsic loss parameters in magnetic hyperthermia materials. Journal of Physics D: Applied Physics. 2014;47(49):495003.

[270] Ng EYK, Kumar SD, et al. Physical mechanism and modeling of heat generation and transfer in magnetic fluid hyperthermia through Néelian and Brownian relaxation: a review. Biomedical engineering online. 2017;16(1):36.

[271] Hiergeist R, Andrä W, Buske N, Hergt R, Hilger I, Richter U, et al. Application of magnetite ferrofluids for hyperthermia. Journal of Magnetism and Magnetic Materials. 1999;201(1-3):420–422.

[272] Wang X, Gu H, Yang Z. The heating effect of magnetic fluids in an alternating magnetic field. Journal of Magnetism and Magnetic Materials. 2005;293(1):334–340.

[273] Li CH, Hodgins P, Peterson G. Experimental study of fundamental mechanisms in inductive heating of ferromagnetic nanoparticles suspension (Fe3O4 Iron Oxide Ferrofluid). Journal of Applied Physics. 2011;110(5):054303.

[274] Kashevsky BE, Kashevsky SB, Prokhorov IV. Dynamic magnetic hysteresis in a liquid suspension of acicular maghemite particles. Particuology. 2009;7(6):451–458.

[275] Lacroix LM, Malaki RB, Carrey J, Lachaize S, Respaud M, Goya G, et al. Magnetic hyperthermia in single-domain monodisperse FeCo nanoparticles: evidences for Stoner–Wohlfarth behavior and large losses. Journal of Applied Physics. 2009;105(2):023911.

[276] Dennis CL, Krycka KL, Borchers JA, Desautels RD, Van Lierop J, Huls NF, et al. Internal magnetic structure of nanoparticles dominates time-dependent relaxation processes in a magnetic field. Advanced Functional Materials. 2015;25(27):4300–4311.

[277] Conde-Leboran I, Baldomir D, Martinez-Boubeta C, Chubykalo-Fesenko O, del Puerto Morales M, Salas G, et al. A single picture explains diversity of hyperthermia response of magnetic nanoparticles. The Journal of Physical Chemistry C. 2015;119(27):15698–15706.

[278] Bae S, Lee SW, Takemura Y, Yamashita E, Kunisaki J, Zurn S, et al. Dependence of frequency and magnetic field on self-heating characteristics of NiFe $_2$ O $_4$ nanoparticles for hyperthermia. IEEE transactions on magnetics. 2006;42(10):3566–3568.

[279] Kallumadil M, Tada M, Nakagawa T, Abe M, Southern P, Pankhurst QA. Suitability of commercial colloids for magnetic hyperthermia. Journal of Magnetism and Magnetic Materials. 2009;321(10):1509–1513.

[280] Blanco-Andujar C, Ortega D, Southern P, Pankhurst Q, Thanh N. High performance multi-core iron oxide nanoparticles for magnetic hyperthermia: microwave synthesis, and the role of core-to-core interactions. Nanoscale. 2015;7(5):1768–1775.

[281] Lima Jr E, De Biasi E, Mansilla MV, Saleta ME, Granada M, Troiani HE, et al. Heat generation in agglomerated ferrite nanoparticles in an alternating magnetic field. Journal of Physics D: Applied Physics. 2012;46(4):045002.

[282] Verde E, Landi GT, Carrião M, Drummond AL, Gomes J, Vieira E, et al. Field dependent transition to the non-linear regime in magnetic hyperthermia experiments: Comparison between maghemite, copper, zinc, nickel and cobalt ferrite nanoparticles of similar sizes.

Aip Advances. 2012;2(3):032120.

[283] Verde EL, Landi GT, Gomes JdA, Sousa MH, Bakuzis AF. Magnetic hyperther-mia investigation of cobalt ferrite nanoparticles: Comparison between experiment, lin-ear response theory, and dynamic hysteresis simulations. Journal of Applied Physics. 2012;111(12):123902.

[284] Ludwig F, Remmer H, Kuhlmann C, Wawrzik T, Arami H, Ferguson RM, et al. Self-consistent magnetic properties of magnetite tracers optimized for magnetic particle imag-ing measured by ac susceptometry, magnetorelaxometry and magnetic particle spec-troscopy. Journal of magnetism and magnetic materials. 2014;360:169–173.

[285] Niculaes D, Lak A, Anyfantis GC, Marras S, Laslett O, Avugadda SK, et al. Asymmetric Assembling of Iron Oxide Nanocubes for Improving Magnetic Hyperthermia Performance. ACS nano. 2017;11(12):12121–12133.

[286] Cabrera D, Coene A, Leliaert J, Artés-Ibáñez EJ, Dupré L, Telling ND, et al. Dynamical magnetic response of iron oxide nanoparticles inside live cells. ACS nano. 2018;.

[287] Dennis C, Jackson A, Borchers J, Ivkov R, Foreman A, Lau J, et al. The influence of collective behavior on the magnetic and heating properties of iron oxide nanoparticles. Journal of Applied Physics. 2008;103(7):07A319.

[288] Deleu M, Crowet JM, Nasir MN, Lins L. Complementary biophysical tools to investigate lipid specificity in the interaction between bioactive molecules and the plasma membrane: a review. Biochimica et Biophysica Acta (BBA)-Biomembranes. 2014;1838(12):3171–3190.

[289] Ma DD, Wei AQ. Enhanced delivery of synthetic oligonucleotides to human leukaemic cells by liposomes and immunoliposomes. Leukemia research. 1996;20(11-12):925–930.

[290] Wang Y, Miao L, Satterlee A, Huang L. Delivery of oligonucleotides with lipid nanopar-ticles. Advanced drug delivery reviews. 2015;87:68–80.

[291] Slabu I, Roeth A, Engelmann U, Wiekhorst F, Buhl E, Neumann U, et al. Modelling of magnetoliposome uptake in human pancreatic tumor cells in vitro. Nanotechnology: Special issue - focus on personalized medicine and theranostics. 2019;.

[292] Yunis AA, Arimura GK, Russin DJ. Human pancreatic carcinoma (MIA PaCa-2) in contin-uous culture: sensitivity to asparaginase. International journal of cancer. 1977;19(1):128–135.

[293] Edge SB, Compton CC. The American Joint Committee on Cancer: the 7th edition of the AJCC cancer staging manual and the future of TNM. Annals of surgical oncology. 2010;17(6):1471–1474.

[294] Oberlin L. Behandlung von Pankreaskarzinomzelllinien in vitro und in vivo mit einem monoklonalen Antikoerper gegen den Transferrinrezeptor. Universitaetsbibliothek Giessen; Doktorarbeit; 2009.

[295] Tan MH, Nowak NJ, Loor R, Ochi H, Sandberg AA, Lopez C, et al. Characterization of a new primary human pancreatic tumor line. Cancer investigation. 1986;4(1):15–23.

[296] Earle WR, Voegtlin C. A further study of the mode of action of methylcholanthrene on normal tissue cultures. Public Health Reports (1896-1970). 1940;p. 303–322.

[297] Sanford KK, Earle WR, Likely GD. The growth in vitro of single isolated tissue cells. Journal of the National Cancer Institute. 1948;9(3):229–246.

[298] Theerakittayakorn K, Bunprasert T. Differentiation capacity of mouse L929 fibroblastic cell line compare with human dermal fibroblast. World Acad Sci Eng Technol. 2011;50:373–376.

[299] Baca OG, Scott T, Akporiaye E, DeBlassie R, Crissman H. Cell cycle distribution patterns and generation times of L929 fibroblast cells persistently infected with Coxiella burnetii. Infection and immunity. 1985;47(2):366–369.

[300] Wilhelm C, Gazeau F, Roger J, Pons J, Bacri JC. Interaction of anionic superparamagnetic nanoparticles with cells: kinetic analyses of membrane adsorption and subsequent internalization. Langmuir. 2002;18(21):8148–8155.

[301] Chithrani BD, Chan WC. Elucidating the mechanism of cellular uptake and removal of protein-coated gold nanoparticles of different sizes and shapes. Nano letters. 2007;7(6):1542–1550.

[302] Dos Santos T, Varela J, Lynch I, Salvati A, Dawson KA. Quantitative assessment of the comparative nanoparticle-uptake efficiency of a range of cell lines. Small. 2011;7(23):3341–3349.

[303] Limbach LK, Li Y, Grass RN, Brunner TJ, Hintermann MA, Muller M, et al. Oxide nanoparticle uptake in human lung fibroblasts: effects of particle size, agglomeration, and diffusion at low concentrations. Environmental science & technology. 2005;39(23):9370–9376.

[304] Chaudhuri A, Battaglia G, Golestanian R. The effect of interactions on the cellular uptake of nanoparticles. Physical biology. 2011;8(4):046002.

[305] Soukup D, Moise S, Céspedes E, Dobson J, Telling ND. In situ measurement of magnetization relaxation of internalized nanoparticles in live cells. Acs Nano. 2015;9(1):231–240.

[306] Sanz B, Calatayud MP, De Biasi E, Lima Jr E, Mansilla MV, Zysler RD, et al. In silico before in vivo: how to predict the heating efficiency of magnetic nanoparticles within the

intracellular space. Scientific reports. 2016;6:38733.

[307] De Duve C. The lysosome concept. In: Ciba Foundation Symposium-Anterior Pituitary Secretion (Book I of Colloquia on Endocrinology). Wiley Online Library; 1963. p. 1–35.

[308] De Duve C, Wattiaux R. Functions of lysosomes. Annual review of physiology. 1966;28(1):435–492.

[309] McNaught AD, McNaught AD. Compendium of chemical terminology. vol. 1669. Blackwell Science Oxford; 1997.

[310] Chang JS, Chang KLB, Hwang DF, Kong ZL. In vitro cytotoxicitiy of silica nanoparticles at high concentrations strongly depends on the metabolic activity type of the cell line. Environmental science & technology. 2007;41(6):2064–2068.

[311] Kim JA, Åberg C, Salvati A, Dawson KA. Role of cell cycle on the cellular uptake and dilution of nanoparticles in a cell population. Nature nanotechnology. 2012;7(1):62.

[312] Martina MS, Nicolas V, Wilhelm C, Ménager C, Barratt G, Lesieur S. The in vitro kinetics of the interactions between PEG-ylated magnetic-fluid-loaded liposomes and macrophages. Biomaterials. 2007;28(28):4143–4153.

[313] Martina MS, Wilhelm C, Lesieur S. The effect of magnetic targeting on the uptake of magnetic-fluid-loaded liposomes by human prostatic adenocarcinoma cells. Biomaterials. 2008;29(30):4137–4145.

[314] Rappoport J, Preece J, Chipman K, Argatov I, Davies A, Dyson L, et al. „How do Manufactured Nanoparticles Enter Cells. Microporous and Mesoporous Materials. 2011;116:123–130.

[315] Serda RE, Mack A, Van De Ven AL, Ferrati S, Dunner Jr K, Godin B, et al. Logic-Embedded Vectors for Intracellular Partitioning, Endosomal Escape, and Exocytosis of Nanoparticles. Small. 2010;6(23):2691–2700.

[316] Jiang X, Roecker C, Hafner M, Brandholt S, Doerlich RM, Nienhaus GU. Endo-and exocytosis of zwitterionic quantum dot nanoparticles by live HeLa cells. ACS nano. 2010;4(11):6787–6797.

[317] Rascol E, Devoisselle JM, Chopineau J. The relevance of membrane models to understand nanoparticles–cell membrane interactions. Nanoscale. 2016;8(9):4780–4798.

[318] Ding Hm, Ma Yq. Theoretical and computational investigations of nanoparticle–biomembrane interactions in cellular delivery. Small. 2015;11(9-10):1055–1071.

[319] Bahrami AH, Raatz M, Agudo-Canalejo J, Michel R, Curtis EM, Hall CK, et al. Wrapping of nanoparticles by membranes. Advances in colloid and interface science. 2014;208:214–224.

[320] Fenzl C, Genslein C, Domonkos C, Edwards K, Hirsch T, Baeumner A. Investigating non-specific binding to chemically engineered sensor surfaces using liposomes as models. Analyst. 2016;141(18):5265–5273.

[321] Lee RJ, Low PS. Delivery of liposomes into cultured KB cells via folate receptor-mediated endocytosis. Journal of Biological Chemistry. 1994;269(5):3198–3204.

[322] Gao H, Shi W, Freund LB. Mechanics of receptor-mediated endocytosis. Proceedings of the National Academy of Sciences. 2005;102(27):9469–9474.

[323] Osaki F, Kanamori T, Sando S, Sera T, Aoyama Y. A quantum dot conjugated sugar ball and its cellular uptake. On the size effects of endocytosis in the subviral region. Journal of the American Chemical Society. 2004;126(21):6520–6521.

[324] Bartczak D, Nitti S, Millar TM, Kanaras AG. Exocytosis of peptide functionalized gold nanoparticles in endothelial cells. Nanoscale. 2012;4(15):4470–4472.

[325] Bartczak D, Nitti S, Millar TM, Kanaras AG. Exocytosis of peptide functionalized gold nanoparticles in endothelial cells. Nanoscale. 2012;4(15):4470–4472.

[326] Poller WC, Löwa N, Wiekhorst F, Taupitz M, Wagner S, Möller K, et al. Magnetic particle spectroscopy reveals dynamic changes in the magnetic behavior of very small superparamagnetic iron oxide nanoparticles during cellular uptake and enables determination of cell-labeling efficacy. Journal of biomedical nanotechnology. 2016;12(2):337–346.

[327] Dutz S, Hergt R. The role of interactions in systems of single domain ferrimagnetic iron oxide nanoparticles. Journal of Nano- and Electronic Physics. 2012;4(2).

[328] Engelmann U, Buhl EM, Baumann M, Schmitz-Rode T, Slabu I. Agglomeration of magnetic nanoparticles and its effects on magnetic hyperthermia. Current Directions in Biomedical Engineering. 2017;3(2):457–460.

[329] Engelmann UM, Buhl EM, Draack S, Viereck T, Ludwig F, Schmitz-Rode T, et al. Magnetic relaxation of agglomerated and immobilized iron oxide nanoparticles for hyperthermia and imaging applications. IEEE Magnetics Letters. 2018;9:1–5.

[330] Calvet D, Wong JY, Giasson S. Rheological monitoring of polyacrylamide gelation: Importance of cross-link density and temperature. Macromolecules. 2004;37(20):7762–7771.

[331] Engler AJ, Sen S, Sweeney HL, Discher DE. Matrix elasticity directs stem cell lineage specification. Cell. 2006;126(4):677–689.

[332] Pernodet N, Maaloum M, Tinland B. Pore size of agarose gels by atomic force microscopy. Electrophoresis. 1997;18(1):55–58.

[333] Narayanan J, Xiong JY, Liu XY. Determination of agarose gel pore size: Absorbance measurements vis a vis other techniques. In: Journal of Physics: Conference Series.

vol. 28. IOP Publishing; 2006. p. 83.

[334] Roeder L, Reckenthaeler M, Belkoura L, Roitsch S, Strey R, Schmidt A. Covalent ferrohydrogels based on elongated particulate cross-linkers. Macromolecules. 2014;47(20):7200–7207.

[335] Wang J, Ugaz VM. Using in situ rheology to characterize the microstructure in photopolymerized polyacrylamide gels for DNA electrophoresis. Electrophoresis. 2006;27(17):3349–3358.

[336] Khlebtsov B, Khlebtsov N. On the measurement of gold nanoparticle sizes by the dynamic light scattering method. Colloid Journal. 2011;73(1):118–127.

[337] Fissan H, Ristig S, Kaminski H, Asbach C, Epple M. Comparison of different characterization methods for nanoparticle dispersions before and after aerosolization. Analytical Methods. 2014;6(18):7324–7334.

[338] Martens MA, Deissler RJ, Wu Y, Bauer L, Yao Z, Brown R, et al. Modeling the Brownian relaxation of nanoparticle ferrofluids: comparison with experiment. Medical physics. 2013;40(2).

[339] Hrozek J, Nespor D, Bartusek K. Thermal Conductivity and Heat Capacity Measurement of Biological Tissues. In: PIERS Proceedings; 2013. .

[340] Bergman TL, Incropera FP, DeWitt DP, Lavine AS. Fundamentals of heat and mass transfer. John Wiley & Sons; 2011.

[341] Bu-Lin Z, Bing H, Sheng-Li K, Huang Y, Rong W, Jia L. A polyacrylamide gel phantom for radiofrequency ablation. International Journal of Hyperthermia. 2008;24(7):568–576.

[342] Rojas JM, Gavilán H, del Dedo V, Lorente-Sorolla E, Sanz-Ortega L, da Silva GB, et al. Time-course assessment of the aggregation and metabolization of magnetic nanoparticles. Acta biomaterialia. 2017;58:181–195.

[343] Ludwig R, Stapf M, Dutz S, Müller R, Teichgräber U, Hilger I. Structural properties of magnetic nanoparticles determine their heating behavior-an estimation of the in vivo heating potential. Nanoscale research letters. 2014;9(1):602.

[344] Dieckhoff J, Eberbeck D, Schilling M, Ludwig F. Magnetic-field dependence of Brownian and Néel relaxation times. Journal of Applied Physics. 2016;119(4):043903.

[345] Yoshida T, Enpuku K. Simulation and quantitative clarification of AC susceptibility of magnetic fluid in nonlinear Brownian relaxation region. Japanese Journal of Applied Physics. 2009;48(12R):127002.

[346] Fannin P, Charles S. On the calculation of the Neel relaxation time in uniaxial single-domain ferromagnetic particles. Journal of Physics D: Applied Physics. 1994;27(2):185.

[347] Dennis C, Jackson A, Borchers J, Hoopes P, Strawbridge R, Foreman A, et al. Nearly complete regression of tumors via collective behavior of magnetic nanoparticles in hyperthermia. Nanotechnology. 2009;20(39):395103.

[348] Jiang J, Oberdörster G, Biswas P. Characterization of size, surface charge, and agglomeration state of nanoparticle dispersions for toxicological studies. Journal of Nanoparticle Research. 2009;11(1):77–89.

[349] Genck W. Make the most of antisolvent crystallization. Chemical Processing. 2010;.

[350] Tan CY, Huang YX. Dependence of refractive index on concentration and temperature in electrolyte solution, polar solution, nonpolar solution, and protein solution. Journal of Chemical & Engineering Data. 2015;60(10):2827–2833.

[351] Balgavỳ P, Dubničková M, Kučerka N, Kiselev MA, Yaradaikin SP, Uhriková D. Bilayer thickness and lipid interface area in unilamellar extruded 1, 2-diacylphosphatidylcholine liposomes: a small-angle neutron scattering study. Biochimica et Biophysica Acta (BBA)-Biomembranes. 2001;1512(1):40–52.

[352] Lak A, Kraken M, Ludwig F, Kornowski A, Eberbeck D, Sievers S, et al. Size dependent structural and magnetic properties of FeO–Fe 3 O 4 nanoparticles. Nanoscale. 2013;5(24):12286–12295.

[353] Stakhanova M, Karapet'yants MK, Vasil'ev V, Epikhin YA. A comparative study of the specific heats and densities of aqueous electrolyte solutions. Russ J Phys Chem. 1964;38:1306.

[354] Piñeiro-Redondo Y, Bañobre-López M, Pardiñas-Blanco I, Goya G, López-Quintela MA, Rivas J. The influence of colloidal parameters on the specific power absorption of PAA-coated magnetite nanoparticles. Nanoscale research letters. 2011;6(1):383.

[355] Spizzo F, Sgarbossa P, Sieni E, Semenzato A, Dughiero F, Forzan M, et al. Synthesis of ferrofluids made of iron oxide nanoflowers: Interplay between carrier fluid and magnetic properties. Nanomaterials. 2017;7(11):373.

[356] Saville SL, Qi B, Baker J, Stone R, Camley RE, Livesey KL, et al. The formation of linear aggregates in magnetic hyperthermia: Implications on specific absorption rate and magnetic anisotropy. Journal of colloid and interface science. 2014;424:141–151.

[357] Dennis CL, Ivkov R. Physics of heat generation using magnetic nanoparticles for hyperthermia. International Journal of Hyperthermia. 2013;29(8):715–729.

[358] Suto M, Hirota Y, Mamiya H, Fujita A, Kasuya R, Tohji K, et al. Heat dissipation mechanism of magnetite nanoparticles in magnetic fluid hyperthermia. Journal of Magnetism and Magnetic Materials. 2009;321(10):1493–1496.

[359] Franken NA, Rodermond HM, Stap J, Haveman J, Van Bree C. Clonogenic assay of cells in vitro. Nature protocols. 2006;1(5):2315.

[360] Bähring F, Schlenk F, Wotschadlo J, Buske N, Liebert T, Bergemann C, et al. Suitability of viability assays for testing biological effects of coated superparamagnetic nanoparticles. Ieee Transactions on Magnetics. 2013;49(1):383–388.

[361] Wilhelm C, Billotey C, Roger J, Pons J, Bacri JC, Gazeau F. Intracellular uptake of anionic superparamagnetic nanoparticles as a function of their surface coating. Biomaterials. 2003;24(6):1001–1011.

[362] Dickson J, Calderwood S. Temperature range and selective sensitivity of tumors to hyperthermia: a critical review. Annals of the New York Academy of Sciences. 1980;335(1):180–205.

[363] Hensley D, Tay ZW, Dhavalikar R, Zheng B, Goodwill P, Rinaldi C, et al. Combining magnetic particle imaging and magnetic fluid hyperthermia in a theranostic platform. Physics in Medicine & Biology. 2017;62(9):3483.

[364] Tay ZW, Chandrasekharan P, Chiu-Lam A, Hensley DW, Dhavalikar R, Zhou XY, et al. Magnetic particle imaging-guided heating in vivo using gradient fields for arbitrary localization of magnetic hyperthermia therapy. ACS nano. 2018;12(4):3699–3713.

[365] Lim EK, Kim T, Paik S, Haam S, Huh YM, Lee K. Nanomaterials for theranostics: recent advances and future challenges. Chemical reviews. 2014;115(1):327–394.

[366] Rehman M, Asadullah Madni AI, Khan WS, Khan MI, Mahmood MA, Ashfaq M, et al. Solid and liquid lipid-based binary solid lipid nanoparticles of diacerein: in vitro evaluation of sustained release, simultaneous loading of gold nanoparticles, and potential thermoresponsive behavior. International journal of nanomedicine. 2015;10:2805.

[367] Hondow N, Brydson R, Wang P, Holton MD, Brown MR, Rees P, et al. Quantitative characterization of nanoparticle agglomeration within biological media. Journal of Nanoparticle Research. 2012;14(7):977.

[368] Cho EC, Zhang Q, Xia Y. The effect of sedimentation and diffusion on cellular uptake of gold nanoparticles. Nature nanotechnology. 2011;6(6):385.

[369] Park MV, Neigh AM, Vermeulen JP, de la Fonteyne LJ, Verharen HW, Briedé JJ, et al. The effect of particle size on the cytotoxicity, inflammation, developmental toxicity and genotoxicity of silver nanoparticles. Biomaterials. 2011;32(36):9810–9817.

[370] Ross SM. Combined DC and ELF magnetic fields can alter cell proliferation. Bioelectromagnetics: Journal of the Bioelectromagnetics Society, The Society for Physical Regulation in Biology and Medicine, The European Bioelectromagnetics Association. 1990;11(1):27–36.

[371] Fitzsimmons R, Ryaby J, Magee F, Baylink D. Combined magnetic fields increased net calcium flux in bone cells. Calcified tissue international. 1994;55(5):376–380.

[372] Kirson ED, Gurvich Z, Schneiderman R, Dekel E, Itzhaki A, Wasserman Y, et al. Disruption of cancer cell replication by alternating electric fields. Cancer research. 2004;64(9):3288–3295.

C. List of Publications

This thesis is partly based on the following original publications by the writer:

Engelmann UM, Shasha C, Teeman E, Slabu I and Krishnan KM. Predicting size-dependent heating efficiency of magnetic nanoparticles from experiment and stochastic Néel-Brown Langevin simulation. Journal of Magnetism and Magnetic Materials. 2019; 471:450-456. Published online: 2018-10-03.
[251]; used in Chapter 5, Section 5.3.

Slabu I, Roeth AA, Engelmann UM, Wiekhorst F, Buhl EM, Neumann U and Schmitz-Rode T. Modelling of magnetoliposome uptake in human pancreatic tumor cells in vitro. Nanotechnology: Special issue - focus on personalized medicine and theranostics. 2019. Published online: 2019-01-30.
[291]; used in Chapter 6, Sections 6.2 and 6.3.

Engelmann UM, Seifert J, Mues B, Roitsch S, Menager C, Schmidt A and Slabu I. Heating efficiency of magnetic nanoparticles decreases with gradual immobilization in hydrogels. Journal of Magnetism and Magnetic Materials. 2019; 471:486-494. Published online: 2018-09-29.
[264]; used in Chapter 7, Sections 7.1 and 7.2.

Engelmann UM, Buhl EM, Baumann M, Schmitz-Rode T and Slabu I. Agglomeration of magnetic nanoparticles and its effects on magnetic hyperthermia. Current Directions in Biomedical Engineering. 2017; 3(2):457–460. Published online: 2017-09-08.
[328]; used in Chapter 7, Sections 7.3 and 7.4.

Engelmann UM, Buhl EM, Draack S, Viereck T, Ludwig F, Schmitz-Rode T and Slabu I. Magnetic relaxation study of agglomerated and immobilized magnetic iron oxide nanoparticles for hyperthermic and imaging applications. IEEE Magnetics Letters. 2018; 9:1-5. Published online: 2018-11-01.
[329]; used in Chapter 7, Sections 7.3 and 7.4.

Engelmann UM, Roeth AA, Eberbeck D, Buhl EM, Neumann UP, Schmitz-Rode T and Slabu I. Combining bulk temperature and nanoheating enables advanced magnetic fluid hyperthermia efficacy on pancreatic tumor cells. Scientific Reports. 2018; 8(1):13210. Published online: 2018-09-04.
[50]; used in Chapter 2, Section 2.5 and Chapter 8.

D. Acknowledgment

"If I have seen further it is by standing on the shoulders of giants."

Sir Isaac Newton

In diesem Sinne steht diese Arbeit ganz im Zeichen hervorragender Zusammenarbeit und eines regen Austauschs mit anderen, denen ich zum Dank verpflichtet bin und ohne die diese Arbeit nicht gelungen wäre.

Mein größter Dank gilt Dr. Ioana Slabu als meiner direkte Betreuerin, Mentorin und Ideengeberin. Und daher möchte ich ihr auch als Erste danken. Als Forschungsgruppenleiterin hat sie sich unermüdlich für mein wissenschaftliches Vorankommen eingesetzt, mich an ihren Ideen teilhaben lassen und mir einen Aufenthalt als Gastwissenschaftler an der Physikalisch-Technischen Bundesanstalt in Berlin ermöglicht. Vielen Dank für das große Vertrauen und die außergewöhnliche Unterstützung, die vielen Gespräche und die tolle Atmosphäre in der Arbeitsgruppe.

Ebenso hervorragend war die Begleitung durch meine Berichter Professor Dr. Martin Baumann und Professor Dr. Jörg Fitter. Von Ihrer Seite wurde die Arbeit durch die für mich optimale Balance aus Hilfestellung in kniffligen Fragen und Freiheit im Arbeiten bestens begleitet. Danke für die angenehme Betreuung und das in mich gesteckte Vertrauen, diese Promotion trotz anfänglicher finanzieller Engpässe zu begleiten; vielen Dank ebenso für das stets offene Ohr in allen Belangen und die wissenschaftliche Betreuung meiner Studierenden. Besonders danken möchte ich Professor Dr. Jörg Fitter für seinen außergewöhnlichen Einsatz, mir sehr schnelles Feedback zu dieser Arbeit zu geben.

Mein besonderer Dank gilt dem Cusanuswerk für die finanzielle Unabhängigkeit, die diese Promotion durch ein Stipendium erst ermöglicht hat. Über die monetäre Sicherheit hinaus, habe ich besonders den Austausch mit Promovierenden anderer Wissenschaften sehr genossen und danke dem Cusanuswerk, dass insbesondere Auslandskonferenzen und mein Auslandsaufenthalt in Seattle unkompliziert und unbürokratisch unterstützt wurden. Stellvertretend möchte ich besonders Dr. Ingrid Reul für die perfekt durchorganisierten Graduiertentreffen danken und ebenso Dr. Siegfried Kleymann, der mich mehrfach zum Schreiben, Studieren und Diskutieren im Cusanushaus in Mehlem empfangen hat.

Getragen wurde ich durch die gesamte Zeit der Doktorarbeit von der außergewöhnlich guten Arbeitsatmosphäre der Biophysical & Education Engineering Gruppe des Institutes für Angewandte Medizintechnik (AME). Ich danke meinen Arbeitskollegen Dr.-Ing. Andreas Ritter, M.Sc. Benedikt Mues, M.Sc. Michael Gundlach und M.Sc. Max Lindemann für die vielen guten Stunden und erfolgreichen (nicht) Kaffeepausen. Darüber hinaus danke ich allen Studenten, die ich während meiner Zeit am Institut begleiten durfte und deren Arbeiten und Austausch diese Arbeit auf allen Ebenen bereichert haben: M.Sc. Florian Müller, M.Sc. Katharina Kolvenbach, M.Sc. Artur Kessler, B.Sc. Andrea Haack, M.Sc. Simon Lyra, B.Sc. Felix Jiang und B.Sc. Richard Görg.

Die gute Atmosphäre hat aber nicht auf unserem Flur geendet: Ich danke allen Kolleginnen und Kollegen vom Institut für Angewandte Medizintechnik der RWTH Aachen University, allen voran dem Institutsleiter Professor Dr. Thomas Schmitz-Rode, und insbesondere Professor Dr. Ulrich Steinseifer von der Arbeitsgruppe Cardiovascular Engineering, der mich zusammen mit Professor Baumann erfolgreich für das Promotionsstipendium beim Cusanuswerk vorgeschlagen hat. Ich danke M.Sc. Catharina Lierath und M.Sc. Matthias Menne für die temporäre Aufnahme in ihr Büro während meines Messmarathons und Ilona Mager und Judith Maas für die immer-ansprechbare Hilfe in und rund ums Labor; sowie Dr. Nicole Kiesendahl für die Hilfe im Chemielabor.

Ich danke der Klinik für Allgemein-, Viszeral- und Transplantationschirurgie am Uniklinikum Aachen unter Professor Dr. Ulf Neumann und insbesondere Dr. Anjali Röth für die hervorragende Zusammenarbeit zu den Zellversuchen. Ebenso gilt mein Dank dem Team von Professor Dr. Thorsten Cramer, in dessen Labor die Zellversuche durchgeführt wurden. Besonders hervorheben möchte ich die unermüdliche Arbeit von Anne Wernerus, die mir während nächtelanger Versuchsreihen die Grundlagen der Zellversuche beigebracht hat. Danke für diesen großen Beitrag zu dieser Arbeit.

My very special thanks goes to Professor Dr. Kannan M. Krishnan and the Krishnan Group from the University of Washington in Seattle for welcoming me to the group in winter 2017/18. Thank you for introducing me to magnetic relaxation dynamics MC-simulations, thermal decomposition synthesis and MPI. Due to the great freedom and support I experienced, this time broadened my horizon academically, culturally and personally greatly. In particular, I thank M.Sc. Eric Teeman, M.Sc. Ryan Hufschmid and Dr. Vineeth Mohanan Parakkat for their great spirits and support around the lab and M.Sc. Carolyn Shasha for introducing me to her programming code for implementing MC-simulations.

Besonders hervorheben möchte ich die herausragende Kooperation mit dem Institut für Physikalische Chemie der Universität zu Köln, genauer mit der Arbeitsgruppe von Professor Dr. Annette Schmidt: Ihr Wissen um Polymer-Hydrogele und die Durchführung von rheologischen Messungen, sowie freeze-fractured TEM und VSM Analysen von Ferrohydrogel-Proben in ihren Laboren haben diese Arbeit deutlich bereichert. Danke an Dr. Stefan Roitsch für die TEM Aufnahmen. Mein ganz besonderer Dank geht an M.Sc. Julian Seifert, von dem ich voller Überzeugung sagen kann, dass uns eine beispiellos gute Kooperation gelungen ist, auf die ich sehr gerne und auch mit Stolz zurück blicke.

Großer Dank gilt dem Institut für Pathologie der Uniklinik Aachen unter Institutsdirektorin Professor Dr. Ruth Knüchel-Clarke, insbesondere der Elektronenmikroskopischen Einrichtung: Hier wurden unter der Leitung von Dr. Miriam Buhl hervorragende TEM Aufnahmen von Hiltrud

Königs-Werner an Zellen und Partikeln durchgeführt. Für die unkomplizierte Kommunikation und professionelle Kooperation danke ich sehr.

Bedanken möchte ich mich ebenso bei der Physikalisch-Technischen Bundesanstalt Berlin, in deren Abteilung 8.2 Biosignale ich Ende 2015 als Gastwissenschaftler unter Dr. Lutz Trahms an MPS Messungen arbeiten konnte. Vielen Dank stellvertretend an Dr. Dietmar Eberbeck für die ausführlichen Diskussionen zur Auswertung von MPS Daten.

Mit dem Institut für Elektrische Messtechnik und Grundlagen der Elektrotechnik der TU Braunschweig unter Leitung von Professor Dr. Meinhard Schilling blicke ich auf eine sehr gute Zusammenarbeit zurück: Vielen Dank an Professor Dr. Frank Ludwig, Dr. Thilo Viereck und M.Sc. Sebastian Draack für MPS Messungen und ausführliche Diskussionen zur Partikelinteraktion.

Herzlich bedanken möchte ich mich auch beim 1. Physikalischen Institut (IA) der RWTH Aachen University: Danke an Professor Dr. Matthias Wuttig, dass ich in seinen Laboren XRD Messungen durchführen konnte und danke an M.Sc. Stefan Jakobs für die hilfreichen Diskussionen zur Interpretation der XRD Daten.

Mein Dank geht ebenso an das Institut für Anorganische Chemie der RWTH Aachen University, namentlich Professor Dr. Ulrich Simon und seinen Mitarbeiterinnen Dr. Katharina Wiemer und M.Sc. Sabine Eisold, die mir die Nutzung des DLS ermöglicht haben.

Danke an das Rechenzentrum der RWTH Aachen University für die zur Verfügung gestellte Rechenzeit auf dem Rechencluster unter dem Projekt rwth0301. In diesem Zusammenhang danke ich ebenfalls M.Sc. Dennis Többen und M.Sc. Piotr Luczynski vom Institut für Kraftwerkstechnik, Dampf- und Gasturbinen der RWTH Aachen University für ihre Unterstützung beim Aufsetzen der Rechencluster-Umgebung.

Diese Arbeit wurde von Dr. Ari Fogel, M.Sc. Benedikt Mues, M.Sc. Julian Seifert, M.Sc. Carolyn Shasha, M.Sc. Eric Teeman, M.Sc. Ryan Hufschmid, B.Sc. Erin Tang und Brian Amoeni Korrektur gelesen: Vielen Dank an euch / thank you very much for proof-reading!

Meinen wichtigsten Dank habe ich mir für den Schluss aufgehoben; ganz herzlichen Dank an meine Familie, insbesondere durch drei Generationen: meinen Eltern Beate und Martin, meinen Brüder Markus und Georg und meinen Großeltern Maria und Hubert, die immer — auch in den schweren Anfangszeiten ohne finanzielle Sicherheit — fest hinter mir und meinem Wunsch der Promotion gestanden haben. Und meinen herzlichsten Dank für das Durchstehen, Motivieren, Konfrontieren und Ermutigen durch alle Phasen der Promotion hindurch an meine Frau Clara: Es gibt niemanden, mit dem ich dieses große Etappenziel lieber feiern möchte!

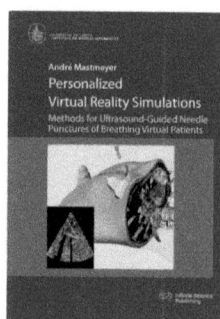